T0190389

Lecture Notes in Computer Science　　12440

More information about this series at http://www.springer.com/series/7409

Shin'ichi Satoh · Lucia Vadicamo ·
Arthur Zimek · Fabio Carrara ·
Ilaria Bartolini · Martin Aumüller ·
Björn Þór Jónsson · Rasmus Pagh (Eds.)

Similarity Search and Applications

13th International Conference, SISAP 2020
Copenhagen, Denmark, September 30 – October 2, 2020
Proceedings

 Springer

Editors
Shin'ichi Satoh
National Institute of Informatics
Tokyo, Japan

Lucia Vadicamo
ISTI-CNR
Pisa, Italy

Arthur Zimek
University of Southern Denmark
Odense M, Denmark

Fabio Carrara
ISTI-CNR
Pisa, Italy

Ilaria Bartolini
University of Bologna
Bologna, Italy

Martin Aumüller
IT University of Copenhagen
Copenhagen, Denmark

Björn Þór Jónsson
IT University of Copenhagen
Copenhagen, Denmark

Rasmus Pagh
IT University of Copenhagen
Copenhagen, Denmark

ISSN 0302-9743   ISSN 1611-3349 (electronic)
Lecture Notes in Computer Science
ISBN 978-3-030-60935-1   ISBN 978-3-030-60936-8 (eBook)
https://doi.org/10.1007/978-3-030-60936-8

LNCS Sublibrary: SL3 – Information Systems and Applications, incl. Internet/Web, and HCI

This Springer imprint is published by the registered company Springer Nature Switzerland AG
The registered company address is: Gewerbestrasse 11, 6330 Cham, Switzerland

Preface

This volume contains the papers presented at the 13th International Conference on Similarity Search and Applications (SISAP 2020), held during September 30 – October 2, 2020. The conference was planned to be hosted by the IT University of Copenhagen, Denmark. Due to the COVID-19 pandemic and international travel restrictions around the globe, however, SISAP 2020 had to be held as an online conference instead.

SISAP is an annual forum for researchers and application developers in the area of similarity data management. It focuses on the technological problems shared by numerous application domains, such as data mining, information retrieval, multimedia, computer vision, pattern recognition, computational biology, geography, biometrics, machine learning, and many others that make use of similarity search as a necessary supporting service.

From its roots as a regional workshop in metric indexing, SISAP has expanded to become the only international conference entirely devoted to the issues surrounding the theory, design, analysis, practice, and application of content-based and feature-based similarity search. The SISAP initiative has also created a repository (http://www.sisap. org/) serving the similarity search community, for the exchange of examples of real-world applications, source code for similarity indexes, and experimental testbeds and benchmark data sets. In addition, SISAP 2020 featured the 2020 edition of the SISAP Doctoral Symposium, for which a technical program was assembled, to give PhD students an opportunity to present their research ideas in an international research venue. The Doctoral Symposium indeed provided a forum that facilitated interactions among PhD students and stimulates feedback from more experienced researchers.

The call for papers welcomed full research papers, short research papers, as well as position and demonstration papers, with all manuscripts presenting previously unpublished research contributions.

We received 50 submissions from authors based in 22 different countries. The Program Committee (PC) was composed of 63 members from 26 countries. Each submission received at least three reviews, and the papers and reviews were thoroughly discussed by the chairs and PC members. Based on the reviews and discussions, the PC chairs accepted 19 full papers and 12 short papers (including 2 demonstration papers and 1 position paper), resulting in an acceptance rate of 38% for the full papers and 62% cumulative for full and short papers. After a separate review by the Doctoral Symposium Program Committee members, two Doctoral Symposium papers, giving a clear sample of emerging topics in similarity search and applications, were accepted for presentation and included in the program and proceedings.

The proceedings of SISAP are published by Springer as a volume in the *Lecture Notes in Computer Science* (LNCS) series. For SISAP 2020, as in previous years, extended versions of selected excellent papers were invited for publication in a special issue of the journal *Information Systems*. The conference also conferred a Best Paper

Award, a Best Student Paper Award, and a Best Doctoral Symposium Paper Award, as judged by the PC co-chairs and the Steering Committee.

Besides the presentations of the accepted papers, the conference program featured three keynote talks from outstanding scientists from industry and academia: Prof. Marcel Worring from University of Amsterdam, The Netherlands, Divesh Srivastava from AT&T Labs-Research, USA, and Ilya Razenshteyn from Microsoft Research, USA.

We would like to thank all the authors who submitted papers to SISAP 2020. We would also like to thank all members of the PC and the external reviewers for their effort and contribution to the conference. We want to extend our gratitude to the members of the Organizing Committee for the enormous amount of work they have done, and our sponsors and supporters for their generosity. Finally, we thank all the participants in the online event, who make up the thriving SISAP community.

September 2020

Shin'ichi Satoh
Lucia Vadicamo
Arthur Zimek
Fabio Carrara
Ilaria Bartolini
Martin Aumüller
Björn Þór Jónsson
Rasmus Pagh

Organization

General Chairs

Martin Aumüller   IT University of Copenhagen, Denmark
Björn Þór Jónsson   IT University of Copenhagen, Denmark
Rasmus Pagh   IT University of Copenhagen, Denmark

Program Committee Chairs

Shin'ichi Satoh   National Institute of Informatics, Japan
Lucia Vadicamo   ISTI-CNR, Italy
Arthur Zimek   University of Southern Denmark, Denmark

Doctoral Symposium Program Committee Chair

Ilaria Bartolini   University of Bologna, Italy

Publication Chair

Fabio Carrara   ISTI-CNR, Italy

Steering Committee

Laurent Amsaleg   CNRS-IRISA, France
Edgar Chávez   CICESE, Mexico
Michael E. Houle   National Institute of Informatics, Japan
Pavel Zezula   Masaryk University, Czech Republic

Program Committee

Giuseppe Amato   ISTI-CNR, Italy
Laurent Amsaleg   CNRS-IRISA, France
Fabrizio Angiulli   University of Calabria, Italy
James Bailey   The University of Melbourne, Australia
Christian Beecks   University of Münster, Germany
Virendra Bhavsar   University of New Brunswick, Canada
Panagiotis Bouros   Johannes Gutenberg University Mainz, Germany
Benjamin Bustos   University of Chile, Chile
Selçuk Candan   Arizona State University, USA
Aniket Chakrabarti   Microsoft AI & Research, India
Edgar Chávez   CICESE, Mexico
Richard Chbeir   University Pau and Pays de l'Adour, France

Richard Connor	University of St Andrews, UK
Petros Daras	Information Technologies Institute, Greece
Alan Dearle	University of St Andrews, UK
Vlastislav Dohnal	Masaryk University, Czech Republic
Vladimir Estivill-Castro	Griffith University, Australia
Andrea Esuli	ISTI-CNR, Italy
Rolf Fagerberg	University of Southern Denmark, Denmark
Fabrizio Falchi	ISTI-CNR, Italy
Claudio Gennaro	ISTI-CNR, Italy
Magnus Lie Hetland	Norwegian University of Science and Technology, Norway
Thi Thao Nguyen Ho	Aalborg University, Denmark
Michael E. Houle	National Institute of Informatics, Japan
Ichiro Ide	Nagoya University, Japan
Kyoung-Sook Kim	National Institute of Advanced Industrial Science and Technology, Japan
Peer Kröger	Ludwig Maximilians University of Munich, Germany
Jakub Lokoč	Charles University, Czech Republic
Rui Mao	Shenzhen University, China
Stephane Marchand-Maillet	University of Geneva, Switzerland
Yusuke Matsui	The University of Tokyo, Japan
Luisa Micó	University of Alicante, Spain
Henning Müller	HES-SO, Switzerland
Chong-Wah Ngo	City University of Hong Kong, Hong Kong
Vincent Oria	New Jersey Institute of Technology, USA
Deepak P.	Queen's University Belfast, UK
Rodrigo Paredes	University of Talca, Chile
Marco Patella	University of Bologna, Italy
Oscar Pedreira	University of A Coruña, Spain
Raffaele Perego	ISTI-CNR, Italy
Miloš Radovanović	University of Novi Sad, Serbia
Nora Reyes	Universidad Nacional de San Luis, Argentina
Marcela Xavier Ribeiro	Federal University of São Carlos, Brazil
Kunihiko Sadakane	The University of Tokyo, Japan
Maria Luisa Sapino	Università di Torino, Italy
Erich Schubert	Technical University of Dortmund, Germany
Matthias Schubert	Ludwig Maximilians University of Munich, Germany
Thomas Seidl	Ludwig Maximilians University of Munich, Germany
Tetsuo Shibuya	The University of Tokyo, Japan
Tomas Skopal	Charles University, Czech Republic
Yasuo Tabei	RIKEN Center for Advanced Intelligence Project, Japan
Joe Tekli	Lebanese American University, Lebanon
Nenad Tomasev	DeepMind, UK
Agma J. M. Traina	University of São Paulo, Brazil
Caetano Traina	University of São Paulo, Brazil

Goce Trajcevski	Iowa State University, USA
Takashi Washio	Osaka University, Japan
Marcel Worring	University of Amsterdam, The Netherlands
Kaoru Yoshida	Sony Computer Science Laboratories, Inc., Japan
Pavel Zezula	Masaryk University, Czech Republic
Kaiping Zheng	National University of Singapore, Singapore
Zhi-Hua Zhou	Nanjing University, China
Andreas Züfle	George Mason University, USA

Doctoral Symposium Program Committee

Selçuk Candan	Arizona State University, USA
Pavel Zezula	Masaryk University, Czech Republic

Special Session Organizers

Giuseppe Amato	ISTI-CNR, Italy
Laurent Amsaleg	CNRS-IRISA, France
Fabio Carrara	ISTI-CNR, Italy
Fabrizio Falchi	ISTI-CNR, Italy
Claudio Gennaro	ISTI-CNR, Italy
Michael E. Houle	National Institute of Informatics, Japan
Sanjiv Kumar	Google Research, USA
Rasmus Pagh	IT University of Copenhagen, Denmark
Anshumali Shrivastava	Rice University, USA

Additional Reviewers

Anna Beer
Fabian Berns
Felix Borutta
Fabio Carrara
Mirela Teixeira Cazzolato
Fabio Fassetti
Luca Ferragina
Tahrima Hashem
Daniyal Kazempour

Oscar Cuadros Linares
Gabriele Lagani
Elio Mansour
Vladimir Mic
Alejandro Moreo Fernández
Ladislav Peska
Phil Sun
Erik Thordsen
Zhaozhuo Xu

Sponsors

IT University of Copenhagen

Google

Springer

Information Systems, Elsevier

Abstracts of Keynotes

Interactive Exploration using Hypergraphs

Marcel Worring

University of Amsterdam, The Netherlands

Abstract. Interactive exploration of a multimedia collection, ranging from search to browsing, requires various tasks to be supported by the system. Categorization, in which each item receives a membership score, provides a unifying framework for many of these tasks that can now, with specialized efficient high-dimensional indexing, interactively be performed even for very large collections. It also provides a proper basis for the notoriously difficult task of evaluating interactive exploration. Categorization is primarily based on the learned features of the items in the collection, possibly implicitly supported by metric learning. It does not explicitly capture the similarity or knowledge-based relations among items in the collection. Hypergraphs generalize graphs by having edges which can connect any number of nodes instead of just two. In doing so they are effectively combining categories and similarity-based relations in one model. Recent advances in graph-convolutional networks bring new opportunities to learning using hypergraphs, predicting a hyperedge membership score that captures both similarity among the elements as well as group membership. In this talk, we highlight progress made in hypergraph learning and how it leads to new opportunities for interactive exploration of multimedia content.

Exploiting Similarity Relationships to Repair Graphs

Divesh Srivastava

AT&T Labs-Research, USA

Abstract. Graphs are a flexible way to represent data in a variety of applications, with nodes representing domain-specific entities (e.g., records in entity resolution, products categories in a taxonomy) and edges capturing a variety of relationships between these entities (e.g., a linkage relationship between records in entity resolution, a category-subcategory relationship between product categories in a taxonomy). Often, the edges in this graph are inferred based on similarity relationships between nodes and are noisy, in that some edges are missing (i.e., real-world relationships that do not have corresponding edges in the graph) and some edges are spurious (i.e., edges in the graph that do not have corresponding real-world relationships). Directly analyzing such graphs can lead to undesirable outcomes, making it important to repair noisy graphs. In this talk, we describe an approach that takes advantage of properties of real-world relationships and their estimated probabilities to ask oracle queries (an abstraction of crowdsourcing) to efficiently repair the noisy graphs. We illustrate this approach for the case of graphs that are unions of cliques (which is the case for entity resolution) and graphs that are tree-structured (which is the case for taxonomies), and present theoretical and empirical results for these cases.

Scalable Nearest Neighbor Search for Optimal Transport

Ilya Razenshteyn

Microsoft Research, USA

Abstract. The Optimal Transport (aka Wasserstein) distance is an increasingly popular similarity measure for structured data domains, such as images or text documents. This raises the necessity for fast nearest neighbor search with respect to this distance, a problem that poses a substantial computational bottleneck for various tasks on massive datasets. In this talk, I will discuss fast tree-based approximation algorithms for searching nearest neighbors with respect to the Wasserstein-1 distance. I will start with describing a standard tree-based technique, known as QuadTree, which has been previously shown to obtain good results. Then I'll introduce a variant of this algorithm, called FlowTree, and show that it achieves better accuracy, both in theory and in practice. In particular, the accuracy of FlowTree is in line with previous high-accuracy methods, while its running time is much faster. The talk is based on a joint work with Arturs Backurs, Yihe Dong, Piotr Indyk, and Tal Wagner. The paper[1] and code[2] is available.

---

[1] https://arxiv.org/abs/1910.04126.
[2] https://github.com/ilyaraz/ot_estimators.

Scalable Nearest Neighbor Searches for Optimal Transport

Contents

High-Dimensional Data and Intrinsic Dimensionality

Clustering

Artificial Intelligence and Similarity

Demo and Position Papers

Doctoral Symposium

Scalable Similarity Search

Accelerating Metric Filtering by Improving Bounds on Estimated Distances

Vladimir Mic$^{(\boxtimes)}$ and Pavel Zezula

Masaryk University, Brno, Czech Republic
xmic@fi.muni.cz

Abstract. Filtering is a fundamental strategy of metric similarity indexes to minimise the number of computed distances. Given a triple of objects for which distances of two pairs are known, the lower and upper bounds on the third distance can be set as the difference and the sum of these two already known distances, due to the *triangle inequality* rule of the metric space. For efficiency reasons, the tightness of bounds is crucial, but as angles within triangles of distances can be arbitrary, the worst case with zero and straight angles must also be considered for correctness. However, in data of real-life applications, the distribution of possible angles is skewed and extremes are very unlikely to occur. In this paper, we enhance the existing definition of bounds on the unknown distance with information about possible angles within triangles. We show that two lower bounds and one upper bound on each distance exist in case of limited angles. We analyse their filtering power and confirm high improvements of efficiency by experiments on several real-life datasets.

Keywords: Metric space · Similarity search · Triangle inequality · Metric filtering · Estimating unknown distance

1 Introduction

Metric spaces are often used to formalise a similarity of complex data objects from various domains. Given a domain of objects D, a metric space is pair (D, d) where $d : D \times D \mapsto \mathbb{R}_0^+$ is a *distance function* which quantifies the dissimilarity of objects. This function must be *non-negative*, *symmetric*, and the distances among three arbitrary objects from D must satisfy the *triangle inequality*. Metric similarity searching has become popular due to its wide applicability, and many metric indexes have been proposed [10,12]. We consider the *query by example* paradigm: having a dataset $X \subseteq D$ and an arbitrary query object $q \in D$, the task is to efficiently find objects $o \in X$ that are close to q according to d.

V. Mic and P. Zezula—This research was supported by ERDF "CyberSecurity, Cyber-Crime and Critical Information Infrastructures Center of Excellence" (No. CZ.02.1.01/0.0/0.0/16_019/0000822).

S. Satoh et al. (Eds.): SISAP 2020, LNCS 12440, pp. 3–17, 2020.
https://doi.org/10.1007/978-3-030-60936-8_1

The simplest type of a similarity query is the *range query*: for a threshold $r \geq 0$ and a query object $q \in D$, its solution is $\{o \in X \mid d(q,o) \leq r\}$. Considering an arbitrary reference object p_i, called *pivot*, and assuming that distances $d(o, p_i)$ and $d(q, p_i)$ are known, the triangle inequalities define the lower bound on $d(q, o)$:

$$d(q,o) \geq |d(o, p_i) - d(q, p_i)|. \tag{1}$$

If the lower bound on distance $d(q, o)$ given by an arbitrary pivot p_i is greater than radius r, o cannot be in the answer of the range query. Accessing o and evaluation of $d(q, o)$ can thus be avoided. By analogy, if the upper bound:

$$d(q,o) \leq d(o, p_i) + d(q, p_i), \tag{2}$$

given by triangle inequalities is smaller than r for an arbitrary p_i, then o is guaranteed to be in the query answer, and the evaluation of $d(q, o)$ can be skipped.

The triangle inequality rule enables to isometrically embed objects q, o, p_i in 2D Euclidean space to form a triangle with sides $d(q, o)$, $d(o, p_i)$, $d(p_i, q)$ [4]. We further focus on angles in this triangle in the Euclidean space. Please notice that findings in this article are valid for all metric spaces thanks to this embedding.

Bounds given by Eqs. 1 and 2 are tight, i.e., the equalities hold, if there are two zero angles and one straight angle within the triangle q, o, p_i with distances $d(q, o)$, $d(p_i, o)$, and $d(p_i, q)$. However, this is an unrealistic case in most of metric spaces that describe the similarity of complex real-life data.

We analyse triangle inequalities under the assumption of limited angles in triangles. We show that the limitation of angles can increase the lower bound given by Eq. 1 even by 66%, and decrease the upper bound given by Eq. 2 by 40% in real scenarios. Moreover, the third bound exists: the lower bound on a side based on a sum of lengths of two other sides in a triangle. These improvements have a dramatic impact on the filtering power of triangle inequalities.

Section 2 contains analysis to enhance triangle inequalities by angles limitation. Section 3 illustrates what happens if the limitation of angles is wrong. Section 4 defines, when are the newly proposed bounds correct. Section 5 provides instructions to set the angles limitation for similarity search. Section 6 presents experimental results and Sect. 7 concludes the paper.

2   Triangle Inequalities with Limited Angles

We define novel lower and upper bounds on distances in this section, which are given by the triangle inequalities that assume the limited range of angles within triangles of distances. Specifically, we consider an arbitrary metric space (D, d) and three objects $q, o, p_i \in D$. The triangle with sides $a, b, c \in \mathbb{R}_0^+$ is given by pairwise distances between these objects, and we assume that the distance $c = d(q, o)$ is unknown and the distances a and b are already evaluated. We denote α, β, γ the angles in triangle $\triangle a, b, c$ that are opposite to respective sides. Since we always focus on just one isolated triangle $\triangle a, b, c$, we can assume[1]

---

[1] The assumption is used just in the last sections of the article, starting from Sect. 4.

Table 1. Notation used throughout this paper

(D, d)	metric space: the domain of objects and the distance function
$X \subseteq D$	the searched dataset
$q \in D, r \in \mathbb{R}_0^+$	the query object and the radius of the query
a, b, c	pairwise distances between arbitrary given objects from D, c is the only unknown distance
α, β, γ	angles in the triangle of distances that are opposite to sides a, b, c
$[\Omega_{min}, \Omega_{max}]$	range of the angles α, β, γ in the metric space, usually limited more than to $[0°, 180°]$ in practice
$f_{sum}(\alpha, \beta), f_{diff}(\alpha, \beta)$	functions of angles used in equations expressing c using the sum and difference of a and b, respectively
$C_{LB\_sum}(\Omega_{min}, \Omega_{max})$, $C_{UB\_sum}(\Omega_{min}, \Omega_{max})$	minimum and maximum possible values of $f_{sum}(\alpha, \beta)$ for given values Ω_{min} and Ω_{max}
$C_{LB\_diff}(\Omega_{min}, \Omega_{max})$, $C_{UB\_diff}(\Omega_{min}, \Omega_{max})$	minimum and maximum possible values of $f_{diff}(\alpha, \beta)$ for given values Ω_{min} and Ω_{max}
$LB_{sum}(a, b, \Omega_{min}, \Omega_{max})$	the lower bound on c based on a sum of a and b, (Eq. 5)
$LB_{diff}(a, b, \Omega_{min}, \Omega_{max})$	the lower bound on c based on a difference of a and b, (Eq. 6)
$UB_{sum}(a, b, \Omega_{min}, \Omega_{max})$	the upper bound on c based on a sum of a and b, (Eq. 8)

$a \leq b$ without a loss of generality for the application of the similarity search[2]. The whole notation is summarised in Table 1.

Two following lemmas form the core of the paper as they allow to define the bounds on c given by triangle inequalities that consider a limited range of angles α, β, γ.

Lemma 1. *For an arbitrary triangle with sides a, b, c and corresponding angles α, β, γ holds:*

$$c = (a + b) \cdot \frac{1 - \cos \gamma}{\cos \alpha + \cos \beta}$$

Proof. All cosines in the fraction can be substituted using the cosine rule to get:

$$(a+b) \cdot \frac{1 - \cos \gamma}{\cos \alpha + \cos \beta} = \frac{(a + b)(1 - \frac{a^2+b^2-c^2}{2ab})}{\frac{b^2+c^2-a^2}{2bc} + \frac{a^2+c^2-b^2}{2ac}} = c\frac{-a^3 + a^2b + ab^2 - b^3 + ac^2 + bc^2}{-a^3 + a^2b + ab^2 - b^3 + ac^2 + bc^2} = c$$

□

---

[2] The swap of distances a and b is achieved by swapping the notation of objects q and o. While this swaps lengths a and b, it preserves the distance $c = d(q, o)$ as distances $d(q, o)$ and $d(o, q)$ are symmetric.

Lemma 2. *For an arbitrary triangle with sides* a, b, c *and corresponding angles* α, β, γ *holds:*

$$c = |a - b| \cdot \frac{1 + \cos \gamma}{|\cos \alpha - \cos \beta|}$$

Proof. The numerator of the fraction is non-negative, and all cosines in the fraction can be substituted using the cosine rule to get:

$$\frac{|a - b| \cdot |1 + \cos \gamma|}{|\cos \alpha - \cos \beta|} = \frac{|(a - b)(1 + \frac{a^2 + b^2 - c^2}{2ab})|}{|\frac{b^2 + c^2 - a^2}{2bc} - \frac{a^2 + c^2 - b^2}{2ac}|} = c \cdot \frac{|a^3 + a^2 b - ab^2 - b^3 - ac^2 + bc^2|}{|a^3 + a^2 b - ab^2 - b^3 - ac^2 + bc^2|} = c$$

$\square$

Lemma 1 expresses c using the sum of a and b and the function of α, β, γ. Notice thus the similarity with Eq. 2. Lemma 2 expresses c using the difference of a and b, similarly as Eq. 1, and another function of α, β and γ.

Real-life metric spaces contain triangles $\triangle a, b, c$ with angles α, β, γ from a more narrow range than $[0°, 180°]$. Let us thus assume bounds Ω_{min}, Ω_{max} on the angles such that $\forall \alpha, \beta, \gamma : \Omega_{min} \leq \alpha, \beta, \gamma \leq \Omega_{max}$. Please notice that bounds Ω_{min} and Ω_{max} are meaningful if and only if $0° \leq \Omega_{min} \leq 60° \leq \Omega_{max} \leq 180°$, since $\alpha + \beta + \gamma = 180°$. The key feature of Lemmas 1 and 2 is that Ω_{min} and Ω_{max} also limit the values of the fractions used in these lemmas. We further denote and alter these fractions as:

$$f_{sum}(\alpha, \beta) = \frac{1 - \cos \gamma}{\cos \alpha + \cos \beta} = \frac{1 - \cos(180° - \alpha - \beta)}{\cos \alpha + \cos \beta} = \frac{1 + \cos(\alpha + \beta)}{\cos \alpha + \cos \beta}, \quad (3)$$

$$f_{diff}(\alpha, \beta) = \frac{1 + \cos \gamma}{|\cos \alpha - \cos \beta|} = \frac{1 + \cos(180° - \alpha - \beta)}{|\cos \alpha - \cos \beta|} = \frac{1 - \cos(\alpha + \beta)}{|\cos \alpha - \cos \beta|} \quad (4)$$

Intuitively, $f_{sum}(\alpha, \beta)$ is a coefficient exploited to express c using the sum of a and b, and $f_{diff}(\alpha, \beta)$ is utilised to express c using the difference of a and b.

We denote $C_{LB\_sum}(\Omega_{min}, \Omega_{max})$ and $C_{LB\_diff}(\Omega_{min}, \Omega_{max})$ the minimum possible values of $f_{sum}(\alpha, \beta)$ and $f_{diff}(\alpha, \beta)$ that are defined for a range of angles $[\Omega_{min}, \Omega_{max}]$. As the notation suggests, these minimum values define two lower bounds on c, since:

$$c = (a + b) \cdot f_{sum}(\alpha, \beta) \geq (a + b) \cdot C_{LB\_sum}(\Omega_{min}, \Omega_{max}),$$

$$c = |a - b| \cdot f_{diff}(\alpha, \beta) \geq |a - b| \cdot C_{LB\_diff}(\Omega_{min}, \Omega_{max})$$

We denote these lower bounds as:

$$LB_{sum}(a, b, \Omega_{min}, \Omega_{max}) = (a + b) \cdot C_{LB\_sum}(\Omega_{min}, \Omega_{max}), \quad (5)$$

Algorithm 1. Algorithm to evaluate value $\mathcal{C}_{LB\_sum}(\Omega_{min}, \Omega_{max})$

Input: $\Omega_{min}, \Omega_{max}$ ▷ bounds on angles α, β, γ; $0° \leq \Omega_{min} \leq 60° \leq \Omega_{max} \leq 180°$
Output: $\mathcal{C}_{LB\_sum}(\Omega_{min}, \Omega_{max})$ ▷ coef. for the lower bound on $\mathfrak{c}$ based on sum of $\mathfrak{a}$, $\mathfrak{b}$
if $180° - 2 \cdot \Omega_{max} \geq \Omega_{min}$ **then**   ▷ Can be two angles in a triangle equal to Ω_{max}?
$\quad \alpha \leftarrow \Omega_{max}$
$\quad \beta \leftarrow \Omega_{max}$
$\quad \gamma \leftarrow 180° - \alpha - \beta$
else
$\quad \alpha \leftarrow \min(\Omega_{max}, (180° - \Omega_{min})/2)$
$\quad \beta \leftarrow \alpha$
$\quad \gamma \leftarrow \max(\Omega_{min}, 180° - 2 \cdot \alpha)$
Substitute values α, β, γ in Eq. 3 to get value $\mathcal{C}_{LB\_sum}(\Omega_{min}, \Omega_{max})$

Algorithm 2. Algorithm to evaluate value $\mathcal{C}_{LB\_diff}(\Omega_{min}, \Omega_{max})$

Input: $\Omega_{min}, \Omega_{max}$ ▷ bounds on angles α, β, γ; $0° \leq \Omega_{min} \leq 60° \leq \Omega_{max} \leq 180°$
Output: $\mathcal{C}_{LB\_diff}(\Omega_{min}, \Omega_{max})$ ▷ coef. for the lower bound on $\mathfrak{c}$ based on diff. of $\mathfrak{a}$, $\mathfrak{b}$
if $180° - \Omega_{max} - \Omega_{min} \leq \Omega_{max}$ **then** ▷ Can be two angles in a tr. Ω_{max} and Ω_{min}?
$\quad \alpha \leftarrow \Omega_{min}$
$\quad \beta \leftarrow \min(\Omega_{max}, 180° - 2 \cdot \Omega_{min})$
$\quad \gamma \leftarrow 180° - \alpha - \beta$
else
$\quad \alpha \leftarrow \max(\Omega_{min}, 180° - 2 \cdot \Omega_{max})$
$\quad \beta \leftarrow 180° - \alpha - \Omega_{max}$
$\quad \gamma \leftarrow \Omega_{max}$
Substitute values α, β, γ in Eq. 4 to get value $\mathcal{C}_{LB\_diff}(\Omega_{min}, \Omega_{max})$

$$LB_{diff}(\mathfrak{a}, \mathfrak{b}, \Omega_{min}, \Omega_{max}) = |\mathfrak{a} - \mathfrak{b}| \cdot \mathcal{C}_{LB\_diff}(\Omega_{min}, \Omega_{max}) \tag{6}$$

Similarly, maximum possible values of $f_{sum}(\alpha, \beta)$ and $f_{diff}(\alpha, \beta)$ for a range $[\Omega_{min}, \Omega_{max}]$, denoted as $\mathcal{C}_{UB\_sum}(\Omega_{min}, \Omega_{max})$ and $\mathcal{C}_{UB\_diff}(\Omega_{min}, \Omega_{max})$, define two upper bounds on $\mathfrak{c}$, since:

$$\mathfrak{c} = (\mathfrak{a} + \mathfrak{b}) \cdot f_{sum}(\alpha, \beta) \leq (\mathfrak{a} + \mathfrak{b}) \cdot \mathcal{C}_{UB\_sum}(\Omega_{min}, \Omega_{max})$$

$$\mathfrak{c} = |\mathfrak{a} - \mathfrak{b}| \cdot f_{diff}(\alpha, \beta) \leq |\mathfrak{a} - \mathfrak{b}| \cdot \mathcal{C}_{UB\_diff}(\Omega_{min}, \Omega_{max}) \tag{7}$$

and we denote just the first one as:

$$UB_{sum}(\mathfrak{a}, \mathfrak{b}, \Omega_{min}, \Omega_{max}) = (\mathfrak{a} + \mathfrak{b}) \cdot \mathcal{C}_{UB\_sum}(\Omega_{min}, \Omega_{max}) \tag{8}$$

A derivation of value $\mathcal{C}_{UB\_diff}(\Omega_{min}, \Omega_{max})$ for given $[\Omega_{min}, \Omega_{max}]$ is simple as it is infinity for all meaningful ranges $[\Omega_{min}, \Omega_{max}]$. This is given by the denominator in Eq. 4, which is zero for $\alpha = \beta$. Equation 7 thus defines a trivial upper bound on $\mathfrak{c}$: infinity, for all meaningful ranges $[\Omega_{min}, \Omega_{max}]$.

A derivation of concrete values of $\mathcal{C}_{LB\_diff}(\Omega_{min}, \Omega_{max})$, $\mathcal{C}_{LB\_sum}(\Omega_{min}, \Omega_{max})$, and $\mathcal{C}_{UB\_sum}(\Omega_{min}, \Omega_{max})$ is slightly complicated as angles α, β, γ are limited not only by Ω_{min} and Ω_{max}, but also by equation $\alpha + \beta + \gamma = 180°$. For this reason,

---

Algorithm 3. Algorithm to evaluate value $\mathcal{C}_{UB\_sum}(\Omega_{min}, \Omega_{max})$

Input: $\Omega_{min}, \Omega_{max}$ ▷ bounds on angles α, β, γ; $0° \leq \Omega_{min} \leq 60° \leq \Omega_{max} \leq 180°$
Output: $\mathcal{C}_{UB\_sum}(\Omega_{min}, \Omega_{max})$ ▷ coef. for the upper bound on $\mathfrak{c}$ based on sum of $\mathfrak{a}, \mathfrak{b}$
if $180° - 2 \cdot \Omega_{min} \leq \Omega_{max}$ **then**   ▷ Can be two angles in a triangle equal to Ω_{min}?
    $\alpha \leftarrow \Omega_{min}$
    $\beta \leftarrow \Omega_{min}$
    $\gamma \leftarrow 180° - 2 \cdot \Omega_{min}$
else
    $\alpha \leftarrow \max(\Omega_{min}, 180° - 2 \cdot \Omega_{max})$
    $\beta \leftarrow 180° - \alpha - \Omega_{max}$
    $\gamma \leftarrow \Omega_{max}$
Substitute values α, β, γ in Eq. 3 to get value $\mathcal{C}_{UB\_sum}(\Omega_{min}, \Omega_{max})$

Table 2. Examples of triangle inequalities for given ranges of angles $\Omega_{min}, \Omega_{max}$

$[\Omega_{min}, \Omega_{max}]$	$UB_{sum}(\mathfrak{a}, \mathfrak{b}, \Omega_{min}, \Omega_{max})$	$LB_{diff}(\mathfrak{a}, \mathfrak{b}, \Omega_{min}, \Omega_{max})$	$LB_{sum}(\mathfrak{a}, \mathfrak{b}, \Omega_{min}, \Omega_{max})$
$[0°, 180°]$	$\mathfrak{c} \leq (\mathfrak{a}+\mathfrak{b}) \cdot 1$	$\mathfrak{c} \geq \lvert\mathfrak{a}-\mathfrak{b}\rvert \cdot 1$	$\mathfrak{c} \geq (\mathfrak{a}+\mathfrak{b}) \cdot 0$
$[60°, 60°]$	$\mathfrak{c} \leq (\mathfrak{a}+\mathfrak{b}) \cdot 0.5$	*undefined*	$\mathfrak{c} \geq (\mathfrak{a}+\mathfrak{b}) \cdot 0.5$
$[20°, 100°]$	$\mathfrak{c} \leq (\mathfrak{a}+\mathfrak{b}) \cdot 0.815$	$\mathfrak{c} \geq \lvert\mathfrak{a}-\mathfrak{b}\rvert \cdot 1.347$	$\mathfrak{c} \geq (\mathfrak{a}+\mathfrak{b}) \cdot 0.174$
$[20°, 80°]$	$\mathfrak{c} \leq (\mathfrak{a}+\mathfrak{b}) \cdot 0.742$	$\mathfrak{c} \geq \lvert\mathfrak{a}-\mathfrak{b}\rvert \cdot 1.532$	$\mathfrak{c} \geq (\mathfrak{a}+\mathfrak{b}) \cdot 0.174$
$[25°, 120°]$	$\mathfrak{c} \leq (\mathfrak{a}+\mathfrak{b}) \cdot 0.869$	$\mathfrak{c} \geq \lvert\mathfrak{a}-\mathfrak{b}\rvert \cdot 1.294$	$\mathfrak{c} \geq (\mathfrak{a}+\mathfrak{b}) \cdot 0.216$
$[25°, 90°]$	$\mathfrak{c} \leq (\mathfrak{a}+\mathfrak{b}) \cdot 0.752$	$\mathfrak{c} \geq \lvert\mathfrak{a}-\mathfrak{b}\rvert \cdot 1.570$	$\mathfrak{c} \geq (\mathfrak{a}+\mathfrak{b}) \cdot 0.216$
$[30°, 100°]$	$\mathfrak{c} \leq (\mathfrak{a}+\mathfrak{b}) \cdot 0.778$	$\mathfrak{c} \geq \lvert\mathfrak{a}-\mathfrak{b}\rvert \cdot 1.580$	$\mathfrak{c} \geq (\mathfrak{a}+\mathfrak{b}) \cdot 0.259$
$[30°, 80°]$	$\mathfrak{c} \leq (\mathfrak{a}+\mathfrak{b}) \cdot 0.684$	$\mathfrak{c} \geq \lvert\mathfrak{a}-\mathfrak{b}\rvert \cdot 1.938$	$\mathfrak{c} \geq (\mathfrak{a}+\mathfrak{b}) \cdot 0.259$
$[0°, 90°]$	$\mathfrak{c} \leq (\mathfrak{a}+\mathfrak{b}) \cdot 1$	$\mathfrak{c} \geq \lvert\mathfrak{a}-\mathfrak{b}\rvert \cdot 1$	$\mathfrak{c} \geq (\mathfrak{a}+\mathfrak{b}) \cdot 0$

we immediately formulate Algorithms 1–3 that evaluate $\mathcal{C}_{LB\_diff}(\Omega_{min}, \Omega_{max})$, $\mathcal{C}_{LB\_sum}(\Omega_{min}, \Omega_{max})$, and $\mathcal{C}_{UB\_sum}(\Omega_{min}, \Omega_{max})$ for given Ω_{min} and Ω_{max}.

Table 2 gives examples of newly derived lower and upper bounds on $\mathfrak{c}$ for selected ranges $[\Omega_{min}, \Omega_{max}]$. We choose these ranges to illustrate several features:

- We limit the angles by trivial values $[\Omega_{min}, \Omega_{max}] = [0°, 180°]$ in the first line, and we get the pure triangle inequalities.
- The second line represents another extreme case: if all angles are $60°$, i.e. the triangle $\triangle \mathfrak{a}, \mathfrak{b}, \mathfrak{c}$ is equilateral, bound $LB_{diff}(\mathfrak{a}, \mathfrak{b}, \Omega_{min}, \Omega_{max})$ is not defined[3], and bounds $LB_{sum}(\mathfrak{a}, \mathfrak{b}, \Omega_{min}, \Omega_{max})$ and $UB_{sum}(\mathfrak{a}, \mathfrak{b}, \Omega_{min}, \Omega_{max})$ are tight. Together, they give the precise value $\mathfrak{c} = 0.5 \cdot (\mathfrak{a} + \mathfrak{b})$.
- The lower bound $LB_{sum}(\mathfrak{a}, \mathfrak{b}, \Omega_{min}, \Omega_{max})$ is zero, and thus ineffective in case of trivially bounded angles $[\Omega_{min}, \Omega_{max}] = [0°, 180°]$. If the angles are more limited, this bound can bring a new effective limitation on $\mathfrak{c}$.
- The last row of the table illustrates that the bounds are not improved beyond pure triangle inequalities when preserving $\Omega_{min} = 0°$ and decreasing Ω_{max} to $90°$.

---

[3] In this case, $LB_{diff}(\mathfrak{a}, \mathfrak{b}, \Omega_{min}, \Omega_{max}) = 0 \cdot \infty$, which is an indefinite expression.

– The table confirms a contribution of angles limitation. For instance, if all angles are guaranteed to be within range $[30°, 80°]$, $UB_{sum}(\mathfrak{a}, \mathfrak{b}, \Omega_{min}, \Omega_{max})$ is decreased by 31.6% to $0.684 \cdot (\mathfrak{a} + \mathfrak{b})$ and $LB_{diff}(\mathfrak{a}, \mathfrak{b}, \Omega_{min}, \Omega_{max})$ is almost doubled to $1.938 \cdot |\mathfrak{a} - \mathfrak{b}|$, in comparison with unlimited angles. Moreover, the lower bound $LB_{sum}(\mathfrak{a}, \mathfrak{b}, \Omega_{min}, \Omega_{max}) = 0.259 \cdot (\mathfrak{a} + \mathfrak{b})$ is established.

Table 3. Examples of angles α, β, γ that do not meet the limitation $[\Omega_{min}, \Omega_{max}] = [30°, 80°]$ and the consequences for the newly proposed bounds on $\mathfrak{c}$ that assume this angles limitation. The wrong assumption may, but does not have to lead to wrong bounds on distances. Wrong coefficients and angles are in red.

1	2	3	Col. 4	Column 5	Col. 6	Column 7	Col. 8	Column 9
α	β	γ	$f_{sum}(\alpha, \beta)$	$C_{UB\_sum}(30°, 80°)$	$f_{diff}(\alpha, \beta)$	$C_{LB\_diff}(30°, 80°)$	$f_{sum}(\alpha, \beta)$	$C_{LB\_sum}(30°, 80°)$
28 °	75°	77°	0.679	0.684	1.963	1.938	0.679	0.259
75°	80°	25°	0.217	0.684	22.382	1.938	0.217	0.259
25°	78°	77°	0.696	0.684	1.754	1.938	0.696	0.259

Please notice that if the maximum permitted angle is e.g. 80°, the sum of two arbitrary angles in a triangle $\triangle\mathfrak{a}, \mathfrak{b}, \mathfrak{c}$ is at most 160°, and thus all angles within triangles are at least 20°. A setting of Ω_{min} smaller than 20° for $\Omega_{max} = 80°$ thus does not play a role as Ω_{min} is effectively at least 20° in this case. Similarly, if e.g. $\Omega_{min} = 30°$, then Ω_{max} is effectively at most 120° as the sum of two smallest angles in a triangle is at least 60°. These features are taken into account by Algorithms 1–3.

3   Impact of Wrong Angles Limitation $[\Omega_{min}, \Omega_{max}]$

Real-life metric space similarity models usually do not guarantee bounds on angles Ω_{min} and Ω_{max}. In these cases, the angles limitation can be set experimentally to be valid for a vast majority of all triangles within a given metric space. Consequences of imprecise bounds $[\Omega_{min}, \Omega_{max}]$ can be of various kinds, as we illustrate by Table 3. Here, we assume limitation $[\Omega_{min}, \Omega_{max}] = [30°, 80°]$, and show three examples of angles α, β, γ within triangles such that they violate the angles limitation.

Examples of the Upper Bound $UB_{sum}(\mathfrak{a}, \mathfrak{b}, \Omega_{min}, \Omega_{max})$: The fourth column of Table 3 contains values $f_{sum}(\alpha, \beta)$ defined by Eq. 3 that are evaluated for actual angles α, β, γ given in columns 1–3. In case of the first and second row of the table, this value is smaller than the value $C_{UB\_sum}(30°, 80°)$ which is provided by the fifth column of the table. Therefore, the upper bound $UB_{sum}(\mathfrak{a}, \mathfrak{b}, \Omega_{min}, \Omega_{max})$ is correct in case of these two rows despite wrong bounds $[\Omega_{min}, \Omega_{max}]$. Specifically, in case of the first row holds:

$$\mathfrak{c} = 0.679 \cdot (\mathfrak{a} + \mathfrak{b}) \quad \leq \quad 0.684 \cdot (\mathfrak{a} + \mathfrak{b}) = UB_{sum}(\mathfrak{a}, \mathfrak{b}, \Omega_{min}, \Omega_{max})$$

and the case of the second row is analogous. In case of the last row, wrong bounds $[\Omega_{min}, \Omega_{max}]$ cause a wrong upper bound $UB_{sum}(\mathfrak{a}, \mathfrak{b}, \Omega_{min}, \Omega_{max})$, since:

$$\mathfrak{c} = 0.696 \cdot (\mathfrak{a} + \mathfrak{b}) \quad > \quad 0.684 \cdot (\mathfrak{a} + \mathfrak{b}) = UB_{sum}(\mathfrak{a}, \mathfrak{b}, \Omega_{min}, \Omega_{max})$$

Wrong limitation of angles $[\Omega_{min}, \Omega_{max}]$ in metric spaces thus can, and does not have to lead to a wrong upper bound $UB_{sum}(\mathfrak{a}, \mathfrak{b}, \Omega_{min}, \Omega_{max})$.

Examples of the Lower Bound $LB_{diff}(\mathfrak{a}, \mathfrak{b}, \Omega_{min}, \Omega_{max})$: Column 6 of Table 3 contains values $f_{diff}(\alpha, \beta)$ defined by Eq. 4 that are evaluated for actual angles α, β, γ given in columns 1–3. In case of first two rows of the table, these values are bigger than $C_{LB\_diff}(30°, 80°)$ which is presented in the seventh column. Therefore, the lower bound $LB_{diff}(\mathfrak{a}, \mathfrak{b}, \Omega_{min}, \Omega_{max})$ is correct in case of corresponding triangles $\triangle \mathfrak{a}, \mathfrak{b}, \mathfrak{c}$ despite wrong bounds on angles $[\Omega_{min}, \Omega_{max}]$. Specifically, in case of the first row holds:

$$\mathfrak{c} = 1.963 \cdot |\mathfrak{a} - \mathfrak{b}| \quad \geq \quad 1.938 \cdot |\mathfrak{a} - \mathfrak{b}| = LB_{diff}(\mathfrak{a}, \mathfrak{b}, \Omega_{min}, \Omega_{max})$$

and the case of the second row is analogous. In case of the last row, wrong bounds $[\Omega_{min}, \Omega_{max}]$ imply a wrong lower bound $LB_{diff}(\mathfrak{a}, \mathfrak{b}, \Omega_{min}, \Omega_{max})$, since:

$$\mathfrak{c} = 1.754 \cdot |\mathfrak{a} - \mathfrak{b}| \quad < \quad 1.938 \cdot |\mathfrak{a} - \mathfrak{b}| = LB_{diff}(\mathfrak{a}, \mathfrak{b}, \Omega_{min}, \Omega_{max})$$

Wrong limitation of angles $[\Omega_{min}, \Omega_{max}]$ in metric spaces thus can, and does not have to lead to a wrong lower bound $LB_{diff}(\mathfrak{a}, \mathfrak{b}, \Omega_{min}, \Omega_{max})$.

Examples of the Lower Bound $LB_{sum}(\mathfrak{a}, \mathfrak{b}, \Omega_{min}, \Omega_{max})$: Examples for the lower bound $LB_{sum}(\mathfrak{a}, \mathfrak{b}, \Omega_{min}, \Omega_{max})$ are provided in columns 8 and 9 of the Table 3. The same reasoning as in the case of lower bound $LB_{diff}(\mathfrak{a}, \mathfrak{b}, \Omega_{min}, \Omega_{max})$ reveals that the lower bound $LB_{sum}(\mathfrak{a}, \mathfrak{b}, \Omega_{min}, \Omega_{max})$ is correct in case of the first and third row of the Table 3, and wrong in case of the second row.

We have also proved that all new bounds on $\mathfrak{c}$: $LB_{sum}(\mathfrak{a}, \mathfrak{b}, \Omega_{min}, \Omega_{max})$, $LB_{diff}(\mathfrak{a}, \mathfrak{b}, \Omega_{min}, \Omega_{max})$ and $UB_{sum}(\mathfrak{a}, \mathfrak{b}, \Omega_{min}, \Omega_{max})$ can be correct at the same time even in case of a triangle that violates the assumption about the range of angles $[\Omega_{min}, \Omega_{max}]$ – example is given by the first row of Table 3. The key question thus is, when are bounds correct, and when are they not.

4   When Are the Bounds on Distances Correct?

Table 2 proves that there exist different values Ω_{min}, Ω_{max} that imply the same value $C_{LB\_sum}(\Omega_{min}, \Omega_{max})$. This is also true for coefficients $C_{LB\_sum}(\Omega_{min}, \Omega_{max})$ and $C_{UB\_sum}(\Omega_{min}, \Omega_{max})$. Identification of bounds Ω_{min}, Ω_{max} that imply a fixed value of each of these coefficients will enable us to formally describe triangles for which are the newly proposed bounds on $\mathfrak{c}$ correct, and for which they are not.

Let us assume a given value of $\mathcal{C}_{LB\_diff}(\Omega_{min}, \Omega_{max})$. Since this is a minimal value of $f_{diff}(\alpha, \beta)$, we can define values α and β that imply the given value of $f_{diff}(\alpha, \beta) = \mathcal{C}_{LB\_diff}(\Omega_{min}, \Omega_{max})$ by an analysis of Eq. 4:

$$\alpha = 2 \cdot \arccos\left(\frac{\mathcal{C}_{LB\_diff}(\Omega_{min}, \Omega_{max}) \cdot \cos\beta + 1}{\sqrt{\mathcal{C}_{LB\_diff}(\Omega_{min}, \Omega_{max})^2 + 2 \cdot \mathcal{C}_{LB\_diff}(\Omega_{min}, \Omega_{max}) \cdot \cos\beta + 1}}\right) - \beta \quad (9)$$

To facilitate an understanding of this equation, we introduce plots as is the one in Fig. 1. It depicts the angles α and β on axes y and x, respectively. Angle $\gamma = 180° - \alpha - \beta$ also exists, despite it is not explicitly shown in the plot. The inequality $\alpha + \beta \leq 180°$ limits the meaningful part of the plot, as well as the assumption $\alpha \leq \beta$ used without a loss of generality for the applications in the similarity searching (see Sect. 2). These limitations are depicted by black lines in the figure, so we consider just the triangular area below these lines in the following.

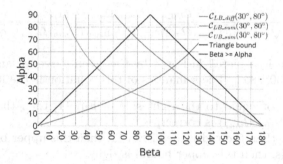

Fig. 1. Functions describing α and β that imply $f_{diff}(\alpha, \beta) = \mathcal{C}_{LB\_diff}(30°, 80°)$; $f_{sum}(\alpha, \beta) = \mathcal{C}_{LB\_sum}(30°, 80°)$; and $f_{sum}(\alpha, \beta) = \mathcal{C}_{UB\_sum}(30°, 80°)$. (Color figure online)

Function given by Eq. 9 for value $\mathcal{C}_{LB\_diff}(30°, 80°)$ is depicted by a blue curve in the Fig. 1. It is easy to verify that points $[\alpha, \beta]$ below this curve imply smaller values $f_{diff}(\alpha, \beta)$ than $\mathcal{C}_{LB\_diff}(30°, 80°)$, and points above the curve imply bigger value $f_{diff}(\alpha, \beta)$ than $\mathcal{C}_{LB\_diff}(30°, 80°)$. Formally:

- if α is smaller than the right side of Eq. 9, then $f_{diff}(\alpha, \beta)$ is smaller than $\mathcal{C}_{LB\_diff}(\Omega_{min}, \Omega_{max})$,
- if α is bigger than the right side of Eq. 9, then $f_{diff}(\alpha, \beta)$ is bigger than $\mathcal{C}_{LB\_diff}(\Omega_{min}, \Omega_{max})$,
- if Eq. 9 holds, then $f_{diff}(\alpha, \beta)$ is equal to $\mathcal{C}_{LB\_diff}(\Omega_{min}, \Omega_{max})$.

Therefore, the lower bound $LB_{diff}(\mathfrak{a}, \mathfrak{b}, \Omega_{min}, \Omega_{max})$ is correct for all triangles with α bigger or equal to the right side of Eq. 9, and wrong for the others. If α equals to the right side of Eq. 9, then the lower bound is tight.

Similarly, we analyse Eq. 3 to reveal, when are the lower and upper bounds on $\mathfrak{c}$ based on a sum of $\mathfrak{a}$ and $\mathfrak{b}$ correct, tight, and wrong, respectively.

The relation between angles α and β that imply a given value of $f_{sum}(\alpha, \beta)$ $= C_{LB\_sum}(\Omega_{min}, \Omega_{max})$ is:

$$\alpha = 2 \cdot \arccos\left(\frac{C_{LB\_sum}(\Omega_{min}, \Omega_{max}) \cdot \sin\beta}{\sqrt{C_{LB\_sum}(\Omega_{min}, \Omega_{max})^2 - 2 \cdot C_{LB\_sum}(\Omega_{min}, \Omega_{max}) \cdot \cos\beta + 1}}\right) - \beta$$
(10)

This function is depicted by the orange curve for angles limitation $[\Omega_{min}, \Omega_{max}] = [30°, 80°]$ in Fig. 1, and its semantics is the following:

- If α is smaller or equal to the right side of Eq. 10, then the lower bound $LB_{sum}(\mathfrak{a}, \mathfrak{b}, \Omega_{min}, \Omega_{max})$ is correct,
- if α is bigger than the right side of Eq. 10, then this lower bound is wrong,
- if Eq. 10 holds, this lower bound is tight.

Finally, the relation between angles α and β that imply a given value of $f_{sum}(\alpha, \beta) = C_{UB\_sum}(\Omega_{min}, \Omega_{max})$ is[4]:

$$\alpha = 2 \cdot \arccos\left(\frac{C_{UB\_sum}(\Omega_{min}, \Omega_{max}) \cdot \sin\beta}{\sqrt{C_{UB\_sum}(\Omega_{min}, \Omega_{max})^2 - 2 \cdot C_{UB\_sum}(\Omega_{min}, \Omega_{max}) \cdot \cos\beta + 1}}\right) - \beta$$
(11)

and this function is depicted in Fig. 1 by the green curve for angles limitation $[\Omega_{min}, \Omega_{max}] = [30°, 80°]$. The semantics of this equation is the following:

- If α is bigger or equal to the right side of Eq. 11, then the upper bound $UB_{sum}(\mathfrak{a}, \mathfrak{b}, \Omega_{min}, \Omega_{max})$ is correct,
- if α is smaller than the right side of Eq. 11, then this upper bound is wrong,
- if Eq. 11 holds, then this upper bound is tight.

Therefore, all three bounds on $\mathfrak{c}$ are correct in case of triangles whose angles α, β are depicted between colour curves in the plot like in Fig. 1.

5   Setting Bounds $[\Omega_{min}, \Omega_{max}]$ for Similarity Search

Test Data. In the experiments that follow, we use three different high dimensional datasets, comprising DeCAF, SIFT and MPEG7 image visual descriptors.

DeCAF descriptors [5] are extracted from the *Profiset image collection*[5]. These descriptors derive from the Alexnet convolutional neural network [6], from which data from the second-last fully connected layer (FC7) is extracted as a 4,096-dimensional array of floating-point values. It has been demonstrated that Euclidean distance applied to the *post-Relu* [9] descriptors gives a good surrogate for semantic similarity over the original images [5].

---

[4] This equation is derived in the same way as Eq. 10, as $C_{UB\_sum}(\Omega_{min}, \Omega_{max})$ is given by the same function as $C_{LB\_sum}(\Omega_{min}, \Omega_{max})$.
[5] http://disa.fi.muni.cz/profiset/.

SIFT descriptors [7] serve us as another real-life example of descriptors of images. Each descriptor comprises 128 floating point values. This dataset is known as *ANN_SIFT1M* dataset[6].

MPEG7 visual descriptors [8] are provided by the CoPhIR data collection [2]. Each of five sub-descriptors is accompanied with a suitable metric function, and all five sub-descriptor spaces are combined into a single metric space (D, d) by a weighted sum of particular distances [1]. In total, this representation can be viewed as a 280-dimensional vector compared by non-Minkowski distance.

Selecting Angles $[\Omega_{min}, \Omega_{max}]$ *for a Good Space Approximation.* A simple way to adjust the lower and upper bounds on c is to sample random triangles of distances from the searched metric data and depict angles α, β in a plot like is the one in Fig. 1. The limitation of angles $[\Omega_{min}, \Omega_{max}]$ should be selected to wrap the points by curves (the green, blue and orange) as tightly as possible.

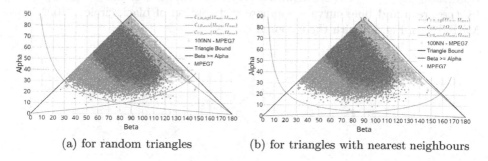

(a) for random triangles   (b) for triangles with nearest neighbours

Fig. 2. Angles limitation $[\Omega_{min}, \Omega_{max}]$ for random triangles and those with NN (Color figure online)

We are, however, interested in the application of newly proposed bounds on c to speed-up the similarity search. We thus have to pay special attention to triangles of distances $\triangle a, b, c$ where c is an extremely small distance considering the data. These extreme cases form a significantly different distribution of angles α, β, and moreover, they are not effectively sampled by random triangles.

Selecting Angles $[\Omega_{min}, \Omega_{max}]$ *for a Similarity Search.* We thus randomly select 1000 objects $q \in D$ and find their 100 nearest neighbours o_{NN} from a random sample of X of size 100,000 objects. We denote $c = d(q, o_{NN})$ and for each o_{NN} select another 100 random pivots $p_i \in X$ to form a triangle $\triangle a, b, c$ where a, b are distances $d(q, p_i)$, $d(o_{NN}, p_i)$ and $b \geq a$. Together, we have 10 million samples of angles α, β for each dataset.

We depict both angles distributions in Figs. 2a and b[7]. Specifically, purple points (MPEG7) and black circles (100NN – MPEG7) depict angles in triangles

---

[6] http://corpus-texmex.irisa.fr/.

[7] We depict just non-overlapping points due to hardware limitations.

sampled in random, and with focus on nearest neighbours, respectively. Distributions are significantly different, and plots for the DeCAF and SIFT descriptors confirm this as well, though not shown due to the limited paper length.

Figure 2a also depicts angles $[\alpha, \beta]$ that imply $f_{sum}(\alpha, \beta)$ and $f_{diff}(\alpha, \beta)$ of the same values as are given by $\mathcal{C}_{LB\_diff}(16°, 110°)$, $\mathcal{C}_{LB\_sum}(16°, 110°)$, and $\mathcal{C}_{UB\_sum}(16°, 110°)$. These curves, shown again in blue, orange and green, tightly embrace angles from randomly sampled triangles (purple points).

Figure 2b illustrates that a distribution of angles $[\alpha, \beta]$ from triangles with a near neighbour makes impossible to select bounds $[\Omega_{min}, \Omega_{max}]$ such that all three curves tightly embrace (the black) sampled points. This is caused by asymmetric semantics of angles α, β, γ in these triangles. We experimentally verified that Ω_{min} cannot be bigger than $4°$ to set properly the orange curve (i.e. coefficient $\mathcal{C}_{LB\_sum}(\Omega_{min}, \Omega_{max})$). Consequently, this Ω_{min} implies the minimum meaningful value of $\Omega_{max} = 88°$ due to equation $\alpha + \beta + \gamma = 180°$, and limitation $[\Omega_{min}, \Omega_{max}] = [4°, 88°]$ defines a very loose embrace of the sampled points by the green and blue curve – compare distribution of black circles "100NN – MPEG7" in Fig. 2b with the coloured curves.

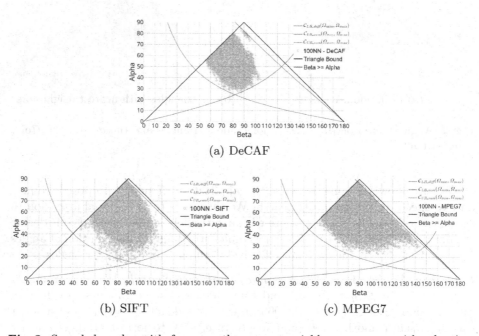

(a) DeCAF

(b) SIFT   (c) MPEG7

Fig. 3. Sampled angles with focus on the nearest neighbours, curves with adaptive limitation $[\Omega_{min}, \Omega_{max}]$ described in Table 4

Therefore, we propose to set bounds $[\Omega_{min}, \Omega_{max}]$ independently for each of the newly proposed bounds $LB_{sum}(\mathfrak{a}, \mathfrak{b}, \Omega_{min}, \Omega_{max})$, $LB_{diff}(\mathfrak{a}, \mathfrak{b}, \Omega_{min}, \Omega_{max})$, $UB_{sum}(\mathfrak{a}, \mathfrak{b}, \Omega_{min}, \Omega_{max})$ on $\mathfrak{c}$ to maximise their tightness. The formal approach to select these values of $[\Omega_{min}, \Omega_{max}]$ forms the future work. For now, we just identify $[\Omega_{min}, \Omega_{max}]$ for all curves and datasets to tightly wrap angles $[\alpha, \beta]$ sampled with a focus on nearest neighbours. These wrappings are illustrated in Fig. 3, and the values $[\Omega_{min}, \Omega_{max}]$ with corresponding bounds on c are provided in Table 4.

6   Searching with New Bounds

We experimentally verify the contribution of proposed bounds on c to the similarity search. We search for $k = 10$ and $k = 100$ closest objects (k *nearest neighbours* – *kNN*) to 1,000 randomly selected query objects $q \in D$ in 1 million datasets, and use a data filtering to speed-up the search. In particular, we select 256 pivots $p_i \in D$ and pre-compute distances $d(o, p_i), o \in X$ for all pivots. When the query object comes, we evaluate all distances $d(q, p_i)$, and then for each $o \in X$ we evaluate the biggest lower bound on $d(q, o)$ provided by an arbitrary pivot p_i – it can be either $LB_{diff}(\mathfrak{a}, \mathfrak{b}, \Omega_{min}, \Omega_{max})$ or $LB_{sum}(\mathfrak{a}, \mathfrak{b}, \Omega_{min}, \Omega_{max})$. If it is bigger than the distance $\mathfrak{r}$ of the current kth nearest object to q, the evaluation of $d(q, o)$ is skipped. Otherwise, the lowest upper bound $UB_{sum}(\mathfrak{a}, \mathfrak{b}, \Omega_{min}, \Omega_{max})$ provided by pivots is computed, and if it is smaller than $\mathfrak{r}$, the evaluation of $d(q, o)$ is skipped and o is added to the query answer. Please notice that the order of the verification of bounds matters, since they are sometimes in a contradiction.

Table 4. Selected angles limitation $[\Omega_{min}, \Omega_{max}]$, and new bounds on c

	$c_{LB\_diff}(\Omega_{min}, \Omega_{max})$	$c_{LB\_sum}(\Omega_{min}, \Omega_{max})$	$c_{UB\_sum}(\Omega_{min}, \Omega_{max})$
DeCAF dataset	$[28°, 90°]$	$[8°, 86°]$	$[30°, 80°]$
	$c \geq 1.664 \cdot |\mathfrak{b} - \mathfrak{a}|$	$c \geq 0.070 \cdot (\mathfrak{b} + \mathfrak{a})$	$c \leq 0.684 \cdot (\mathfrak{b} + \mathfrak{a})$
SIFT dataset	$[12°, 84°]$	$[3°, 88.5°]$	$[20°, 85°]$
	$c \geq 1.264 \cdot |\mathfrak{b} - \mathfrak{a}|$	$c \geq 0.026 \cdot (\mathfrak{b} + \mathfrak{a})$	$c \leq 0.762 \cdot (\mathfrak{b} + \mathfrak{a})$
MPEG7 dataset	$[8°, 86°]$	$[4°, 88°]$	$[40°, 90°]$
	$c \geq 1.162 \cdot |\mathfrak{b} - \mathfrak{a}|$	$c \geq 0.035 \cdot (\mathfrak{b} + \mathfrak{a})$	$c \leq 0.710 \cdot (\mathfrak{b} + \mathfrak{a})$

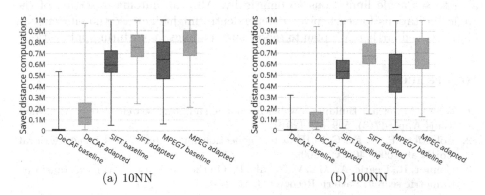

(a) 10NN   (b) 100NN

Fig. 4. Real-life experiments: increase of saved distance computations out of 1M

The proposed bounds on c do not provide any guarantees on their precision. Nevertheless, 999 out of a thousand 10NN queries on each dataset are evaluated with recall 1, i.e. all 10 true nearest neighbours are returned. Query answers may contain more than k objects due to object involvements based on upper-bounds[8]. But despite of this, the median answer on 10NN queries contains 10 objects in case of each examined dataset. Answers to 100NN queries contain all 100 true nearest neighbours in case of 955, 963 and 923 query objects in case of the DeCAF, SIFT, and MPEG7 dataset, respectively. The biggest answers, 105 on median, are returned in case of the MPEG7 dataset.

Numbers of saved distance computations are presented in Fig. 4. Box-plots describe the distribution of values over particular query objects. Dark-grey box-plots form the baseline, i.e. the metric filtering with bounds given by Eqs. 1 and 2. Light-grey box-plots present results achieved by newly proposed bounds. Median numbers of skipped distance computations increase from 0.4% to 11.8% (DeCAF), from 59.4% to 75.2% (SIFT), and from 64.5% to 80.2% (MPEG7) in case of 10NN queries. The results are coherent with 100NN queries.

7   Conclusions and Future Work

We analysed consequences of limited angles within triangles of distances in metric spaces and their impact on the bounds on distances given by triangle inequalities. We derived a new lower bound on a distance $LB_{sum}(\mathfrak{a}, \mathfrak{b}, \Omega_{min}, \Omega_{max})$ which is based on a sum of two opposite sides in a triangle. Our findings have a strong impact on the filtering power of triangle inequalities, which we confirmed by experiments with 3 real-life datasets. Moreover, the proposed enhancement of the filtering is extremely precise, as only 3 out of three thousand 10NN queries did not provide the query answer with the recall 1 in our experiments. The proposed method can be immediately incorporated into metric-based indexes to improve their efficiency, thanks to its simplicity and practically no overhead. In the future work, we would like to clarify the relation of this work to convex transforms of distance functions [3,11]. We also would like to develop algorithms able to set angle limitations automatically. Also, an automatic setting of the angle limitations for each pivot independently might increase the efficiency of filtering even further. We plan to report such findings in our future publications.

References

1. Batko, M., et al.: Building a web-scale image similarity search system. Multimed. Tools Appl. **47**(3), 599–629 (2010)
2. Bolettieri, P., et al.: CoPhIR: a test collection for content-based image retrieval. CoRR abs/0905.4627v2 (2009)
3. Connor, R., Dearle, A., Mic, V., Zezula, P.: On the application of convex transforms to metric search. Pattern Recognit. Lett. **138**, 563–570 (2020)

---

[8] This is also caused by a gradual shrinking of a search radius to evaluate kNN queries.

4. Connor, R., Vadicamo, L., Cardillo, F.A., Rabitti, F.: Supermetric search. Inf. Syst. **80**, 108–123 (2019)
5. Donahue, J.: DeCAF: a deep convolutional activation feature for generic visual recognition. In: Proceedings of International Conference on Machine Learning, ICML, Beijing, pp. 647–655 (2014)
6. Krizhevsky, A., Sutskever, I., Hinton, G.E.: ImageNet classification with deep convolutional neural networks. In: Advances in Neural Information Processing Systems 25, pp. 1097–1105. Curran Associates, Inc. (2012)
7. Lowe, D.G.: Object recognition from local scale-invariant features. In: Proceedings of the International Conference on Computer Vision, Kerkyra, Corfu, Greece, 20–25 September 1999, pp. 1150–1157 (1999)
8. MPEG7: Multimedia content description interfaces. Part 3: Visual (2002)
9. Nair, V., Hinton, G.E.: Rectified linear units improve restricted Boltzmann machines. In: Proceedings of the 27th International Conference on International Conference on Machine Learning. ICML 2010, pp. 807–814. Omnipress, USA (2010)
10. Novak, D., Kyselak, M., Zezula, P.: On locality-sensitive indexing in generic metric spaces. In: Third International Workshop on Similarity Search and Applications. SISAP 2010, 18–19 September 2010, Istanbul, Turkey, pp. 59–66. ACM (2010)
11. Skopal, T.: Unified framework for fast exact and approximate search in dissimilarity spaces. ACM Trans. Database Syst. **32**(4), 29 (2007)
12. Zezula, P., Amato, G., Dohnal, V., Batko, M.: Similarity Search - The Metric Space Approach, Advances in Database Systems, vol. 32. Springer, Heidelberg (2006)

Differentially Private Sketches for Jaccard Similarity Estimation

Martin Aumüller$^{(\boxtimes)}$ (iD), Anders Bourgeat, and Jana Schmurr

IT University of Copenhagen, Copenhagen, Denmark
{maau,anfh,jansc}@itu.dk

Abstract. This paper describes two locally-differential private algorithms for releasing user vectors such that the Jaccard similarity between these vectors can be efficiently estimated. The basic building block is the well known MinHash method. To achieve a privacy-utility trade-off, MinHash is extended in two ways using variants of Generalized Randomized Response and the Laplace Mechanism. A theoretical analysis provides bounds on the absolute error and experiments show the utility-privacy trade-off on synthetic and real-world data. A full version of this paper is available at http://arxiv.org/abs/2008.08134.

1 Introduction

Privacy of user data is becoming an ever increasing need for organizations and users alike. Multiple large-scale privacy breaches in the last years showed how critical and vulnerable most of today's infrastructure is [8]. In particular, there is dispute about the concept of a *trusted data curator* to whom users send their original data, and who uses this data to build models for different tasks such as targeted advertisement. As Kearns and Roth put it in their recent book about ethical algorithms [10], "[to] make sure that the effect of these models respect the societal norms that we want to maintain, we need to learn how to design these goals directly into our algorithms." In pursue of this goal, the present paper studies how we can implicitly incorporate privacy into a similarity search system.

The concept of differential privacy as introduced by Dwork et al. in [7] defines privacy in a precise mathematical way that often allows the design of efficient randomized algorithms. In the case of an untrusted data curator, the concept can be extended to *local differential privacy*, where users themselves run randomized algorithms to make their data private before sending it to an untrusted curator.

This paper proposes two randomized mechanisms when users have a collection of items and are interested in finding their similarity with other users under the *Jaccard similarity* in a private manner. The proposed algorithms build upon the papers [4,11,17] and a precise account of the relation will be given in the related work section at the end of this paper. In a nutshell, each user starts by applying MinHash as introduced by Broder [1] with the range compression of Li and König [12] (Sect. 3.1) to produce a sketch of their data. It is well known

© Springer Nature Switzerland AG 2020
S. Satoh et al. (Eds.): SISAP 2020, LNCS 12440, pp. 18–32, 2020.
https://doi.org/10.1007/978-3-030-60936-8_2

that these sketches can be used to efficiently estimate the original Jaccard similarity. Now, each user applies a local randomization to their sketch to satisfy the notion of differential privacy as introduced in the next section. One randomization mechanism is based on the concept of randomized response (Sect. 4), the other mechanism uses the concept of Laplacian noise (Sect. 5). We provide probabilistic bounds on the estimation error of these mechanisms as Theorems 2 and 3. A running example of our setting and the mechanisms is provided in the full version of this paper. The mechanisms will be evaluated in a real-world setting in Sect. 6. There we will see that they allow for precise similarity estimations if user vectors do not contain too few elements.

We hope that the proposed methods will help in building privacy-preserving similarity search systems with good utility and precise privacy guarantees.

2   (Local) Differential Privacy

Differential privacy conveys a precise mathematically definition of privacy. It says that a randomized algorithm is private if for two "neighboring" databases, there must be a "good enough" probability that the algorithm produces the same output. Here, a clean definition of *neighboring* is a key criterion and we will introduce our notion in the next section. While differential privacy usually works with a trusted data curator, the notion of *local differential privacy* describes the setting in which the user apply the randomized algorithm themselves. Thus, the curator never sees the original data.

Definition 1 (Sect. 12.1 [8]). *Let $\varepsilon, \delta \geq 0$. Let $\mathcal{A}$ be a randomized algorithm with output space $\mathcal{R}$. $\mathcal{A}$ satisfies (ε, δ)-local differential privacy $((\varepsilon, \delta)$-LDP$)$, if and only if for any neighboring input x and y we have: $\forall v \in \mathcal{R}: \Pr[\mathcal{A}(x) = v] \leq e^{\varepsilon} \Pr[\mathcal{A}(y) = v] + \delta$.*

Note that $\mathcal{A}$ is run by each individual user.

3   Basic Setup

Let $\mathcal{U}$ be a collection of n users and $\mathcal{I}$ be a collection of m items. Each user has a subset of the items. Formally, user $u \in \mathcal{U}$ is associated with a bit vector $x_u = (X_1, \ldots, X_m) \in \{0,1\}^m$, where $X_i = 1$ means that item i is present in the user's item set. From a practical point of view, such a representation is often obtained from a real-valued vector $(X'_1, \ldots, X'_m) \in \mathbb{R}^m$ by setting $X_i = 1$ iff $X'_i \geq t$, for some chosen threshold $t \in \mathbb{R}$.

This paper will focus on the similarity of user's item sets with regard to their *Jaccard similarity*. For two vectors $x, y \in \{0,1\}^m$, the Jaccard similarity $J(x,y) = |x \cap y|/|x \cup y|$ is the fraction of positions with a common one over the number of positions with at least a one.

We want to release the matrix $M = (x_u)_{u \in \mathcal{U}} \in \{0,1\}^{n \times m}$ in a locally differential private way. This means that each user locally produces a differential private version $\hat{x}_u$ of x_u such that if two vectors x_u and y_u do not differ by much, there is a good chance that they map to the same output. Sending all $\hat{x}_u$ to an untrusted curator, we obtain a matrix $\hat{M} = (\hat{x}_u)_{u \in \mathcal{U}}$ that can be published. The utility of this mapping $M \mapsto \hat{M}$ is the ability to recover from any two vectors $\hat{x}$ and $\hat{y}$ their original similarity $J(x, y)$. Since the mapping introduced random noise to preserve privacy, custom similarity estimation algorithms are required to solve this task.

Neighboring Notion. Throughout this paper, we will often make the assumption that each user vector has at least $\tau \geq 1$ items, i.e., at least τ bits are set. We say that two vectors x and y in $\{0,1\}^m$ are neighboring if they differ in at most α positions. In this case, $J(x, y) \geq 1 - \alpha/\tau$.

Basic Building Blocks of Differential Privacy. We review the Laplace mechanism [8, Chap. 3.3] to produce differential privacy mechanisms in our context. The ℓ_1-sensitivity $\Delta(f)$ of a function $f: \{0,1\}^m \to \mathbb{R}^K$ is defined as $\Delta(f) = \max \|f(x) - f(y)\|_1$, where the maximum is taken over all neighboring bitstrings x, y. Given $f(x) \in \mathbb{R}^K$ and a privacy budget ε, the Laplace mechanism returns the value $f(x) + (Y_1, \ldots, Y_K)$, where each Y_i is drawn independently from the Laplace distribution with shape parameter $\Delta(f)/\varepsilon$ and mean 0.

Theorem 1 (Thm 3.6 [8]). *The Laplace mechanism preserves* $(\varepsilon, 0)$*-LDP.*

Another way of preserving $(\varepsilon, 0)$-LDP is via generalized randomized response [16]. The variant used in this paper will be described in Lemma 2.

3.1 Jaccard Similarity Estimation via MinHash

MinHash. Our approach relies on the *MinHash algorithm* that was first described by Broder in [1]. Choosing a MinHash function $h: \{0,1\}^m \to [m] := \{1, \ldots, m\}$ amounts to choosing a random permutation π over $[m]$. The hash value of $x \in \{0,1\}^m$ is the position of the first 1 in x under π. MinHash has the property that for any pair $x, y \in \{0,1\}^m$, we have $\Pr[h(x) = h(y)] = J(x, y)$ where the probability is taken over the random choice of h. Repeating this construction K times results in an output $(h_1(x), \ldots, h_K(x)) \in [m]^K$. By linearity of expectation, the value $\frac{1}{K} \sum_{i=1}^{K} [h_i(x) = h_i(y)]$ is an unbiased estimator of $J(x, y)$.

b-bit MinHash. Li and König described in [12] the following twist to the standard MinHash approach. For an integer $B \geq 2$, choosing a *range-B MinHash* function amounts to choosing a MinHash function $h_{\min}: \{0,1\}^m \to [m]$ and a universal hash function [2] $h_{\mathrm{uni}}: [m] \to [B]$. The range-$B$ MinHash function is $h := h_{\mathrm{uni}} \circ h_{\min}: \{0,1\}^m \to [B]$. This mapping has the property that $\Pr[h(x) = h(y)] = (1 - J(x, y))1/B + J(x, y)$, since with probability $J(x, y)$ the MinHash value is identical—which yields a collision—and with probability $1 - J(x, y)$ the MinHash value is different but the random mapping generates a collision. [12] discussed the case $B = 2^b$ for $b \geq 1$ where the hash function gives a b-bit value. In this paper we will use their approach for general $B \geq 2$.

3.2 Generalized Randomized Response for Close Vectors

To have a chance for good utility of our mechanisms, we will use the additive δ summand available in LDP (cf. Definition 1) to collect cases where the mapping h maps two neighboring user vectors far away from each other. We then provide ε-LDP on the remaining cases. We will need two technical lemmata.

Lemma 1. *Let $x, y \in \{0,1\}^m$ such that $J(x,y) \geq 1 - \alpha/\tau$. Let $\delta > 0$. Let $h_1, \ldots, h_K$ be a collection of K random range-B MinHash functions. Let $x^* = (h_1(x), \ldots, h_K(x))$ and $y^* = (h_1(y), \ldots, h_K(y))$. With probability at least $1 - \delta$, the number of positions where x^* and y^* differ is at most $K(\alpha/\tau)\left(1 - \frac{1}{B}\right) + \sqrt{3\ln(1/\delta)\left(1 - \frac{1}{B}\right)K\alpha/\tau}$.*

Proof. For each $i \in [K]$, define the random variable $X_i = [h_i(x) \neq h_i(y)]$. Let $X = \sum_{i=1}^K X_i$ denote the number of differences between x^* and y^*. Since all X_i are independent and $\Pr(X_i = 1) = (1 - J(x,y))\left(1 - \frac{1}{B}\right) \leq \alpha/\tau\left(1 - \frac{1}{B}\right)$, we have $\mathrm{E}[X] \leq K\alpha/\tau\left(1 - \frac{1}{B}\right)$. Using the Chernoff bound $\Pr(X > (1+\beta)\mathrm{E}[X]) \leq \exp\left(-\beta^2/3\mathrm{E}[X]\right)$ [5, Theorem 1.1] with $\beta = \sqrt{3\ln(1/\delta)/\mathrm{E}[X]}$ proves the lemma.

The next lemma shows that we can avoid loosing a factor K in the privacy budget[1] when using generalized randomized response [16] on vectors with few differences.

Lemma 2. *Fix $\varepsilon > 0$. Let $x, y \in [B]^K$ be two arbitrary vectors that differ in at most L positions. Let $\varepsilon' = \varepsilon/L$. Let $\mathcal{A}$ be generalized randomized response mapping from $z \in [B]^K$ to $z^* \in [B]^K$ such that with probability $e^{\varepsilon'}/(e^{\varepsilon'} + B - 1)$ we have that $z_i^* = z_i$, and otherwise z_i^* is uniformly picked from $[B] - \{z_i\}$. Then $\mathcal{A}$ is ε-differentially private.*

Proof. Fix an arbitrary $v \in [B]^K$. We have to show that $\frac{\Pr[A(x)=v]}{\Pr[A(y)=v]} \leq e^\varepsilon$. Let the set $I_{x,v}$ collect all positions in which $x_i = v_i$, and let $N_{x,v}$ collect all positions in which $x_i \neq v_i$. We observe that $\Pr[A(x) = v] = \prod_{i \in I_{x,v}} \frac{e^{\varepsilon'}}{e^{\varepsilon'}+B-1} \cdot \prod_{i \in N_{x,v}} \frac{1}{e^{\varepsilon'}+B-1}$. The expression for $\Pr[A(y) = v]$ follows analogously. Let $D = \{i \mid x_i \neq y_i\}$ denote all positions where x and y differ. Because all terms where x and y are identical cancel out, we may conclude that

$$\frac{\Pr[A(x) = v]}{\Pr[A(y) = v]} = \frac{\prod_{i \in I_{x,v} \cap D} \frac{e^{\varepsilon'}}{e^{\varepsilon'}+B-1} \prod_{i \in N_{x,v} \cap D} \frac{1}{e^{\varepsilon'}+B-1}}{\prod_{i \in I_{y,v} \cap D} \frac{e^{\varepsilon'}}{e^{\varepsilon'}+B-1} \prod_{i \in N_{y,v} \cap D} \frac{1}{e^{\varepsilon'}+B-1}} \leq \prod_{i \in D} \frac{\frac{e^{\varepsilon'}}{e^{\varepsilon'}+B-1}}{\frac{1}{e^{\varepsilon'}+B-1}} \leq e^{\varepsilon' L} = e^\varepsilon.$$

4   LDP Sketches via Generalized Randomized Response

This section introduces an (ε, δ)-locally differential private algorithm to produce the user vectors $\hat{x}_u$ using generalized randomized response.

---

[1] Traditionally, a standard application of the composition theorem [8] shows that the composition of K ε-DP mechanisms satisfies $(K\varepsilon)$-DP.

The idea of the following algorithm is that each user receives the description of K range-B MinHash functions that map from $[m]$ to $[B]$ for $B \geq 2$. Each user applies the range-B MinHash functions and perturbs the hash value using a variant of generalized randomized response [16]. We proceed to describe the RRMinHash approach. An example is given in the full version of this paper.

Preprocessing. Each user accesses $K \geq 1$ range-B MinHash functions $h_1, \ldots, h_K$ shared among all users. Each user u applies $h_1, \ldots, h_K$ to their vector x_u to obtain $x_u^* \in [B]^K$. Now, each position of x_u^* is perturbed using generalized randomized response (Lemma 2) with an individual privacy budget of $\varepsilon' = \varepsilon/L$ to generate the response $\hat{x}_u$, where L is an upper bound on the number of differences between neighboring user vectors as in Lemma 1. $\hat{x}_u$ is the public response of user u.

Lemma 3. *The randomized mechanism $x \mapsto \hat{x}$ is (ε, δ)-LDP.*

Proof. Fix $\varepsilon, \delta > 0$ and let $x, y \in \{0,1\}^m$ such that they differ in at most α positions. By Lemma 1, with probability at least $1 - \delta$, the vectors x^* and y^* differ in at most $L = \lceil K(\alpha/\tau)\left(1 - \frac{1}{B}\right) + \sqrt{3\ln(1/\delta)\left(1 - \frac{1}{B}\right)K\alpha/\tau} \rceil$ positions. If x^* and y^* differ in at most L positions, Lemma 2 guarantees that the mapping $x^* \mapsto \hat{x}$ is ε-differential private.

Similarity Estimation. Given two responses $\hat{x} \in [B]^K$ and $\hat{y} \in [B]^K$, count collisions to obtain $p_{\mathrm{col}} = \sum [\hat{x}_i = \hat{y}_i]/K$. Given p_{col}, B, and $p^* = e^{\varepsilon'}/(e^{\varepsilon'}+B-1)$, we estimate the Jaccard similarity of x and y as

$$\hat{J}_{RR}(\hat{x}, \hat{y}) = \frac{(B-1)(B \cdot p_{\mathrm{col}} - 1)}{(B \cdot p^* - 1)^2} \tag{1}$$

Lemma 4. $\hat{J}_{RR}(\hat{x}, \hat{y})$ *is an unbiased estimator of* $J(x, y)$.

Proof. We proceed in two parts. First, we calculate the probability of the event "$\hat{x}_i = \hat{y}_i$". Next, we connect this probability to the estimation given above.

To compute the collision probability, we split up the probability space in two stages. In the first stage, we condition on the events "$x_i^* = y_i^*$" and "$x_i^* \neq y_i^*$", i.e. on whether the range-B MinHash values collide or not. In the second stage, we calculate the probability that the perturbed responses collide. As discussed in Sect. 3.1, a random range-B MinHash function has the property that $\Pr[x_i^* = y_i^*] = \frac{(B-1)J(x,y)+1}{B}$.

Given that $x_i^* = y_i^*$, we observe $\hat{x}_i = \hat{y}_i$ if both keep their answer, or if both change their answer to the same of the other $B - 1$ possible responses. Since both pick a choice uniformly at random, this means that $\Pr[\hat{x}_i = \hat{y}_i \mid x_i^* = y_i^*] = (p^*)^2 + \frac{(1-p^*)^2}{B-1}$. Consider that the event $x_i^* \neq y_i^*$ happened. In this case, we observe a collision of the perturbed values in the following cases: (i) one response is truthful, the other is changed and picks the truthful response as answer, and

(ii) both responses are obtained by changing the answer, and they both choose the same answer at random. Computing these probabilities, we conclude that $\Pr[\hat{x}_i = \hat{y}_i \mid x_i^* \neq y_i^*] = 2p^*(1-p^*)\frac{1}{B-1} + (1-p^*)^2\left(1-\frac{1}{B-1}\right)^2\frac{1}{B-2}$. The last term is obtained by first conditioning that neither choice picks the other's truthful answer, and then using the random choice of the remaining $B-2$ buckets.

Putting everything together, we obtain

$$\Pr[\hat{x}_i = \hat{y}_i] = \frac{(B-1)J(x,y)+1}{B}\left((p^*)^2 + \frac{(1-p^*)^2}{B-1}\right)$$

$$+ \left(1 - \frac{(B-1)J(x,y)+1}{B}\right)\left(2p^*(1-p^*)\frac{1}{B-1} + (1-p^*)^2\left(1-\frac{1}{B-1}\right)^2\frac{1}{B-2}\right).$$

Simplifying this formula by collecting terms yields

$$\Pr[\hat{x}_i = \hat{y}_i] = \frac{J(x,y) + BJ(x,y)p^*(Bp^*-2) + B - 1}{B(B-1)}. \tag{2}$$

Solving (2) for $J(x,y)$ and using linearity of expectation to connect p_{col} to $\Pr[\hat{x}_i = \hat{y}_i]$ results in (1).

Utility Analysis. Next we will discuss probabilistic bounds on the absolute error that the similarity estimation algorithm achieves on the private vectors. This means that we want upper bound the value $|\hat{J}_{RR}(\hat{x},\hat{y}) - J(x,y)|$. In the following, we will consider the absolute error in the case $\hat{J}_{RR}(\hat{x},\hat{y}) > J(x,y)$. The case $\hat{J}_{RR}(\hat{x},\hat{y}) < J(x,y)$ follows by symmetry.

Lemma 5. *With probability at least* $1 - \delta$,

$$|\hat{J}_{RR}(\hat{x},\hat{y}) - J(x,y)| \leq \sqrt{\frac{3\ln(1/\delta)B^3(1+p^*(Bp^*-2))}{K(Bp^*-1)^4}}. \tag{3}$$

Proof. Fix x and y. We let X_i be the indicator variable for the event "$\hat{x}_i = \hat{y}_i$". Define $X = X_1 + \cdots + X_K$. By (2) we know that X_i is Bernoulli-distributed with $q := \Pr[X_i = 1] = \frac{J(x,y) + BJ(x,y)p^*(Bp^*-2) + B - 1}{B(B-1)}$. Again using a Chernoff bound, we see that with probability at least $1 - \delta$, $X \leq E[X] + \sqrt{E[X]3\ln(1/\delta)}$. Assume from here on that this inequality holds. From (1), we start by observing that

$$\hat{J}_{RR}(\hat{x},\hat{y}) = \frac{(B-1)(B \cdot X/K - 1)}{(B \cdot p^* - 1)^2} \leq \frac{(B-1)(B(E[X] + \sqrt{E[X]3\ln(1/\delta)})/K - 1)}{(B \cdot p^* - 1)^2}$$

$$\stackrel{\text{Lem } 4}{=} J(x,y) + \frac{(B-1)B\sqrt{E[X]3\ln(1/\delta)}}{K(Bp^*-1)^2} < J(x,y) + \sqrt{\frac{3\ln(1/\delta)B^3(1+p^*(Bp^*-2))}{K(Bp^*-1)^4}}.$$

Theorem 2. *Fix* $\varepsilon, \delta_{DP}, \delta_{fail} > 0$. *There exists B and K such that with probability at least* $1 - \delta_{fail}, |\hat{J}_{RR}(\hat{x},\hat{y}) - J(x,y)| = O(\sqrt{\alpha/(\tau \cdot \varepsilon)})$. *The constant hidden in the big-Oh notation depends on δ_{DP} and δ_{fail}.*

Proof. Lemma 5 tells us that for every choice of B and K, with probability at least $1 - \delta_{\text{fail}}$ it holds that

$$|\hat{J}_{RR}(\hat{x}, \hat{y}) - J(x, y)| \leq \sqrt{\frac{3\ln(1/\delta)B^3(1 + p^*(Bp^* - 2))}{K(Bp^* - 1)^4}}. \qquad (4)$$

Since $p^* \leq 1$, we continue to bound the right-hand side of (4) by $\sqrt{\frac{3\ln(1/\delta_{\text{fail}})}{K}}$. $B^2/(Bp^* - 1)^2$. Assume that $Bp^* \geq 2$, which means that $\varepsilon' \geq \ln(2(B - 1)/(B - 2))$ and $\varepsilon \geq L \cdot \varepsilon'$. Since $1/(x - 1) \leq 2/x$ for $x \geq 2$, we continue to bound the absolute error from above by $\sqrt{\frac{3\ln(1/\delta_{\text{fail}})}{K}} \cdot B^2(2/(Bp^*))^2 \leq \sqrt{\frac{48\ln(1/\delta_{\text{fail}})}{K}}(1 + (B - 1)/e^{\varepsilon'})^2$. Now, we may set $B = 3$ since it makes the numerator as small as possible. ($B = 2$ is no valid choice because of the assumption $Bp^* \geq 2$.) Choosing $K = \Theta(\tau\varepsilon/\alpha)$, the absolute error is bounded by $\Theta(\sqrt{\alpha/(\tau\varepsilon)})$ for all $\varepsilon \geq L\ln 4$. $\qquad \square$

5   LDP Sketches via the Laplace Mechanism

This section introduces an (ε, δ)-LDP protocol for generating private user vectors $\hat{x}_u$ using the Laplace Mechanism. As before, for fixed integers B and K, we use K range-B MinHash functions such that each user produces a sketch in $[B]^K$.

Let x and y be neighboring vectors and let x^* and y^* be the two sketches in $[B]^K$. As before, with probability at least $1 - \delta$, we can assume that x^* and y^* differ in at most $L = K(\alpha/\tau)\left(1 - \frac{1}{B}\right) + \sqrt{3\ln(1/\delta)\left(1 - \frac{1}{B}\right)K\alpha/\tau}$ positions.

Before we can use Theorem 1, we have to compute the sensitivity Δ of the local sketches under the assumption that neighboring vectors differ in at most L positions. Since each coordinate in which the two vectors differ contributes at most $B - 1$ to the ℓ_1 norm, the sensitivity is at most $\Delta := L(B - 1)$. According to Theorem 1, adding Laplace noise with scale Δ/ε to x^* to produce $\hat{x}$ guarantees $(\varepsilon, 0)$-differential privacy as long as the number of differences is at most L. With probability at most δ_{DP}, there are more than L differences. We are now ready to describe the NoisyMinHash approach, with an example provided in the full version.

Preprocessing. Let K, B, α, and τ be integers, and let $\varepsilon > 0$ and $\delta > 0$ be the privacy budget. Choose K range-B MinHash functions $h_1, \dots, h_K$ and distribute them to the users. Each user with vector x returns $\hat{x} = (h_1(x) + N_{x,1}, \dots, h_K(x) + N_{x,K}) \in \mathbb{R}^K$, where each $N_{x,i} \sim \text{Lap}(\Delta/\varepsilon)$ with $\Delta = (B - 1)\left(K(\alpha/\tau)\left(1 - \frac{1}{B}\right) + \sqrt{3\ln(1/\delta)\left(1 - \frac{1}{B}\right)K\alpha/\tau}\right)$.

Similarity Estimation. Given $\hat{x}$ and $\hat{y}$ from $\mathbb{R}^K$, return

$$\hat{J}_{Lap}(\hat{x}, \hat{y}) = \frac{(B^2 - 1)K - 6\sum_{i=1}^{K}(\hat{x}_i - \hat{y}_i)^2 + 24K(\Delta/\varepsilon)^2}{(B - 1)(B + 1)K}. \qquad (5)$$

Notably, the estimation algorithm just computes the squared Euclidean distance and adjusts for the noise added.

Lemma 6. $\hat{J}_{Lap}(\hat{x}, \hat{y})$ *is an unbiased estimator for* $J(x, y)$.

Proof. Given x and y from $\{0, 1\}^m$, apply `NoisyMinHash` to compute $\hat{x}, \hat{y} \in \mathbb{R}^K$. Using linearity of expectation, we proceed as follows:

$$
\mathrm{E}\left[\sum_{i=1}^{K}(\hat{x}_i - \hat{y}_i)^2\right] = \sum_{i=1}^{K} E[(\hat{x}_i - \hat{y}_i)^2] = K\mathrm{E}[((h_1(x) - h_1(y)) + (N_{x,1} - N_{y,1}))^2]
$$

$$
= K\mathrm{E}[(h_1(x) - h_1(y))^2 + 2(h_1(x) - h_1(y))(N_{x,i} - N_{y,i}) + (N_{x,i} - N_{y,i})^2]
$$

$$
\overset{(1)}{=} K(\mathrm{E}[(h_1(x) - h_1(y))^2] + \mathrm{E}[(N_{x,i} - N_{y,i})^2])
$$

$$
\overset{(2)}{=} K\mathrm{E}[(h_1(x) - h_1(y))^2] + 2K\mathrm{Var}[N_{x,i}] = K\mathrm{E}[(h_1(x) - h_1(y))^2] + 4K(\Delta/\varepsilon)^2.
$$

In our calculations, both (1) and (2) used that $N_{x,i}$ and $N_{y,i}$ are independently chosen, $\mathrm{E}[N_{x,i}] = \mathrm{E}[N_{y,i}] = 0$, and $\mathrm{Var}[N_{y,i}] = 2(\Delta/\varepsilon)^2$.

Let $x^* = h_1(x)$ and $y^* = h_1(y)$. We continue by calculating $\mathrm{E}[(x^* - y^*)^2]$ as

$$
\sum_{j=1}^{B-1} j^2 \Pr[|x^* - y^*| = j] = \sum_{j=1}^{B-1} j^2 \frac{(B-1)(1 - J(x, y))}{B} \Pr[|x^* - y^*| = j \mid x^* \neq y^*]
$$

$$
\overset{(1)}{=} \sum_{j=1}^{B-1} j^2 \frac{(B-1)(1 - J(x, y))}{B} 2(B - j)/B = \frac{2(B-1)(1 - J(x, y))}{B^2} \sum_{j=1}^{B-1} j^2(B - j)
$$

$$
= \frac{2(B-1)(1 - J(x, y))}{B^2} \cdot \frac{B^2}{12}(B + 1)(B - 1) = \frac{(B-1)^2(B+1)(1 - J(x, y))}{6},
$$

where (1) is obtained by noticing that for a fixed x^* (with B choices), there are $B - 2j$ choices with two y^* such that $|x^* - y^*| = j$, and there are $2j$ choices with only one choice y^*. Putting everything together, we summarize

$$
\mathrm{E}\left[\sum_{i=1}^{K}(\hat{x}_i - \hat{y}_i)^2\right] = \frac{K(B-1)^2(B+1)(1 - J(x, y))}{6} + 4K(\Delta/\varepsilon)^2.
$$

The result is obtained by rearranging terms.

Utility Analysis

Theorem 3. *Fix* $\varepsilon, \delta_{DP}, \delta_{fail} > 0$. *There exists* B *and* k *such that with probability at least* $1 - \delta_{fail}$, $|\hat{J}_{Lap}(\hat{x}, \hat{y}) - J(x, y)| = \tilde{O}\left((\alpha/\tau)^{4/5} \cdot \varepsilon^{-2/5}\right)$. *The constant hidden in the big-Oh notation depends on* δ_{DP} *and* δ_{fail}, *and the tilde notation suppresses polylogarithmic factors.*

Proof. We first describe and analyze the two events which constitute the failure probability δ_{fail}. Next we proceed to analyze the estimation error under the condition that none of these events occur. We will only analyze the case that $\hat{J}_{Lap}(\hat{x}, \hat{y})$ is larger than $J(x, y)$. The other case follows by symmetry.

First, we assume that the number of differences between x^* and y^* (among the K functions) does not differ by more than a value L' from its expectation. This is true for $L' = \sqrt{3\ln(2/\delta_{\text{fail}})\left(1 - \frac{1}{B}\right)K\alpha/\tau}$ by Lemma 1 for failure probability $\delta_{\text{fail}}/2$. Second, we use Theorem 3.8 in [8] (reproduced in the full version of this paper) that says that with probability at least $1 - \delta_{\text{fail}}/4$, the maximum absolute difference in a coordinate of $\hat{x}$ compared to x^* is at most $D = \ln(4K/\delta_{\text{fail}})\Delta/\varepsilon$.

By a union bound, with probability at least $1 - \delta_{\text{fail}}$ none of these events occur, i.e., we observe a deviation of at most L' in the number of differences of two vectors x^* and y^* from their expectation, and the Laplace noise added to both x^* and y^* keeps all coordinates within D in their absolute value. Under this condition, we will study the value $|X - \mathrm{E}[X]|$ for the random variable $X = \sum_{i=1}^{K}(\hat{x}_i - \hat{y}_i)^2$. If this value is at most t, the absolute estimation error is at most $\frac{6t}{(B-1)(B+1)K}$, cf. (5).

As in the proof of Lemma 6, we split up $\hat{x}_i$ into x_i^* and N_i to calculate $\sum_{i=1}^{K}(\hat{x}_i - \hat{y}_i)^2 = \sum_{i=1}^{K}\left((x_i^* - y_i^*)^2 + 2(x_i^* - y_i^*)(N_i - N_j) + (N_i - N_j)^2\right)$. By our second condition, we may assume that $|N_i - N_j| \leq 2D$, which means that the last summand is at most $4KD^2$ over the whole sum. For the first summand, we use the first condition that says that the number of observed differences is within L' from its expectation. Since each individual term in the sum contributes at most $(B-1)^2$, the deviation from the expectation over the whole sum is not more than $(B-1)^2L'$. Lastly, using both conditions, the contribution of the middle term over the whole sum is bounded by $4D(K\alpha/\tau(1 - 1/B) + L')$.

Using $P := K\alpha/\tau(1 - 1/B)$ and rewriting $L' = \sqrt{3P\ln(2/\delta_{\text{fail}})}$, we can put the observations from above together and conclude that with probability at least $1 - \delta_{\text{fail}}$ the estimation error is

$$O\left(\frac{(B-1)^2\sqrt{3P\ln(2/\delta_{\text{fail}})} + KD^2 + D(P + \sqrt{3P\ln(2/\delta_{\text{fail}})})}{(B-1)(B+1)K}\right).$$

Comparing the second and the third term of the sum, we notice that $D > (B-1)(P + \sqrt{P})$ for $\varepsilon > 1$, so the second term is always larger than the third and we may bound the estimation error by $O\left(\frac{(B-1)^2\sqrt{3P\ln(2/\delta_{\text{fail}})} + KD^2}{(B-1)(B+1)K}\right)$. The function $(x-1)^2/((x-1)(x+1))$ is monotonically increasing for $x \geq 1$, so the choice $B = 2$ minimizes the expression above. Now, observe that the first term is $O(\sqrt{\alpha/(K\tau)})$ and the second term is $\tilde{O}((\alpha K/\tau\varepsilon)^2)$, where the tilde notation suppresses the logarithmic dependence on K. To balance the estimation error, we set these terms in relation to each other and solve for K. This shows that the asymptotic minimum is achieved for $K = \varepsilon^{4/5}(\tau/\alpha)^{3/5}$. Using this value to bound the estimation error results in the bound stated in the theorem.

Comparing `RRMinHash` and `NoisyMinHash`. Comparing Theorems 2 to 3, both analyses provide bounds on the absolute error in terms of the length τ of individual vectors, the neighboring notion α, and the privacy budget ε.

Since the value α/τ is between 0 and 1, the contribution of $(\alpha/\tau)^{4/5}$ to the error of NoisyMinHash (Theorem 3) is smaller than the term $(\alpha/\tau)^{1/2}$ for RRMinHash. However, the $\varepsilon^{-1/2}$ dependence of RRMinHash is better than $\varepsilon^{-2/5}$ for $\varepsilon \geq 1$. This should mean that while NoisyMinHash might guarantee smaller error for small epsilon settings, the error decreases faster for RRMinHash.

In both mechanisms, the preprocessing time to generate a private vector is $O(K\tau)$ for a vector with τ set bits. It consists of evaluating K range-B MinHash functions (each taking time $O(\tau)$) and sampling $O(K)$ values from a uniform (RRMinHash) or Laplace distribution (NoisyMinHash). The similarity estimation of two vectors takes time $O(K)$. A private vector for RRMinHash consists of K bits and the similarity estimation uses the Hamming distance, while NoisyMinHash uses K floating point values and uses Euclidean distance as basis for similarity estimation. Given the difficulty of correctly implementing the Laplace Mechanism [13], RRMinHash has a simpler basis for a correct implementation.

6   Experimental Evaluation

All algorithms described in this paper where implemented in Python 3. The code, raw results, and evaluation notebooks can be accessed at https://github. com/maumueller/ldp-jaccard. Due to space restrictions, we only present a few selected results. See the Jupyter notebook at the web page for additional plots and tables.

Experimental Setting. We conduct experiments in two different directions.

First, we create artificial vectors and test how well the algorithms estimate Jaccard similarity for a fixed privacy budget. We use the *mean absolute error* as our quality measure, which is defined as $\frac{1}{\ell}\sum_{i=1}^{\ell}|d_i - e_i|$ for true similarities $d_1, \ldots, d_\ell$ and their estimates $e_1, \ldots, e_\ell$ returned by the algorithm. In the experiment, we create user vectors x with $\tau \in \{20, 50, 100, 250, 500, 1000, 2000\}$ entries. For each such x, we create vectors y' with τ entries and Jaccard similarity in $\{0.1, 0.5, 0.9\}$ to x. The number K of hash functions considered is chosen from $\{10, 20, 30, \ldots, 500\}$. For each algorithm, we vary the privacy budget and internal parameters such as the range B of the MinHash functions. All runs were repeated 100 times with random hash functions.

Second, we study how well these algorithms work on real-world datasets. Following [17], we chose the MovieLens and Last.FM dataset available at https:// grouplens.org/datasets/hetrec-2011. We obtain a set representation by collecting all movies rated at least 4 (MovieLens, $m = 65\,536$) and the top-20 artists (Last.FM, $m = 18\,739$). The average set size is 178.1 ($\sigma = 187.5$) and 19.8 ($\sigma = 1.78$), respectively. To account for the influence of the size of the user vectors, we create different versions of these datasets. From the MovieLens dataset, we make three versions containing all users that have at least 50, 100, and 500 entries, respectively. This results in datasets with 1636, 1205, and 124 users. From the Last.FM dataset, we collect all users that have at least 20 entries which amounts to 1860 users. For each dataset, we take 50 query points at random

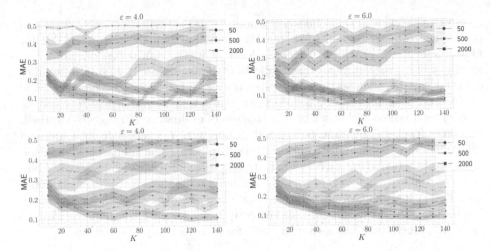

Fig. 1. Results on synthetic vectors with $\tau \in \{50, 500, 2000\}$, privacy budget $\varepsilon \in \{4, 6\}$, and vectors with Jaccard similarity of 0.5. Blue, red, green lines represent runs with choice of 2, 3, 5 for B, respectively; top: `RRMinHash`, bottom: `NoisyMinHash`. (Color figure online)

among all data points for which the 10-th nearest neighbor has at least Jaccard similarity 0.1. As quality measure, we use recall@k (R@k) which measures the average number of times that the (index of the) true nearest neighbor is found among the first k nearest neighbors in the private vectors. (Note that the true vectors are not revealed, so there cannot be a re-ranking step as is tradition in nearest neighbor search.) Moreover, we report on the approximate similarity ratio, which is defined as the ratio of the sum of similarities to the 10 original nearest neighbors, and the sum of similarities to the 10 nearest neighbors among the private vectors computed with their original similarities.

For the whole evaluation, we will use range-2 MinHash (i.e., 1-bit Min-Hash [12]) with $K \leq 100$ as a baseline for comparison. For all experiments, we set $\alpha = 1$, i.e. we allow for a single item change. Results for other values can be read off the plots by looking at different τ values. For example, a combination $(\alpha = 1, \tau = 500)$ is identical to $(\alpha = 10, \tau = 50)$ since all bounds depend on the ratio of α and τ. For all private mechanisms, we use $\delta = 0.0001$.

6.1   Result Discussion on Artificial Data

Figure 1 visualizes the mean absolute error (with standard deviation as error bars) for runs of `RRMinHash` (top) and `NoisyMinHash` (bottom) for privacy budgets of $\varepsilon = 4$ (left) and $\varepsilon = 6$ (right) and choices 2, 3, 5 of the B parameter (blue/red/green lines). With respect to `RRMinHash` and a privacy budget of $\varepsilon = 4$, we notice that for each choice of τ the trend is that smaller B values produce smaller absolute error, which is in accordance with our analysis in Sect. 4. (Larger B values can be found on the supplemental website; they performed much worse.) For vectors of 50 elements, the smallest MAE error is achieved with the smallest choice of K, resulting in an MAE of around 0.35. The error shrinks to around 0.15 for 500 elements (with K of around 20), and 0.05 for vectors with 2 000 elements (with K around 80). The linear increase of K with τ further motivates the choice of K in Theorem 2. Increasing the privacy budget to $\varepsilon = 6$ further decreases the error but results in the same trends. We note that a growing privacy budget also corresponds to a larger K choice, again as motivated in Theorem 2. Increasing K will sometimes result in worse error because of integer constraints in Lemma 2. From a practical point of view, one should choose K as large as possible before this increase occurs. The trends are identical with regard to NoisyMinHash, but it is much clearer that a smaller choice of B is preferable (as motivated in the proof of Theorem 3). We achieve an MAE of around 0.43, 0.18, 0.1 for vectors of size 50, 500, 2000 and $\varepsilon = 4$, respectively, slightly worse than RRMinHash.

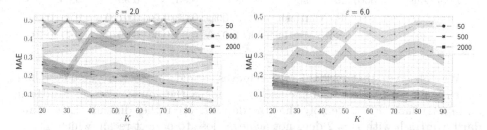

Fig. 2. Results on synthetic vectors with $\tau \in \{50, 500, 2000\}$, privacy budget $\varepsilon \in \{2, 6\}$, and vectors with Jaccard similarity of 0.5. Blue, red, green lines represent runs with RRMinHash, NoisyMinHash, Range-2 MinHash (non-private), respectively. There is only one line for MinHash because its error is independent of the vector size. (Color figure online)

Figure 2 sets our two mechanisms in relation to MinHash with $B = 2$ and a privacy budget of 2 (left) and 6 (right). For $\varepsilon = 2$, we need large vectors to guarantee an error that is roughly a factor of two larger than that achieved by MinHash. For $\varepsilon = 6$, both larger vectors allow for an estimation vector that is nearly as small as MinHash. Again, RRMinHash achieves smaller error than NoisyMinHash, in particular for larger privacy budgets.

We conclude that RRMinHash with $B = 2$ is a good choice in all considered experiments on artificial data. For small privacy budget, large user vectors are needed to get small estimation errors. A larger privacy budget allows to accommodate smaller vectors.

6.2   Results on Real-World Data

Table 1 summarizes the observed results for runs on the Last.FM and MovieLens datasets. Again, we set MinHash in relation to RRMinHash and NoisyMinHash. Motivated by the observations above we only discuss the case $B = 2$.

Table 1. Results on real-world datasets for different quality measures and privacy budget ε of 4 and 8 (split up via "/" in individual cells).

Dataset	Algorithm	R@10	R@50	R@100	Approx
Last.FM ($\tau = 20$)	MinHash	0.42	0.72	0.82	0.55
	RRMinHash	0.04/0.16	0.15/0.38	0.25/0.51	0.19/0.35
	NoisyMinHash	0.03/0.06	0.11/0.19	0.19/0.31	0.16/0.23
MovieLens ($\tau = 50$)	MinHash	0.13	0.34	0.47	0.61
	RRMinHash	0.02/0.05	0.09/0.17	0.18/0.28	0.49/0.53
	NoisyMinHash	0.01/0.02	0.05/0.11	0.11/0.21	0.49/0.50
MovieLens ($\tau = 100$)	MinHash	0.31	0.62	0.75	0.72
	RRMinHash	0.04/0.07	0.12/0.24	0.22/0.36	0.52/0.57
	NoisyMinHash	0.04/0.04	0.09/0.15	0.18/0.27	0.52/0.54
MovieLens ($\tau = 500$)	MinHash	0.58	0.93	0.99	0.83
	RRMinHash	0.19/0.31	0.65/0.76	0.91/0.96	0.72/0.76
	NoisyMinHash	0.14/0.25	0.63/0.69	0.90/0.92	0.71/0.73

We observe that `RRMinHash` achieves equal or better quality than `NoisyMinHash` in all measurements, so we focus the comparison on MinHash and `RRMinHash`. First, we note that the datasets are rather difficult. Even standard MinHash with $B = 2$ does not achieve close to perfect recall, which means that all vectors are rather close to each other. The Last.FM dataset provides very small user vectors. Accordingly there is a big difference between the quality achieved by the two algorithms. For a privacy budget of $\varepsilon = 4$, the quality is between a factor of around 10 (R@10) and of around 3 (R@100, Approx) worse if solving the similarity search task on private vectors. For a privacy budget of $\varepsilon = 8$, these factors shrink to 1.5–3. With regard to MovieLens, we observe that it is difficult for MinHash to achieve high recall values for $\tau = 50$. Results for `RRMinHash` are again a factor 3–6 worse for privacy budget 4, with the exception of the relative approximation that is rather close (0.49 vs. 0.61). Quality increases slowly from 50 to 100 items, and rapidly for 500 items (because of its small size).

We summarize that there is a clear trade-off between the utility and privacy of the proposed mechanisms. The results on artificial and real-world data show that to ensure good utility under a small privacy budget, user vectors have to contain many items, say in the 100s. Many of the theoretical choices translated well into practice. Most interestingly, while the upper bounds in the theory section painted an unclear picture about the utility at a fixed privacy budget, our empirical analysis clearly suggests that `RRMinHash` is both easier to implement and achieves higher utility for the same privacy budget.

7   Related Work

The paper by Kenthapadi [11] shows how to estimate vector differences under the ℓ_2 norm in a differentially private setting in the centralized model of differential privacy. More precisely, their algorithm has privacy guarantees with respect to a single element change (i.e., one user changes one item). In very recent work, Dhaliwal et al. [4] show how to achieve the same guarantees when the privacy-guarantees are over the change of a fraction of a user vector in the central model. Both approaches apply a Johnson-Lindenstrauss transform [9] and add noise of a certain scale to the resulting matrix. Our `NoisyMinHash` approach can be seen as a natural generalization of their method, but there are some stand-alone features such as the mapping to B buckets.

With respect to similarity estimation under Jaccard similarity, the paper by Riazi et al. [15] describes a privacy-preserving approach for similarity estimation both for inner product similarity (using SimHash [3]) and Jaccard similarity (using MinHash). Their privacy notion does not satisfy differential privacy.

The paper by Yan et al. [17] is closest to our approach. It discusses an LDP approach based on MinHash by selecting certain hash values in a differentially private manner using the exponential mechanism. As we argue in the full version of this paper, their approach does not provide the guarantees they state and quickly degrades to a basic MinHash approach without noise addition.

Concurrent to our work, Pagh and Stausholm [14] describe LDP sketches for approximating the number of items in a set. Their sketches are *linear*, which allows them to approximate the size of the union and the intersection of two sets, and thus their Jaccard similarity. In contrast to our bounds, their bounds rely on the universe size of set elements. It would be interesting to compare their mechanism to ours in a practical setting, in particular because their lower-order error terms [14, Theorem 1] suggest that they need much larger vectors than the ones considered in our empirical study in Sect. 6.

Finally, this paper studied the privacy/utility-tradeoff achievable with our proposed methods. While an important issue, it does not discuss (un)desirable privacy budgets, which will be application-specific and lack consensus [6].

References

1. Broder, A.Z.: On the resemblance and containment of documents. In: Proceedings of the Compression and Complexity of Sequences 1997, pp. 21–29. IEEE (1997)
2. Carter, J.L., Wegman, M.N.: Universal classes of hash functions. J. Comput. Syst. Sci. **18**(2), 143–154 (1979)
3. Charikar, M.: Similarity estimation techniques from rounding algorithms. In: Proceedings of the 34th ACM Symposium on Theory of Computing (STOC), pp. 380–388 (2002)
4. Dhaliwal, J., So, G., Parker-Wood, A., Beck, M.: Utility preserving secure private data release. CoRR abs/1901.09858 (2019). http://arxiv.org/abs/1901.09858
5. Dubhashi, D.P., Panconesi, A.: Concentration of Measure for the Analysis of Randomized Algorithms. Cambridge University Press, Cambridge (2009)

6. Dwork, C., Kohli, N., Mulligan, D.: Differential privacy in practice: Expose your epsilons!. J. Priv. Confid. **9**(2) (2019)
7. Dwork, C., McSherry, F., Nissim, K., Smith, A.: Calibrating noise to sensitivity in private data analysis. In: Halevi, S., Rabin, T. (eds.) TCC 2006. LNCS, vol. 3876, pp. 265–284. Springer, Heidelberg (2006). https://doi.org/10.1007/11681878_14
8. Dwork, C., Roth, A., et al.: The algorithmic foundations of differential privacy. Found. Trends Theor. Comput. Sci. **9**, 211–407 (2014)
9. Johnson, W.B., Lindenstrauss, J.: Extensions of Lipschitz mappings into a Hilbert space. Contemp. Math. **26**(189–206), 1 (1984)
10. Kearns, M., Roth, A.: The Ethical Algorithm. Oxford University Press, Oxford (2019)
11. Kenthapadi, K., Korolova, A., Mironov, I., Mishra, N.: Privacy via the Johnson-Lindenstrauss transform. J. Priv. Confid. **5**(1), 39–71 (2013)
12. Li, P., König, A.C.: b-Bit minwise hashing. In: WWW 2010, pp. 671–680 (2010)
13. Mironov, I.: On significance of the least significant bits for differential privacy. In: ACM CCS, pp. 650–661 (2012)
14. Pagh, R., Stausholm, N.M.: Efficient differentially private F_0 linear sketching (2020)
15. Riazi, M.S., Chen, B., Shrivastava, A., Wallach, D.S., Koushanfar, F.: Sublinear privacy-preserving near-neighbor search with untrusted server on large-scale datasets (2016). arXiv:1612.01835
16. Wang, T., Blocki, J., Li, N., Jha, S.: Locally differentially private protocols for frequency estimation. In: USENIX Security, pp. 729–745 (2017)
17. Yan, Z., Wu, Q., Ren, M., Liu, J., Liu, S., Qiu, S.: Locally private Jaccard similarity estimation. Concurr. Comput. Pract. Exp. **31**, e4889 (2018)

Pivot Selection for Narrow Sketches by Optimization Algorithms

Naoya Higuchi[1], Yasunobu Imamura[2], Vladimir Mic[3], Takeshi Shinohara[1($\boxtimes$)], Kouichi Hirata[1], and Tetsuji Kuboyama[4]

[1] Kyushu Institute of Technology, Kawazu 680-4, Iizuka 820-8502, Japan
nac24nh@gmail.com, shino.kyutech@gmail.com, hirata@ai.kyutech.ac.jp
[2] Third Inc., Shinjuku, Tokyo 160-0004, Japan
imamura.kit@gmail.com
[3] Masaryk University, Brno, Czech Republic
xmic@fi.muni.cz
[4] Gakushuin University, Mejiro 1-5-1, Toshima, Tokyo 171-8588, Japan
ori-sisap2020@tk.cc.gakushuin.ac.jp

Abstract. Sketches are compact bit strings that are considered as products of an LSH for high-dimensional data. We use them in filtering for narrowing down solution candidates in similarity search. We propose a pivot selection method for narrow sketches with a length such as 16-bits by optimization algorithms with the accuracy of filtering itself as the objective function.

1 Introduction

The purpose of our research is to establish fast similarity search techniques for large-scale high-dimensional data. Such techniques can contribute to pattern recognition, data mining, knowledge discovery, etc. In this paper, we aim to improve the accuracy and speed of approximate similarity search using sketches.

The *sketch* [2,8,10,12,15] is a compact bit string that represents high-dimensional data. We can regard a mapping from data to sketch as a type of locality-sensitive hash (LSH) that can maintain some similarity of data. We use sketches for filtering to narrow down the solution candidates at the first stage of similarity search. The similarity search using sketches is only an approximation, and improving accuracy is one of the most important issues. Since it is costly to measure the accuracy of filtering by sketches, we have conventionally been designing the sketching based on the optimization by using an evaluation that is different from the accuracy such as the collision probability and the entropy. However, there is a limit to optimize with such an indirect evaluation [7].

We have established a very fast search technique using relatively short and narrow such as 16-bit sketches [4,5]. In this paper, we propose an efficient method to evaluate a set of pivots P by utilizing the fast search principle using narrow sketches. The method gives an evaluation measure directly related to the filtering performance by the sketches defined by P. The effectiveness of the proposed

© Springer Nature Switzerland AG 2020
S. Satoh et al. (Eds.): SISAP 2020, LNCS 12440, pp. 33–46, 2020.
https://doi.org/10.1007/978-3-030-60936-8_3

method is confirmed by experiments using two real-life datasets consisting of a million to several million data. The proposed method enables faster and more accurate nearest neighbor searches than the optimization of collision probability.

There exist two important issues on sketches, the *quality* and *indexing* [11]. For higher quality, sketches are proposed to well approximate the original search space. For faster search, the indexing has been studied. These are usually independently treated with each other. Our approach addresses both issues together.

2　Fast Similarity Search Using Narrow Sketches

Here, we prepare concepts necessary for similarity search using sketches and later discussions, and introduce the results of our research leading up to this paper.

2.1　Similarity Search Using Sketches

We assume that the data to be searched is given as points in a metric space with a coordinate system. We use one of the basic similarity search query types, the k *nearest neighbor queries* (k-NN, for short). Their goal is to acquire the k closest points to the point given as the query. Here, $k \geq 1$, and this paper focuses mainly to search for one nearest neighbor, i.e. $k = 1$. When there are multiple nearest neighbor solutions, we consider any one of them as a correct answer.

Data points in the dataset are indexed by natural numbers (referred to as data IDs) from 0 to $n-1$. In other words, the dataset consisting of n points is set $ds = \{x_0, \ldots, x_{n-1}\} \subseteq U$, where U is the data space. The dissimilarity between two data points x and y is defined by the distance $D(x, y)$. The *nearest neighbor search* (NN search, for short) for a *query* $q \in U$ is to find a data point $x \in ds$ such that $D(q, x) \leq D(q, y)$ for all $y \in ds$. The main symbols and notations are summarized in Table 1.

The k-NN search for a query q using sketches is performed by the following two stages, where $k' \geq k \geq 1$ is a parameter of the number of *candidates* obtained in the first stage. The *priority* of sketches is conventionally determined by the Hamming distance. We, however, use asymmetric distance measures between the query point and a sketch, in this paper.

1. The first stage (the filtering using the priority by sketches):
   Select k' candidates $x_{i_0}, \ldots, x_{i_{k'-1}}$ whose sketches $\sigma(x_{i_0}), \ldots, \sigma(x_{i_{k'-1}})$ have top k' priorities for q.
2. The second stage (the k-NN search using distances $D(q, x_{i_0}), \ldots, D(q, x_{k'-1})$):
   Select the k-NN data points from the candidates $x_{i_0}, \ldots, x_{i_{k'-1}}$.

The *recall* of the k-NN search using sketches is given by the ratio of the number of correct answers included in k' candidates selected by the first stage filtering with respect to k. We say that the filtering by sketches has a high *accuracy* when the search recall is high. Clearly, the larger the k', the higher the filtering accuracy is. The accuracy of filtering is also influenced by the used priority, as well as by the sketching transformation.

Table 1. Notations

Notation	Description
$\mathcal{U}$	The data space
m	Dimensionality of data space
$x, y, x_0, \ldots$	Data points in $\mathcal{U}$
$D(x, y)$	The distance between x and y
ds	Dataset $\{x_0, x_1, \ldots, x_{n-1}\}$ indexed by numbers
n	Number of data points in ds
k	Number of data points in the answer by the k-NN search
k'	Number of candidates obtained in the 1st stage filtering
$q \in \mathcal{U}$	A query point
(p, r)	A pivot for BP, a pair of the center $p \in \mathcal{U}$ and the radius r
w	The width (length) of sketches
P	A w-tuple of pivots $((p_0, r_0), (p_1, r_1), \ldots, (p_{w-1}, r_{w-1}))$
$\sigma(P, x), \sigma(x)$	The sketch of x using P and the sketch of x if P is omitted
ς	A sketch of unspecified data point
$score_p(q, \varsigma)$	L_p-like asymmetric distance between a query q and a sketch ς
Q	Sample of queries $\{q_0, q_1, \ldots, q_{|Q|-1}\}$
$Q' \subseteq Q$	The resampling of queries
$\mathcal{P}$	The space of w-tuple of pivots
$E(P, Q')$	The evaluation value of P for $Q' \subseteq Q$
$Nb(P) \subseteq \mathcal{P}$	The neighborhood of P
$t \in \mathbb{N}$	The time step $(0, 1, 2, \ldots)$
$size(t) \leq |Q|$	The resampling size at t (monotonically increasing)

2.2 Space Partitioning for Sketching Transformation

Ball partitioning (BP, for short) is used to define bit-string sketches of data points. BP assigns a data point x bit 0 if x is in the sphere specified by a pair (p, r) with the center $p \in \mathcal{U}$ and the radius r, and bit 1 otherwise. Formally:

$$BP_{(p,r)}(x) = \begin{cases} 0, & \text{if } D(p, x) \leq r, \\ 1, & \text{otherwise.} \end{cases}$$

The pair (p, r) is called a *pivot*. The length of a sketch is called *width*. A sketch of width w is defined as a concatenation of w bits assigned by BP using a w-tuple of pivots $P = ((p_0, r_0), \ldots, (p_{w-1}, r_{w-1}))$ as follows:

$$\sigma(P, x) = BP_{(p_{w-1}, r_{w-1})}(x) \cdots BP_{(p_0, r_0)}(x)$$

The arrangement of bits in sketches is from right to left since we apply bit operators to them like "right-shift." The sketch of x is also denoted as $\sigma(x)$ by omitting P if clear from the context. A sketch of unspecified data point is denoted by using ς.

2.3 More Accurate Priorities than Hamming Distance

Traditionally, the Hamming distance between sketches, which is the number of mismatched bits between sketches, has been used as the priority in the first stage filtering. Dong et al. [2] introduced asymmetric distances between data point and sketch which give more accurate priorities than the Hamming distance. We introduced similar asymmetric distances, such as $score_\infty$, $score_1$ and $score_2$, as the priorities by sketches [3]. While we use them for relatively narrow sketches [4, 5] based on the viewpoint that sketches can be considered as quantized images of a dimension reduction called Simple-Map [14], Dong et al. use long sketches as a kind of data compression to improve filtering performance.

The minimum distance from a query q to the partition boundary of $BP_{(p_i,r_i)}$ is given by the following:

$$e_i(P, q) = |D(p_i, q) - r_i|$$

Using this, a lower bound on $D(q, x)$ is obtained by using $\varsigma = \sigma(P, x)$:

$$b_i(q, \varsigma) = \begin{cases} e_i(P, q), & \text{if } BP_{(p_i,r_i)}(q) \neq (\varsigma \gg i) \& 1, \\ 0, & \text{otherwise,} \end{cases}$$

where $\gg$ and $\&$ are bit operators of *logical right-shift* and *bitwise AND*, respectively. Therefore, $(\varsigma \gg i) \& 1$ is the i-th bit of ς from the right. Let x be any point included in the opposite side to a query q of the partition boundary by $BP_{(p_i,r_i)}$. Then, the corresponding bits of sketches for q and x disagree, and the distance between q and x is at least $e_i(P, q)$. Thus, we can prove that b_i gives a lower bound. By aggregating these lower bounds like Minkowski's L_p distance, we define the following asymmetric distances.

$$score_1(q, \varsigma) = \sum_{i=0}^{w-1} b_i(q, \varsigma), \quad score_2(q, \varsigma) = \sqrt{\sum_{i=0}^{w-1} b_i(q, \varsigma)^2},$$

$$score_\infty(q, \varsigma) = \max_{i=0}^{w-1} b_i(q, \varsigma)$$

The only $score_\infty(q, \varsigma)$ is guaranteed to be a true lower bound on distance $D(q, x)$.

We experimentally confirmed that all of these provide more accurate filtering than the Hamming distance [3]. Here, we should notice that the $score_p(q, \varsigma)$ requires no access to the data point x whose sketch is ς, and therfore, it is less expensive to evaluate than the distance $D(q, x)$. These asymmetric distances made us notice that the required search accuracy may be obtained even when using narrower sketches than previously thought.

2.4 Fast Filtering by Sketch Enumeration

Here, we explain the main points of the fast search method introduced in [4,5].

We assume that the size of the dataset is several million and the width of sketches is 16-bits. Since the number of 16-bit patterns is only 2^{16}, the data

```
1  function SEARCH(q, ς, NN, nearest, checked)
2  │  for i = f[ς] to f[ς] + num[ς] − 1 do
3  │  │  if D(q, x[id[i]]) ≤ nearest then
4  │  │  │  (NN, nearest) ← (id[i], D(q, x[id[i]]));
5  │  │  checked ← checked + 1;
6  │  │  if checked ≥ k' then
7  │  │  │  return (NN, nearest, checked);
8  │  return (NN, nearest, checked);
9  function SEARCHBYSKETCHENUM(q)
10 │  Setup the sketch enumeration in the ascending order of score_p for q;
11 │  (j, NN, nearest, checked) ← (0, "none", ∞, 0);
12 │  while checked ≤ k' do
13 │  │  ς ← the next sketch in the enumeration;
14 │  │  (NN, nearest, checked) ←SEARCH(q, ς, NN, nearest, checked);
15 │  return NN;
```

Algorithm 1: NN Search by Sketch Enumeration

points can be managed by the bucket method with the sketch as the key, where data points with a sketch ς are immediately accessed if ς is given. As for the bucket we use, we will explain details shortly.

Let $\varsigma_0, \varsigma_1, \ldots$ be the sketch enumeration in the priority order for the query q. For each $j = 0, 1, \ldots$, the number of data points x with $\sigma(x) = \varsigma_j$ is expected to be more than 15, because the average number of data points for each sketch is larger than $1,000,000 \div 2^{16} \fallingdotseq 15$. Therefore, sketches to be enumerated are only a small part of approximately 66,000 sketches. By using an algorithm that enumerates sketches *one by one* in the ascending order of the priority, it is possible to speed up the first stage search without collating sketches so that the cost is practically ignored. We call a method of the first stage filtering based on sketch enumeration a *filtering by sketch enumeration*. On the other hand, many conventional search methods are based on the first stage filtering by sequential search on all sketches of data points. We call such a method a *sequential filtering*.

Algorithm 1 outlines the NN search method that utilizes the sketch enumeration in the ascending order of $score_p$ for a query. For details of the enumeration, refer to papers [4,5]. Let data points in the dataset be stored in the array, $x_i = x[i]$ for $i = 0, \ldots, n - 1$. We use a bucket with the sketch as key for the dataset. The bucket is assumed to be represented by three arrays id, f, and num that satisfy the following conditions.

· $id[j]$ = the ID of the j-th data point in the sketch order. Data points are not necessarily sorted on memory. Indirect sorting via id is enough.
· $f[\varsigma]$ = the first position in the sketch order of data points with sketch ς.
· $num[\varsigma]$ = the number of data points in the dataset whose sketches are ς.

Data points can be considered to be sorted in the sketch order as follows:

$$\sigma(x[id[0]]) \leq \cdots \leq \sigma(x[id[n - 1]]).$$

```
1  procedure LS
2  │  P ← any tuple of pivots in P;
3  │  for t = 0 to ∞ do
4  │  │  P' ← randomly selected tuple of pivots from Nb(P);
5  │  │  if E(P', Q) is better than E(P, Q) then
6  │  │  │  P ← P'
```

```
7  procedure AIR
8  │  P ← any tuple of pivots in P;
9  │  for t = 0 to ∞ do
10 │  │  P' ← randomly selected tuple of pivots from Nb(P);
11 │  │  Q' ← randomly selected resampling from Q such that |Q'| = size(t);
12 │  │  if E(P', Q') is better than E(P, Q') then
13 │  │  │  P ← P'
```

Algorithm 2: Local Search and Annealing by Increasing Resampling

Also, we can get the following data IDs whose sketches are ς without accessing the array x:

$$id[f[\varsigma]], \ldots, id[f[\varsigma] + num[\varsigma] - 1].$$

Using this sequence, for each data point whose sketch is ς, function SEARCH in Algorithm 1 compares the distance $D(q, x[id[i]])$ with the current nearest distance (see line 3 in Algorithm 1).

3   Pivot Selection by Optimization Algorithms

To improve the accuracy of filtering by sketches, we consider optimization algorithms to select pivots using a dataset and a set of queries as a training example. As the methods, we use the local search (LS, for short) and the simulated annealing. The objective function is the accuracy of filtering by sketches, that is, the search recall using the sketches. As an efficient alternative of the simulated annealing, we use the AIR (Annealing by Increasing Resampling) [6,7].

Algorithm 2 formalizes the LS and AIR specialized for the pivot selection based on the evaluation $E(P, Q)$ of a tuple of pivots P using a sample Q of queries. Both the LS and AIR iterate trials to improve the tuple of pivots P. The AIR uses a small resampling of the sample to evaluate pivots at the beginning iterations, and then, gradually increases the resampling size. The number of iterations in the LS and AIR is infinity, which is replaced with an integer specified by parameter N-*trials* in realistic setting. We regard the final P as the result obtained by the LS and AIR.

Let us consider optimization using the search recall itself for a sample consisting of 10,000 queries as the objective function to *maximize*. In this paragraph, we use the NN search speeds reported in [4,5]. They include a speed fast enough to remind us of the idea of using the search recall itself as the objective function for optimization. Using the sequential filtering on 32-bit sketches with Hamming distance, the time required for the NN search with recall = 91.5% is 139 ms per query (presented in Table 5 in [5] and Table 1 in [4]). If the filtering by 18-bit sketch enumeration in $score_1$ order is used, it takes 4.97 ms per query (presented in Table 3 of [4]), which is 28 times faster than the sequential filtering Using the sequential filtering on 32-bit sketches, one measurement requires 139 ms × 10,000 = 1,390 s. If optimization is performed using the LS, it is necessary to repeat trials at least about 10,000 times, which is a rough estimation from our experiments. Since the LS measures the search recall in each trial, it takes 13.9 million seconds, 5 months and more, in total.

Thus, according to our best knowledge, no one even tried to optimize the actual search recall as a direct objective function evaluated by using the conventional search method based on sequential filtering. On the other hand, filtering by sketch enumeration enables extremely fast search, so even in a naive method using SEARCHBYSKETCHENUM to evaluate the search recall it is possible to perform the optimization within a week.

3.1   Fast Evaluation of the Search Recall

We present a method to further speed-up the evaluation of the search recall. First, we prepare the correct query answers that include the IDs of data points. We use these answers in the training phase, in which we evaluate the search recall. The approximate query answers are evaluated just using the sketches, without the actual distance calculations: we compare the answer data IDs with those obtained from the enumerated sketches based on the high-speed search principle of narrow sketches.

SEARCHBYSKETCHENUM uses sketch enumeration in increasing order of $score_p$. For each enumerated sketch ς, the actual distance $D(q, x[id[i]])$ between the query q and the data point $x[id[i]]$ is calculated (see line 3 in Algorithm 1). However, using the correct answer information about data ID of the nearest neighbor we can obtain in advance, to calculate the search recall, it is sufficient to confirm whether or not the data ID $id[i]$ appears in k' candidates (see line 5 in Algorithm 3). Therefore, the actual distance calculation can be omitted. Algorithm 3 formalizes the improved evaluation of the search recall with the function EVALRECALLBYSKETCHENUM.

4   Experiments

We propose to use the LS or AIR to select a tuple of pivots for narrow sketches, where EVALRECALLBYSKETCHENUM is used to compute the evaluation $E(\cdot, \cdot)$ of pivots as the objective function of optimization. The AIR treats the set of queries in the training example as the *sample* and uses the resampling for queries. On the other hand, the AIR uses the fixed dataset throughout in the optimization process. To confirm the effectiveness of the proposed method, we run experiments using the following two datasets, which are referred to as the Image dataset and the DeCAF dataset in the rest, respectively.

```
// Q[0],..., Q[|Q| − 1]: queries
// ans[0],..., ans[|Q| − 1]: answer (array of data ID)
1  function CHECK(ans, ς, checked)
2  |   i ← f[ς];
3  |   while i < f[ς] + num[ς] and checked < k' do
4  |   |   checked ← checked + 1;
5  |   |   if ans = id[i] then return (true, checked);
6  |   |   i ← i + 1;
7  |   |   ;
8  |   return (false, checked);

9  function EVALRECALLBYSKETCHENUM(P, Q, ds)
10 |   Prepare arrays id, f, num of bucket for ds using pivots P;
11 |   success ← 0;
12 |   for i = 0 to |Q| − 1 do
13 |   |   Setup the sketch enumeration in the ascending order of score_p for Q[i];
14 |   |   (found, checked) ← (false, 0);
15 |   |   while checked < k' and not found do
16 |   |   |   ς ← the next sketch in the enumeration;
17 |   |   |   (found, checked) ← CHECK(ans[i], ς, checked);
18 |   |   |   if found then success ← success + 1;
19 |   |   |   ;
20 |   return success/|Q|;
```

Algorithm 3: Evaluation of Search Recall by Sketch Enumeration

1. Image features (also used in papers [3–5]):
   - 6.9 million 64-dimensional vectors extracted as the 2D frequency spectrum of image frames of video movies.
   - Each of 64 axis values is represented by an 8-bit unsigned integer from 0 to 255.
   - The distance between vectors is measured by the L_1 distance function.
2. DeCAF descriptors:
   - 1 million 4,096-dimensional DeCAF descriptors [1,13].
   - Data conversion: To save the size in memory, we use converted axis values into 8-bit unsigned integers after multiplied by 10. Almost no difference in search results is derived by the conversion.
   - The distance between vectors is measure by the L_2 (Euclidean) distance function.

We use a PC with AMD Ryzen 7 3700X 3.59 GHz and memory 16 GBytes. We use OpenMP only for pivot selection in training phase, to speed-up optimization by parallel computing. We measure the speed of the NN search with no parallel computing in testing phase. All the data are stored in the main memory.

4.1   A Baseline: Pivot Selection Based on Collision Minimization

When the sketches match for different data points x and y, we say that a *collision* occurs. If the tuple of pivots P is selected so that the collision probability becomes small, it is expected that the accuracy of the filtering by sketches will be improved to some extent, as long as the width w of sketches is relatively short for the dimensionality m of data space.

We introduced *quantization ball partitioning* (QBP, for short) [3] for pivot selection based on collision minimization. QBP assumes that each coordinate of the data space has the minimum and maximum values, MIN and MAX. For example, in experiments in this paper, we use 8-bit unsigned integers where $MIN = 0$ and $MAX = 255$. For floating point values, we may use the lower and upper bound on the dataset as MIN and MAX. QBP selects any point from the dataset, quantizes it to MIN or MAX using the median of the dataset as a threshold, and makes it the candidate of the center of pivot.

QBP limits the center of pivot to be one of the *corners* of the space, data points whose axis values are MIN or MAX. It has been experimentally confirmed that QBP provides lower collision probability and more accurate filtering than BP which uses an arbitrary point as the center. This is very similar to the fact reported by Mic et al. [9], where BP with *outliers* as the center candidates achieves highly accurate filtering.

We run the *precise NN* search with 10,000 queries for the Image dataset and 1,000 queries for the DeCAF dataset, using the sequential evaluation of all distances. Then, we run SEARCHBYSKETCHENUM using pivots found by SELECT-PIVOTQBP [3]. We use sketches of width $w = 18$ for the Image dataset, and $w = 22$ for the DeCAF dataset, since these widths lead to the fastest search. We use $score_1$ for the Image dataset and $score_2$ for the DeCAF dataset, to achieve the best filtering accuracy. These search speeds are summarized in Table 2 with required k'/n and achieved recall. We will later compare these with the search speed using improved sketches.

4.2   Training and Test

Our goal is to select a tuple of pivots for sketches so that the filtering accuracy is high. We use a pair of a dataset and a set of queries to find an optimized tuple of pivots by the LS and AIR, using the search recall itself as the objective function. Such a pair should be regarded as a "training example" in terms of so-called machine learning. Therefore, the quality of the selected tuple of pivots should be evaluated using a "test case" different from the training example.

Table 2. Speed of nearest neighbor search before proposed optimization

Dataset	$n(\times 10^6)$	m	Precise NN search	SEARCHBYSKETCHENUM	k'/n	Recall
Image	6.9	64	114 (ms/query)	3.0 (ms/query)	1.5%	90%
DeCAF	1.0	4,096	625 (ms/query)	48 (ms/query)	4.5%	90%

We create a dataset for the training example of the Image dataset as a uniform sample of each 16th data point. As for the queries of the training example, a set of 10,000 queries synthesized by weighted averaging of two data points randomly selected from the dataset, in the same manner as in papers [3–5], is used. As the dataset of the test case, the whole dataset is used. A different set of 10,000 queries from the training example is used for the test case.

For the DeCAF dataset, we decided to use one eighth of 1 million data points as the dataset of the training example, and other 5,000 data points as queries. For the test case, we use the entire 1 million dataset and 1,000 queries provided by Profiset, which are disjoint from the queries of the training example.

Table 3 summarizes the training example and the test case for each dataset.

4.3   Experiments on Image Dataset

LS and AIR algorithms use the neighborhood $Nb(P)$ of the tuple of pivots P. Using our experience with the QBP, we choose one of the 2^m corners of the data space with dimensionality m as the center of pivot, and determine the radius by the median of the distances between the center and data points. Therefore, we regard that P has $m \cdot w$ *parameters* for sketches with width w. It is natural that a neighborhood of P is obtained by inverting only one parameter between *MIN* and *MAX*. However, to reduce the number of iterations of LS and AIR, we invert 10 parameters at once in the initial trials. The number of parameters to be inverted is gradually decreasing and only one is inverted at the final stage.

We use the AIR as a kind of simulated annealing, where *cooling schedule* is very important. In the AIR, we use a monotonically decreasing function $size(t)$ that specifies the size of resampling at the time t instead of temperature. According to [6], we can determine $size(t)$ so that the AIR's resampling schedule is compatible with the cooling schedule of the ordinal simulated annealing (SA, for short).

Let T_0 be the initial temperature and T_r the ratio of the current temperature T with respect to T_0, that is:

$$T = T_0 \cdot T_r \qquad (0 < T_r < 1).$$

Table 3. Training examples and test cases

Dataset	Training example		Test case	
	n	$|Q|$	n	$|Q|$
Image	6.9 million $\times \dfrac{1}{16}$	10,000	6.9 million	10,000
DeCAF	1 million $\times \dfrac{1}{8}$	5,000	1 million	1,000

Here, we adopt the most standard cooling schedule in which the temperature decreases linearly. Let *Progress* be the ratio of the number of the trials already done with the total number of trials, which is corresponding to $\dfrac{t}{N\text{-}trials}$ in AIR of Algorithm 2. Then,

$$T_r = 1 - Progress.$$

As similarly to selecting T_0 for the initial temperature, we have to select s_0 for the initial resampling size. Let $s_0 > 0$ be arbitrarily selected for the initial resampling size, that is, $size(0) = s_0$. Then, we can determine the resampling size for $t > 0$ by the following:

$$size(t) = \frac{|Q|}{\left(\frac{|Q|-s_0}{s_0}\right) T_r^2 + 1},$$

where $|Q|$ is the number of queries in the sample Q.

The choice of the initial resampling size may affect not only computing cost but also the optimization quality when $N\text{-}trials$ is not large enough. The smaller the initial resampling size is, the smaller the computing cost required for the optimization is. This property is one of the key benefits of the AIR over the SA. However, too small initial resampling size corresponds to a very high initial temperature in the SA, where stable and better solutions cannot be obtained unless the number of transition trials is increased. In our experiments, $|Q| = 10,000$ (Image), $|Q| = 5,000$ (DeCAF) and $N\text{-}trials \geq 10^5$, we choose $s_0 = 10$.

Table 4 summarizes the results of pivot selection by the LS and AIR, where the computational efficiency is observed from the elapsed time of pivot selection in the row of Time, as well as the quality of optimization from the ratios k'/n of the number of candidates k' to the dataset size n needed to achieve the respective search recall. All values are averages over 5 runs using different random seeds. Note that the LS and AIR randomly select a neighbor from $Nb(P)$ and the AIR randomly selects a resampling. We omit the NN search time by SEARCH-BYSKETCHENUM using selected tuple of pivots, which is determined only by k' when w is fixed. Later, we present the shortest NN search times using the best tuples of pivots.

Table 4. Pivot selection for image dataset

Optimization method	$N\text{-}trials$	Optimization time	Filtering performance: k'/n to achieve recall				
			75%	80%	85%	90%	95%
LS	5,000	7.4 (min)	0.142%	0.256%	0.466%	0.924%	2.27%
	10,000	15 (min)	0.138%	0.244%	0.450%	0.910%	2.20%
	15,000	23 (min)	0.136%	0.240%	0.446%	0.900%	2.21%
AIR	500,000	91 (min)	0.128%	0.222%	0.404%	0.832%	2.02%
	1,000,000	181 (min)	0.118%	0.210%	0.384%	0.762%	1.89%

By comparing results in Table 2 and Table 4, we can observe that both methods can find better tuples of pivots than SELECTPIVOTQBP. For example, to achieve 90% recall, SEARCHBYSKETCHENUM using the tuple of pivots obtained by SELECTPIVOTQBP needs $k'/n = 1.5\%$ (see Table 2), while using the tuple of pivots optimized by the AIR with N-$trials = 1,000,000$ needs only $k'/n = 0.762\%$.

When using the LS, the results are being improved just until N-$trials = 15,000$, because the number of parameters of the tuple of pivots for the Image dataset is $m \cdot w = 1,152$, and the LS gets stuck in a local optimum after approximately 20,000 trials. On the other hand, the AIR can find better tuples of pivots, but needs more computing cost. Here we point out that the running time of the AIR with N-$trials = 500,000$ is approximately 12 times bigger than the running time of the LS with N-$trials = 5,000$. Since the SA usually uses the whole training example to evaluate the objective function, the SA needs optimization time almost same as the LS. Therefore, when N-$trials = 500,000$, the AIR is expected to be about 8 times faster than the ordinal SA.

4.4 Experiments on DeCAF Dataset

Since we use sketches of width $w = 22$ for the DeCAF dataset which consists of vectors of length $m = 4,096$, we have to carefully set the neighborhood for the tuple of pivots since the number of parameters $w \cdot m = 90,112$ provides many possibilities. If about 10 parameters are inverted at once, similarly as in the case of the Image dataset, a very large number of N-$trials$ is required until a satisfactory solution is obtained for both LS and AIR. Also, simply increasing the number of inverting parameters does not work because it reduces the chance of moving to a better solution in the optimization process. Therefore, when the LS is used, we replace one of the w pivots with the one obtained by QBP, and use it as the neighborhood when searching for pivots. When the AIR is used, we need a slight modification on the neighborhood. In each trial, we make the AIR to choose a pivot with the smallest difference to the current one from 50 pivots obtained by QBP.

The results are summarized in Table 5. Increasing N-$trials$ from 5,000 to 15,000 for the LS does not significantly improve the tuple of pivots. The AIR with N-$trials = 500,000$ selects better tuple of pivots than the AIR with N-$trials = 100,000$.

4.5 Speed Tests by the Best Tuples of Pivots

Finally, we summarize the shortest NN search times achieved by SEARCHBY-SKETCHENUM using the best tuples of pivots. For Image dataset, the best tuple of pivots is selected by AIR with N-$trials = 1,000,000$. For the DeCAF dataset, it is selected by AIR with N-$trials = 500,000$. The shortest NN search times to achieve the recall of 75%, 80%, 85%, 90%, 95% on average are provided in Table 6. For instance, the NN search with recall = 90% takes 1.7 ms per query in case of Image dataset. This is by 43 % less in comparison with 3 ms per query required by SEARCHBYSKETCHENUM using the tuple of pivots selected by SELECTPIVOTQBP (see Table 2). In case of the DeCAF dataset, we reduce the NN search time from 48 to 27 ms, again when fixing the average recall of the searching to 90 %. Thus proposed method for pivot selection by AIR using accelerated evaluation algorithm EVALRECALLBYSKETCHENUM can provide a significantly better tuple of pivots, by which the more accurate and faster NN search is achieved.

We also performed 100-NN queries using the DeCAF dataset to verify the robustness of our method. The number of candidates $k'/n = 3.0\%$ imply the query execution time 40 ms per query, and the recall just 75 %.

Table 5. Pivot selection for DeCAF dataset

Optimization method	N-trials	Optimization time	Filtering performance: k'/n to achieve recall				
			75%	80%	85%	90%	95%
LS	5,000	25 (min)	1.02%	1.42%	2.19%	3.39%	6.66%
	10,000	50 (min)	1.00%	1.39%	2.09%	3.33%	6.67%
	15,000	74 (min)	1.01%	1.43%	2.15%	3.31%	6.74%
AIR	100,000	98 (min)	1.02%	1.41%	2.01%	3.17%	6.06%
	500,000	490 (min)	0.95%	1.33%	1.99%	2.99%	5.42%

Table 6. NN search time using the best tupples of pivots

Dataset	NN search time (ms/query) for recall				
	75%	80%	85%	90%	95%
Image	0.27	0.47	0.85	1.7	4.2
DeCAF	8.3	12	18	27	53

5 Conclusion

We proposed a method of pivot selection for sketches by optimization algorithms that use the recall of the NN search as the objective function. By experiments, we confirmed that optimized sketches reduce of the NN search time by 40% for both of two real-life datasets keeping the recall = 90%. As a future work, we plan to use the recall of the k-NN search as the objective function for $k > 1$. In this paper, we run all the experiments with the data stored in the memory. Therefore, we want to focus on the efficiency of the similarity search in more realistic setting for very large-scale datasets where data does no longer fit into the main memory.

Acknowledgments. This research was partly supported by ERDF "CyberSecurity, CyberCrime and Critical Information Infrastructures Center of Excellence" (No. CZ.02.1.01/0.0/0.0/16_019/0000822), and also by JSPS KAKENHI Grant Numbers 17H00762, 19K12125, 19H01133, and 20H00595.

References

1. Budikova, P., Batko, M., Zezula, P.: Evaluation platform for content-based image retrieval systems. In: Gradmann, S., Borri, F., Meghini, C., Schuldt, H. (eds.) TPDL 2011. LNCS, vol. 6966, pp. 130–142. Springer, Heidelberg (2011). https://doi.org/10.1007/978-3-642-24469-8_15

2. Dong, W., Charikar, M., Li, K.: Asymmetric distance estimation with sketches for similarity search in high-dimensional spaces. In: Proceedings of the ACM SIGIR 2008, pp. 123–130 (2008)

3. Higuchi, N., Imamura, Y., Kuboyama, T., Hirata, K., Shinohara, T.: Nearest neighbor search using sketches as quantized images of dimension reduction. In: Proceedings of the ICPRAM 2018, pp. 356–363 (2018)

4. Higuchi, N., Imamura, Y., Kuboyama, T., Hirata, K., Shinohara, T.: Fast filtering for nearest neighbor search by sketch enumeration without using matching. In: Liu, J., Bailey, J. (eds.) AI 2019. LNCS (LNAI), vol. 11919, pp. 240–252. Springer, Cham (2019). https://doi.org/10.1007/978-3-030-35288-2_20

5. Higuchi, N., Imamura, Y., Kuboyama, T., Hirata, K., Shinohara, T.: Fast nearest neighbor search with narrow 16-bit sketch. In: Proceedings of the ICPRAM 2019, pp. 540–547 (2019)

6. Higuchi, N., Imamura, Y., Shinohara, T., Hirata, K., Kuboyama, T.: Annealing by increasing resampling. In: De Marsico, M., Sanniti di Baja, G., Fred, A. (eds.) ICPRAM 2019. LNCS, vol. 11996, pp. 71–92. Springer, Cham (2020). https://doi.org/10.1007/978-3-030-40014-9_4

7. Imamura, Y., Higuchi, N., Kuboyama, T., Hirata, K., Shinohara, T.: Pivot selection for dimension reduction using annealing by increasing resampling. In: Proceedings of the LWDA 2017, pp. 15–24 (2017)

8. Mic, V., Novak, D., Zezula, P.: Improving sketches for similarity search. In: Proceedings of the MEMICS 2015, pp. 45–57 (2015)

9. Mic, V., Novak, D., Zezula, P.: Designing sketches for similarity filtering. In: ICDMW 2016, pp. 655–662 (2016)

10. Mic, V., Novak, D., Zezula, P.: Speeding up similarity search by sketches. In: Proceeding of the SISAP 2016, pp. 250–258 (2016)

11. Mic, V.: Binary Sketches for Similarity Search. Doctoral thesis, Masaryk University, Faculty of Informatics, Brno (2020)

12. Müller, A., Shinohara, T.: Efficient similarity search by reducing i/o with compressed sketches. In: Proceedings of the SISAP 2009, pp. 30–38 (2009)

13. Novak, D., Batko, M., Zezula, P.: Large-scale image retrieval using neural net descriptors. In: Proceedings of the SIGIR 2015, pp. 1039–1040 (2015)

14. Shinohara, T., Ishizaka, H.: On dimension reduction mappings for approximate retrieval of multi-dimensional data. In: Arikawa, S., Shinohara, A. (eds.) Progress in Discovery Science. LNCS (LNAI), vol. 2281, pp. 224–231. Springer, Heidelberg (2002). https://doi.org/10.1007/3-540-45884-0_14

15. Wang, Z., Dong, W., Josephson, W., Lv, Q., Charikar, M., Li, K.: Sizing sketches: a rank-based analysis for similarity search. In: Proceedings of the ACM SIGMETRICS 2007, pp. 157–168 (2007)

mmLSH: A Practical and Efficient Technique for Processing Approximate Nearest Neighbor Queries on Multimedia Data

Omid Jafari$^{(\boxtimes)}$ (iD), Parth Nagarkar (iD), and Jonathan Montaño (iD)

New Mexico State University, Las Cruces, USA
{ojafari,nagarkar,jmon}@nmsu.edu

Abstract. Many large multimedia applications require efficient processing of nearest neighbor queries. Often, multimedia data are represented as a collection of important high-dimensional feature vectors. Existing Locality Sensitive Hashing (LSH) techniques require users to find top-k similar feature vectors for each of the feature vectors that represent the query object. This leads to wasted and redundant work due to two main reasons: 1) not all feature vectors may contribute equally in finding the top-k similar multimedia objects, and 2) feature vectors are treated independently during query processing. Additionally, there is no theoretical guarantee on the returned multimedia results. In this work, we propose a practical and efficient indexing approach for finding top-k approximate nearest neighbors for multimedia data using LSH called *mmLSH*, which can provide theoretical guarantees on the returned multimedia results. Additionally, we present a buffer-conscious strategy to speed up the query processing. Experimental evaluation shows significant gains in performance time and accuracy for different real multimedia datasets when compared against state-of-the-art LSH techniques.

Keywords: Approximate nearest neighbor search · High-dimensional spaces · Locality Sensitive Hashing · Multimedia indexing

1 Introduction

Finding nearest neighbors in high-dimensional spaces is an important problem in several multimedia applications. In multimedia applications, content-based data objects, such as images, audio, videos, etc., are represented using high-dimensional feature vectors. Locality Sensitive Hashing (LSH) [8] is one of the most popular solutions for the approximate nearest neighbor (ANN) problem in high-dimensional spaces. Since it was first introduced in [8], many variants of LSH have been proposed [4,7,9,13] that mainly focused on improving the search accuracy and/or the search performance of the given queries. LSH is known for two main advantages: its sub-linear query performance (in terms of

© Springer Nature Switzerland AG 2020
S. Satoh et al. (Eds.): SISAP 2020, LNCS 12440, pp. 47–61, 2020.
https://doi.org/10.1007/978-3-030-60936-8_4

the data size) and theoretical guarantees on the query accuracy. While the original LSH index structure suffered from large index sizes (in order to obtain a high query accuracy), state-of-the-art LSH techniques [7,9] have alleviated this issue by using advanced methods such as *Collision Counting* and *Virtual Rehashing*. Thus, owing to their small index sizes, fast index maintenance, fast query performance, and theoretical guarantees on the query accuracy, we propose to build *mmLSH* upon existing state-of-the-art LSH techniques.

Motivation of Our Work: Drawbacks of LSH on Multimedia Data
Popular feature extraction algorithms, such as SIFT, SURF (for images), Marsyas (for audio), etc., extract multiple features that collectively represent the object of interest for improved accuracy during retrieval. Hence, if a user wants to find similar objects to a given query object, nearest-neighbor queries have to be performed for every individual feature vector representing the query object (and then these intermediate results are aggregated to find the final object results (Sect. 4)). Existing techniques treat these individual feature vectors as independent of each other, and hence cannot leverage common elements between these feature vector queries for improved query performance. Most importantly, existing techniques can only give theoretical guarantees on the accuracy of the individual feature vector queries, but not on the final object results, unlike our proposed index structure, *mmLSH*.

Contributions of this Paper: In this paper, we propose a practical and efficient indexing approach for finding top-k approximate nearest neighbors for multimedia data using LSH, called *mmLSH*. To the best of our knowledge, we are the first work to provide a rigorous theoretical analysis for answering approximate nearest neighbor queries on high-dimensional multimedia data using LSH. Our main contributions are:

- *mmLSH* can efficiently solve approximate nearest neighbor queries for multimedia data while providing rigorous theoretical analysis and guarantees on the accuracy of the query result.
- Additionally, we present an advanced buffer-conscious strategy to speedup the processing of a multimedia query.
- Lastly, we experimentally evaluate *mmLSH*, on diverse real multimedia datasets and show that *mmLSH* can outperform the state-of-the-art solutions in terms of performance efficiency and query accuracy.

2   Related Work

LSH was originally proposed in [8] for the Hamming distance and then later extended to the popular Euclidean distance [6]. C2LSH [7] introduced two main concepts of *Collision Counting* and *Virtual Rehashing* that solved the two main drawbacks of E2LSH [6]. QALSH [9] used these two concepts to build query-aware hash functions such that the hash value of the query object is considered as the anchor bucket during query processing. [19] proposes an efficient distributed LSH implementation which includes a cache-conscious hash table generation (to avoid cache misses to improve the index construction time). Our proposed cache-conscious optimization is to improve the efficiency of the query processing (and hence very different).

Query Workloads in High-Dimensional Spaces: Until now, only two works [10,15] have been proposed that focus on efficient execution of query workloads in high-dimensional spaces. Neither of these two works provide a rigorous theoretical guarantees on the accuracy of the final result. In [15], the authors propose to efficiently execute set queries using a two-level index structure. The problem formulation, which is quite restrictive compared to our work, states that a point will be considered in the result set only if it satisfies a certain user-defined percentage of the queries in the query workload. In [10], the authors build a model based on the cardinality and dimensionality of the high-dimensional data to efficiently utilize the cache. The main drawback of these two approaches is that they require prior information that is found by analyzing past datasets. Hence the accuracy and efficiency of the index structures is determined by the accuracy of the models. Our proposed work is very different from these previous works: *mmLSH* does not require any training models and additionally, we provide a theoretical guarantee on the accuracy of the returned results.

3   Key Concepts and Problem Specification

A *hash function family* H is (R, cR, p_1, p_2)-sensitive if it satisfies the following conditions for any two points x and y in a d-dimensional dataset $D \subset \mathbb{R}^d$: if $|x - y| \leq R$, then $Pr[h(x) = h(y)] \geq p_1$, and if $|x - y| > cR$, then $Pr[h(x) = h(y)] \leq p_2$. Here, p_1 and p_2 are probabilities and c is an approximation ratio. LSH requires $c > 1$ and $p_1 > p_2$. In the original LSH scheme for Euclidean distance, each hash function is defined as $h_{a,b}(x) = \lfloor \frac{a.x+b}{w} \rfloor$, where a is a d-dimensional random vector and b is a real number chosen uniformly from $[0, w)$, such that w is the width of the hash bucket [6]. **C2LSH** [7] showed that two close points x and y collide in at least l hash layers (out of m) with a probability $1 - \delta$.

Given a multidimensional database $\mathcal{D}$, $\mathcal{D}$ consists of n d-dimensional points that belongs to $\mathbb{R}^d$. Each d-dimensional point x_i is associated with an object X_j s.t. multiple points are *associated* with a single object. There are S objects in the database $(1 \leq S \leq n)$, and for each object X_j, $set(X_j)$ denotes the set of points that are associated with X_j. Thus, $n = \sum_{j=1}^{S} |X_j|$.

Our goal is to provide a k-NN version of the c-approximate nearest neighbor problem for multidimensional objects. For this, we propose a notion of distance between multidimensional objects called Γ-*distance* (defined in Sect. 4.1) and that depends on a percentage parameter that we denote by Γ.

Let us denote the Γ-distance between two objects X_1 and X_2 by $\Gamma dist(X_1, X_2)$. For a given query object Q, an object X_j is a Γ-*c-approximate nearest neighbor of* Q if the Γ-distance between Q and X_j is at most c times the Γ-distance between Q and its true (or exact) nearest neighbor, X_j^*, i.e. $\Gamma dist(Q, X_j) \leq c \times \Gamma dist(Q, X_j^*)$, where $c > 1$ is an *approximation ratio*.

Similarly, the Γk-NN version of this problem states that we want to find k objects that are respectively the Γ-c-approximate nearest neighbors of the exact k-NN objects of Q.

4   mmLSH

The Borda Count method [17] (along with other aggregation techniques [16]) are popular existing techniques to aggregate results of multiple point queries to find similar objects in multimedia retrieval [2]. In order to find top-k nearest neighbor *objects* of multimedia object query Q, the existing methods find the top-k' nearest neighbor *points* for each query *point* q_i, where $1 \leq i \leq |set(Q)|$, k is the number of desired results by the user, and k' is an arbitrarily chosen number such that $k' >> k$ [2]. Once the top-k' nearest neighbors of each query point q_i is found, an overall score is assigned to each multimedia object X_j based on the depth of the points (associated with X_j) in the top-k' results for each of the point queries q_i of Q. **Drawbacks of this approach:** 1) there is no theoretical guarantee for the accuracy of the returned top-k result objects, and 2) all query points q_i of the query object Q are executed independently of each other. Hence, if a query point takes too long to execute as compared to others, then the overall processing time is negatively affected. Our proposed method, *mmLSH*, solves both these drawbacks as explained in the next sections.

4.1   Key Definitions of mmLSH

Justification for Using R-Object Similarity and Γ-distance: In order to define two *Nearby Objects*, we first define a similarity/distance measure between two objects in the context of ANN search. Note that, there have been several works that have defined voting-based similarity/distance measures between two multimedia objects, especially images [11,22,23]. Also, region-based algorithms have been explored in the past, whose main strategy is to divide the query object in regions and compare these with regions of the dataset objects via some region distance, and then aggregate the resulting distances [1]. In this work we define the Γ-distance as a way to measure distances between objects as a whole. Our definition follows the naive strategy of comparing all pairs of features of the objects but it uses a percentage parameter Γ to ensure two identical objects have a near zero distance. Another key advantage of the proposed distance is that it allows us to provide theoretical guarantees for our results. In order to do so, we leverage the theoretical guarantees of LSH in our design.

Definition 1 (R-Object Similarity). *Given a radius R, the R-Object Similarity between two objects Q and X_j, that consists of $set(Q)$ and $set(X_j)$ d-dimensional feature vectors respectively, is defined as:*

$$sim(Q, X_j, R) = \frac{|\{q \in set(Q), x_i \in set(X_j) : \|q - x_i\| \leq R\}|}{|set(Q)|.|set(X_j)|} \quad (1)$$

Note that, $0 \leq sim(Q, X_j, R) \leq 1$. $sim(Q, X_j, R)$ will be equal to 1 if every point of Q is a distance at most R to every point of X_j (e.g. if you are comparing an entirely green image with another green image - and assuming the feature vectors were based on the color of the pixel. But if you are comparing two identical images, then $sim(Q, X_j, R) < 1$ if R is less than the largest among $||q - x_i||$). Since the number of points associated with two objects can be different, we normalize the similarity w.r.t the points associated with Q and X_j.

Definition 2 (Γ-distance). *Given a two objects Q and X_j, the Γ-distance between Q and X_j is defined as:*

$$\Gamma dist(Q, X_j) = \inf\{R \mid sim(Q, X_j, R) \geq \Gamma\} \tag{2}$$

In order to find points that are within R distance, we use the *Collision Counting* method that is introduced in C2LSH [7].

We define a *Collision Index* (denoted by $ci(Q, X_j)$) that determines how close two objects are based on the number of points between the two objects that are considered *close* (i.e. the collision counts between the points of the two objects is greater than the collision threshold l).

Definition 3 (Collision Index of Two Objects). *Given two objects Q and X_j, the collision index of X_j with respect to Q is defined as:*

$$ci(Q, X_j) = \frac{|\{q \in set(Q), x_i \in set(X_j): cc(q, x_i) \geq l\}|}{|set(Q)| . |set(X_j)|} \tag{3}$$

The *Collision Index* between two objects depends on how many nearby points are considered as candidates between the two objects. Thus, in turn, the accuracy of the collision index depends on the accuracy of the collision counting process (which is shown to be very high [7,9]). Hence we define an object X_j to be a Γ-*candidate* if the collision index between them is greater than or equal to $(1 - \varepsilon)\Gamma$, where $\varepsilon > \delta$ is an approximation factor which we set to 2δ.

Definition 4 (Γ-candidate Objects). *Given an object query Q and an object X_j, we say that X_j is a Γ-candidate with respect to Q if $ci(Q, X_j) \geq (1 - \varepsilon)\Gamma$.*

Additionally, we define an object to be a Γ-*false positive* if it is a Γ-candidate but its Γ-distance to the object query is too high.

Definition 5 (Γ-False Positives). *Given an object query Q and an object X_j, we say X_j is a Γ-false positive with respect to Q if we have $ci(Q, X_j) \geq \Gamma + \frac{\beta}{2}$ but $\Gamma dist(Q, X_j) > cR$.*

4.2 Design of mmLSH

During query processing, instead of executing the query points of Q independently, we execute them one at a time in each projection (Lines 5–6 in Algorithm 1). The function $CountCollisions(q_i)$ (Line 7), an existing function from

---

Algorithm 1. k-Nearest Neighbor Object

1: **while** TRUE **do**
2:     **if** $|\{X_j|X_j \in \mathcal{CL} \wedge \Gamma dist(Q, X_j) \leq cR\}| \geq k$ **then**
3:         return the top-k objects from $\mathcal{CL}$;
4:     **end if**
5:     **for** $g = 1;\ g \leq m;\ g + +$ **do**
6:         **for** $i = 1;\ i \leq |set(Q)|;\ i + +$ **do**
7:             $CountCollisions(q_i)$;
8:             $\forall_{X_j \in S}$ Update $ci(Q, X_j)$;
9:         **end for**
10:        **if** $|\mathcal{CL}| \geq k + \beta S$ **then**
11:            return the top-k objects from $\mathcal{CL}$;
12:        **end if**
13:    **end for**
14:    $R = c^{numIter}$;
15:    $numIter + +$;
16: **end while**

C2LSH, is responsible for counting collisions of query points and points in the database. The *Buffer-conscious Optimizer* module (Sect. 4.3) is responsible for finding an effective strategy to utilize the buffer to speed up the query processing. This module decides which query and the hash bucket should be processed next. The Γ-*Analyzer* module is in charge of calculating the *collision indexes* (Sect. 4.1) for objects in the database and for checking/terminating the process if the terminating conditions are met.

Terminating Conditions for mmLSH: The existing solution (Sect. 4) finds top-k' candidates for each query point in Q and then terminates. Instead, *mmLSH* stops when top-k objects are found. These conditions guarantee that Γ-c^2-approximate NN are found with constant probability (Sect. 4.4):

$\mathcal{T}1$) At certain point at level-R, at least $k + \beta S$ Γ-candidates have been found, where βS is the allowed number of false positives. (Line 14, Algorithm 1)
$\mathcal{T}2$) At the end of level-R, there exists at least k Γ-candidates whose Γ-distance to Q is at most R. (Line 6, Algorithm 1)

4.3   Buffer-Conscious Optimization for Faster Query Processing

Another goal of *mmLSH* is to improve the processing speed by efficiently utilizing a given buffer space. In order to explain our strategy, we first analyze the two expensive operations and the two naive strategies for solving the problem. The two main dominant costs in LSH-based techniques are the *Algorithm time* (which is the time required to find the candidate points that collide with the given query point) and the *Index IO time* (which is the time needed to bring the necessary index files from the secondary storage to the buffer).

Due to space limitations, we do not present a formal cost model for this process. Our main focus is on minimizing the above mentioned two dominant costs: *algorithm time* and *index IO time*. We want to store the most important hash buckets from the cache to maximize total number of buffer hits.

Table 1. Performance comparison of naive strategies NS1 and NS2 (in sec)

	Total	AlgTime	IndexIOTime
NS1: LRU	263.6	117.6	146.0
NS2: Per-Bucket	279.5	260.8	18.7

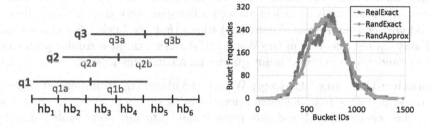

Fig. 1. (a) Query split strategy, (b) Comparison of exact and approx. Frequencies on a real and random query

Naive Strategy 1: Using LRU Eviction Strategy on a Given Buffer. Given Q, we first find the hash bucket locations for each of point queries of Q. In order to make the LRU (Least Recently Used) eviction strategy more effective, in each hash function m, we order the execution of point queries of Q according to the hash bucket locations from left to right. During query processing, we evict the LRU index files from the buffer when the buffer gets full.

Naive Strategy 2: Using a Per-bucket Execution Strategy. Since one of our goals is to reduce the *indexIOCost*, we also consider a Per-bucket execution strategy. Given a query object Q, we bring each useful hash bucket, hb, into the buffer, and for every q in Q that requires hb, we perform *Collision Counting* to find the candidate nearest neighbor points to the point query. In Fig. 1(a), this strategy would bring in hb_1 (then solve for q1), then bring hb_2 (and then solve for q1 and q2, since both queries using hb_2) and so on.

As seen from Table 1, NS1, due to its simplicity, has a lot smaller *AlgTime* than NS2, but its *IndexIOTime* is a lot more than NS2. NS2 needs to find the queries that require the particular hash bucket brought into the main memory. While this process can be sped up with more index structures, it is still an expensive operation to check for all queries. In each projection , since a hash bucket is brought into the buffer only once for NS2, *IndexIOTime* is the lowest.

Hence we propose an efficient and effective buffer-conscious strategy that reduces the *IndexIOTime* of NS1 without adding significant overhead to the *AlgTime*. Instead of using LRU, our eviction strategy is to evict a bucket based on the following three intuitive criteria. **Criterion 1:** if the bucket was not added to the buffer *very recently*. When a bucket is added to the buffer, then there is a high likelihood that another query might use it in the near future. **Criterion 2:** if the bucket is *far away* from the current query. It is more beneficial to evict another bucket that is far apart in the projection than the position of the current query. **Criterion 3:** if the number of queries that still require this bucket (called *frequency* of the bucket) is the lowest *after* the first two criteria are satisfied. Criterion 3 ensures that a bucket needed by a lot of queries is not evicted.

The main challenge is that the main criterion (Criterion 3) requires *mmLSH* to know the frequencies of each bucket to decide which bucket to evict. This is an unfair expensive requirement to have in online query processing. Across different multimedia datasets, we *observed* that the frequencies of buckets on collection of queries associated with an object showed a behavior very similar to a collection of randomly chosen queries. Figure 1(b) shows that the bucket frequencies for a randomly chosen query from the Wang [21] dataset exhibit a similar pattern for a set of randomly generated point queries on a single projection.

Projection-Dividing Strategy: We use the above stated important observation to estimate the frequencies of buckets during offline processing. The following is the overview: 1) *We divide a projection into different regions.* Too few divisions will result in a high error between the estimated and actual frequencies. Too many divisions will also result in a high error because if the frequency behavior is slightly deviated than the random queries' behavior, then we assign same frequencies as that of the random queries. For this paper, we empirically decide the total number of divisions (set to 10). 2) *We calculate the average frequencies for the random point queries for each region*, and assign the region's frequency to each bucket in that particular region. 3) For each projection, *we assign approximate frequencies to all buckets in each projection*.

Query-Splitting Strategy: In order to utilize the buffer more effectively, we split the queries into multiple sub-queries and reorder the execution of the queries based on these new set of queries. In Fig. 1(a), the query execution order will change from $q1, q2, q3$ to $q1a, q2a, q1b, q3a, q2b, q3b$ to utilize the buffer more effectively. Note that too many splits is still detrimental due to the increase in the overall Algorithm time (like Naive Strategy 2). We also try different number of splits and get the overall times of 173, 169, 171, and 176 s for 5, 10, 15, and 20 splits respectively. In this work, we empirically find a good split (that is found during the indexing phase, and set to 10).

4.4   Theoretical Analysis

Guarantees on the Stopping Conditions. The goal of this section is to prove the following theorem which provides a theoretical guarantee to *mmLSH*. For simplicity we perform the theoretical analysis for the case $k = 1$, the general case follows similarly after simple adaptations.

Theorem 1. *Let Q be a query object and let $L = \min\{|X| \mid X \in \mathcal{D}\}$. If*

$$\Gamma \geq \sqrt{\max\left\{\frac{\ln\frac{1}{\delta}}{(\varepsilon - \delta)^2 |Q|L}, \frac{2\ln\frac{2}{\beta}}{\beta^2 |Q|L}\right\}},$$

then mmLSH finds a Γ-c^2-approximate NN with constant high probability.

For the proof of this theorem we need the following lemma. For this, we consider the following two properties for a given query object Q and level R:

$\mathcal{P}1)$ If X is an object such that $\Gamma dist(Q, X) \leq R$ then X is a Γ-candidate.
$\mathcal{P}2)$ The number of Γ-false positives is at most βS.

In the next lemma, we show that the above properties hold with high probability.

Lemma 1. *Let δ be the probability defined in Sect. 3 and $\varepsilon > \delta$ as defined in Sect. 4.1, then if Γ satisfies the inequality in Theorem 1 we have $Pr[\mathcal{P}1] \geq 1 - \delta$ and $Pr[\mathcal{P}2] > \frac{1}{2}$.*

Proof. For $q_i \in set(Q)$ and $x_j \in set(X)$, let A be the condition $cc(q, x_j) \geq l$, B be $\|q - x_j\| \leq R$, and C be $\|q - x_j\| > cR$. From the proof of Lemma 1 in [7] we know the following inequalities hold:

$$Pr[A|B] \geq 1 - \delta \quad \text{and} \quad Pr[\neg A|C] \geq (1 - \exp(-2(\alpha - p_2)^2 m)) \geq (1 - \frac{\beta}{2}). \quad (4)$$

We proceed to prove inequality $Pr[\mathcal{P}1] \geq 1 - \delta$. Assume $\Gamma dist(Q, X) \leq R$, which is equivalent to $Pr[\|q_i - x_j\| \leq R] \geq \Gamma$, where $q_i \in set(Q)$ and $x_j \in set(X)$. Therefore, $p = Pr[A] \geq Pr[A \wedge B] = Pr[A|B]Pr[B] \geq (1 - \delta)\Gamma$, where the last inequality follows from the left hand side inequality in Eq. (4).

For every $1 \leq i \leq |Q|$ and $1 \leq j \leq |X|$, let $Y_{i,j} \sim Ber(1 - p)$ be a Bernoulli random variable which is equal to 1 if $cc(q_i, x_j) < l$. Then

$$Pr[ci(Q, X) \geq (1 - \varepsilon)\Gamma] = 1 - Pr[\sum_{i,j} Y_{i,j} \geq (1 - (1 - \varepsilon)\Gamma)|Q||X|]$$

$$\geq 1 - \exp(-2(\varepsilon - \delta)^2 \Gamma^2)|Q||X|,$$

where the inequality follows from Hoeffding's Inequality. Therefore for the given range of Γ we have $Pr[\mathcal{P}1] = Pr[ci(Q, X) \geq (1 - \varepsilon)\Gamma] \geq 1 - \delta$.

We continue with the proof of $Pr[\mathcal{P}2] > \frac{1}{2}$. For this, we assume $\Gamma dist(Q, X) > cR$. Which is equivalent to $Pr[\|q_i - x_j\| > cR] \geq 1 - \Gamma$. Then

$$1 - p = Pr[\neg A] \geq Pr[\neg A \wedge C] = Pr[\neg A|C]Pr[C] \geq (1 - \frac{\beta}{2})(1 - \Gamma)$$

where the last inequality follows from the right hand side inequality in Eq. (4). Therefore, $p \leq \Gamma + \frac{\beta}{2} - \frac{\beta\Gamma}{2}$. For every $1 \leq i \leq |Q|$ and $1 \leq j \leq |X|$, let $Y_{i,j} \sim Ber(1 - p)$ be a Bernoulli random variable defined as above. Thus

$$Pr[ci(Q, X) \geq \Gamma + \frac{\beta}{2}] = Pr[\sum_{i,j} Y_{i,j} \leq (1 - \Gamma - \frac{\beta}{2} - \Delta)|Q||X|]$$

for some $\Delta > 0$. Thus, from Hoeffding's Inequality it follows that

$$q = Pr[ci(Q,X) \geq \Gamma + \frac{\beta}{2}] < \exp(-2(\Gamma + \frac{\beta}{2} - p)^2|Q||X|) \leq \exp(-2(\frac{\beta}{2})^2\Gamma^2|Q||X|).$$

Let FP be the set of false positives, that is $FP = \{X \in \mathcal{D} \mid ci(Q,X) \geq \Gamma + \frac{\beta}{2}$ and $\Gamma dist(Q,X) > cR\}$, then $Pr[\mathcal{P}2] = Pr[|FP| \leq \beta S]$. Therefore, it suffices to show the latter is larger than $\frac{1}{2}$.

Let $X_1,\ldots,X_S$ denote the elements of $\mathcal{D}$. For every $1 \leq i \leq S$ let $Z_i \sim Ber(q)$ be the Bernoulli random variable which is equal to one if $X_i \in FP$. Then the expected value of the size of FP satisfies

$$E(|FP|) = E(\sum_i Z_i) = \sum_i E(Z_i) = S \cdot q < S \cdot \exp(-2(\frac{\beta}{2})^2\Gamma^2|Q||X|).$$

Therefore, from Markov's Inequality it follows that

$$Pr[|FP|] \leq \beta S]1- \geq \frac{E[|FP|]}{\beta S} > 1 - \frac{1}{\beta}\exp(-2(\frac{\beta}{2})^2\Gamma^2|Q||X|) \geq \frac{1}{2},$$

where the last inequality holds by the assumption on Γ. This finishes the proof.

We are now ready to prove the theorem.

Proof (of Theorem 1). By Lemma 1 properties $\mathcal{P}1$ and $\mathcal{P}2$ hold with constant high probability. Therefore, we may assume these properties hold simultaneously.

Let r be the smallest Γ-distance between Q and an object of $\mathcal{D}$. Set $t = \lceil \log_c r \rceil$ and $R = c^t$.

Assume first that the algorithm finishes with terminating condition $\mathcal{T}1$, that is at level R at least $1 + \beta S$ Γ-candidates have been found. By property $\mathcal{P}2$ at most βS of these are false positives. Let X be the object returned by the algorithm, then we have $\Gamma dist(Q,X) \leq cR \leq c^2 r$.

Now, if the algorithm does not finish with $\mathcal{T}1$, then property $\mathcal{P}1$ guarantees it finishes with $\mathcal{T}2$ at the end of level R. Let X be the object returned by the algorithm, then we have $\Gamma dist(Q,X) \leq R \leq cr < c^2 r$. This finishes the proof.

5 Experimental Evaluation

In this section, we evaluate the effectiveness of $mmLSH$ on four real multimedia data sets. All experiments were run on the nodes of the Bigdata cluster[1] with: two Intel Xeon E5-2695, 256 GB RAM, and CentOS 6.5 operating system. We used the state-of-the-art C2LSH [7] as our base implementation.[2] All codes were written in C++11 and compiled with gcc v4.7.2 with the -O3 optimization flag. For existing state-of-the-art algorithms (C2LSH and QALSH), we used the Borda Count process (Sect. 4) to aggregate the results of the point queries to find the

---

[1] Supported by NSF Award #1337884.
[2] $mmLSH$ can be implemented over any state-of-the-art LSH technique.

nearest neighbor objects. Additionally, since the accuracy and the performance of the aggregation is affected by the chosen number of top-k' results of the point queries, we choose a varying k' for Linear, C2LSH, and QALSH for a fair comparison: $k' = 25, 50, 100$. We also implement an LRU buffer for the indexes in C2LSH and QALSH to show a fair comparison with our results. We compare our work with the following alternatives:

- **LinearSearch-Borda:** In this alternative, the top-k' results of the point queries are found using a brute-force linear search. This method does not utilize the buffer since it does not have any indexes.
- **C2LSH-Borda:** top-k' results of point queries are found using C2LSH [7].
- **QALSH-Borda:** top-k' results of point queries are found using QALSH [9].

5.1 Datasets

We use the following four real multimedia datasets to evaluate *mmLSH*:

- **Caltech** [3] This dataset consists of 3,767,761 32-dimensional points that were created using BRIEF on 28,049 images belonging to 256 categories.
- **Corel** [5] This dataset consists of 1,710,725 64-dimensional points that were created using SURF on 9,994 images belonging to 100 categories.
- **MirFlicker** [14] This dataset consists of 12,004,143 32-dimensional points that were created using ORB on 24,980 images.
- **Wang** [21] This dataset consists of 695,672 128-dimensional SIFT descriptors belonging to 1000 images. These images belong to 10 different categories.

5.2 Evaluation Criteria and Parameters

We evaluate the execution time and accuracy using the following criteria:

- **Time:** Two main dominant costs in LSH-based techniques are the algorithm time and the index IO time. We observed that index IO times were not consistent (i.e. running the same query multiple times, would return drastically different results, mainly because of disk cache and instruction cache issues). Thus, the overall execution time is modeled for an HDD where an average disk seek requires 8.5 ms and an average data read rate is 0.156 MB/ms [18].
- **Accuracy:** Similar to the ratio defined in earlier works [7,9], we define an object ratio to calculate the accuracy of the returned top-k objects as following: $OR_\Gamma(Q) = \frac{1}{k} \sum_{i=1}^{k} \frac{\Gamma dist(Q, X_i)}{\Gamma dist(Q, X_i^*)}$ where $X_1, \ldots, X_k$ denote the top-k objects returned from the algorithm and $X_1^*, \ldots, X_k^*$ denote the real objects found from the ground truth. $\Gamma dist$ is computed using Eq. 2. Object Ratio of 1 means 100% accuracy and as it increases, the accuracy decreases.

We do not report the index size or the index construction cost, since they would be the same as the underlying LSH implementation that we use (C2LSH [7]).

We choose $\delta = 0.1$, $\beta = \frac{25}{S}$, $\varepsilon = 0.2$, $w = 2.184$ [9] for C2LSH and mmLSH, $w = 2.7191$ [9] for QALSH. We randomly chose 10 multimedia objects as queries from each dataset and report the average of the results.

5.3   Discussion of the Results

In this section, we analyze the execution time and accuracy of *mmLSH* using the criteria explained in Sect. 5.2 against its alternatives. We note that QALSH gives drastically worse times than C2LSH, and hence when comparing the effectiveness for varying parameters, we only compare with C2LSH.

Effect of Buffer Size: Figure 2(a) shows the benefit of our eviction strategy (Sect. 4.3) when compared with C2LSH + LRU for varying buffer sizes. It is evident from this figure that our three criterion are helpful in evicting less useful

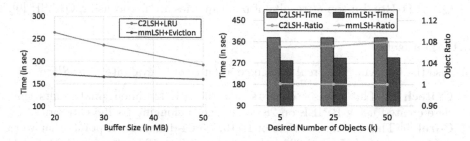

Fig. 2. Effect of (a) Buffer size on time, (b) Varying k on time and accuracy

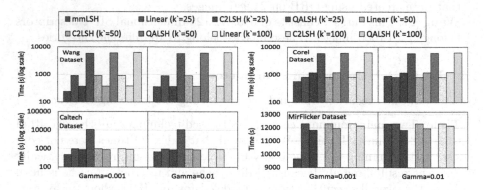

Fig. 3. Comparison of time of *mmLSH* against alternatives

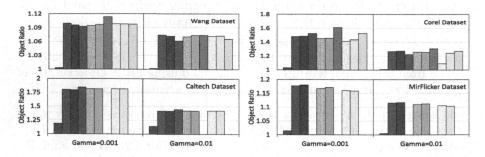

Fig. 4. Comparison of accuracy of *mmLSH* against alternatives

index files. The small overhead in Algorithm time is offset by a significant reduction in the number of index IOs. We found out that the highest L3 cache size is 24.75 MB for desktop processors and 60MB for server processors. Therefore, we decided to choose the buffer sizes between 20 MB and 50 MB. We use 30 MB as the default cache size in the following experiments.

Effect of Number of Desired Objects: Figure 2(b) shows the execution time and accuracy of *mmLSH* against C2LSH for varying number of desired objects (k). This figure shows that *mmLSH* has better time and object ratio for different k values. Additionally, it shows that *mmLSH* is scalable for a large number of desired objects as well. Moreover, although the object ratio of *mmLSH* stays the same by increasing k, the object ratio of C2LSH increases. We use $k = 25$ as the default for the following experiments.

Comparison of mmLSH vs. State-of-the-art Methods. Figures 3 and 4 show the time and accuracy of *mmLSH*, LinearSearch-Borda, C2LSH-Borda, QALSH-Borda for 4 multimedia datasets with varying characteristics. The Borda count process is done after query processing and takes very negligible time. Note that, In our work, we consider all feature-vectors that are extracted by a feature-extraction algorithm. Several works [12, 20] have been proposed that cluster these points with the purpose of finding a representative point to reduce the complexity and overall processing time of the problem. Our work is orthogonal to those approaches and hence are not included in this paper.

For the Caltech and MirFlickr datasets, QALSH did not finish the experiments due to their slow execution and hence are not included in the charts. The slow execution is mainly due to the use of the B+-tree index structures to find the nearest neighbors in the hash functions. *mmLSH* always returns a higher accuracy than the alternatives while being much faster than all three alternatives. This is because *mmLSH* is able to leverage the common elements between queries and improve cache utilization along with being able to stop earlier than the state-of-the-art algorithms. For future work, we plan on investigating the application of *mmLSH* to other distance measures and compare with other feature vector aggregation techniques [11].

6   Conclusion

In this paper, we presented a novel index structure for efficiently finding top-k approximate nearest neighbors for multimedia data using LSH, called *mmLSH*. Existing LSH-based techniques can give theoretical guarantees on these individual high-dimensional feature vector queries, but not on the multimedia object query. These techniques also treat each individual feature vector belonging to the object as independent of one another. In *mmLSH*, novel strategies are used that improve execution time and accuracy of a multimedia object query. Additionally, we provide rigorous theoretical analysis and guarantees on our returned results. Experimental evaluation shows the benefit of *mmLSH* in terms of execution time and accuracy compared to state-of-the-art algorithms. Additionally, *mmLSH* can give theoretical guarantees on the final results instead of the individual point queries.

References

1. Bartolini, I., Ciaccia, P., Patella, M.: Query processing issues in region-based image databases. Knowl. Inf. Syst. **25**, 389–420 (2010). https://doi.org/10.1007/s10115-009-0257-4
2. Arora, A., Sinha, S., Kumar, P., Bhattacharya, A.: Hd-index: pushing the scalability-accuracy boundary for approximate kNN search. In: VLDB (2018)
3. Caltech dataset. http://www.vision.caltech.edu/Image_Datasets/Caltech256
4. Christiani, T.: Fast Locality-sensitive hashing frameworks for approximate near neighbor search. In: Amato, G., Gennaro, C., Oria, V., Radovanović, M. (eds.) SISAP 2019. LNCS, vol. 11807, pp. 3–17. Springer, Cham (2019). https://doi.org/10.1007/978-3-030-32047-8_1
5. Corel dataset. http://www.ci.gxnu.edu.cn/cbir/Dataset.aspx
6. Datar, M., Immorlica, N., Indyk, P., Mirrokni, V.S.: Locality-sensitive hashing scheme based on p-stable distributions. In: SOCG (2004)
7. Gan, J., Feng, J., Fang, Q., Ng, W.: Locality-sensitive hashing scheme based on dynamic collision counting. In: SIGMOD (2012)
8. Gionis, A., Indyk, P., Motwani, R.: Similarity search in high dimensions via hashing. In: VLDB (1999)
9. Huang, Q., Feng, J., Zhang, Y., Fang, Q., Ng, W.: Query-aware locality-sensitive hashing for approximate nearest neighbor search. In: VLDB (2015)
10. Jafari, O., Ossorgin, J., Nagarkar, P.: qwLSH: cache-conscious indexing for processing similarity search query workloads in high-dimensional spaces. In: ICMR (2019)
11. Jégou, H., Douze, M., Schmid, C.: Improving bag-of-features for large scale image search. Int. J. Comput. Vis. **87**, 316–336 (2010). https://doi.org/10.1007/s11263-009-0285-2
12. Križaj, J., Štruc, V., Pavešić, N.: Adaptation of SIFT features for robust face recognition. In: Campilho, A., Kamel, M. (eds.) ICIAR 2010. LNCS, vol. 6111. Springer, Heidelberg (2010). https://doi.org/10.1007/978-3-642-13772-3_40
13. Liu, W., Wang, H., Zhang, Y., Wang, W., Qin, L.: I-LSH: I/O efficient c-approximate nearest neighbor search in high-dimensional space. In: ICDE (2019)
14. MirFlicker dataset. http://press.liacs.nl/mirflickr
15. Nagarkar, P., Candan, K.S.: PSLSH: an index structure for efficient execution of set queries in high-dimensional spaces. In: CIKM (2018)
16. Perez, C.A., Cament, L.A., Castillo, L.E.: Methodological improvement on local Gabor face recognition based on feature selection and enhanced Borda count. Pattern Recogn. **44**, 951–963 (2011)
17. Reilly, B.: Social choice in the south seas: electoral innovation and the Borda count in the Pacific Island countries. IPSR **23**, 355–372+467 (2002)
18. Seagate ST2000DM001 Manual. https://www.seagate.com/files/staticfiles/docs/pdf/datasheet/disc/barracuda-ds1737-1-1111us.pdf
19. Sundaram, N., et al.: Streaming similarity search over one billion tweets using parallel locality-sensitive hashing. In: VLDB (2013)
20. Tao, C., Tan, Y., Cai, H., Tian, J.: Airport detection from large IKONOS images using clustered SIFT keypoints and region information. In: GRSL (2011)

21. Wang, J.Z., Li, J., Wiederhold, G.: Simplicity semantics-sensitive integrated matching for picture libraries. TPAMI **23**, 947–963 (2001)
22. Wu, Z., Ke, Q., Isard, M., Sun, J.: Bundling features for large scale partial-duplicate web image search. In: CVPR (2009)
23. Zhou, W., Li, H., Lu, Y., Tian, Q.: Large scale image search with geometric coding. In: MM 2011 (2011)

Parallelizing Filter-Verification Based Exact Set Similarity Joins on Multicores

Fabian Fier[1(✉)], Tianzheng Wang[2], Erkang Zhu[3],
and Johann-Christoph Freytag[1]

[1] Humboldt-Universität zu Berlin, Berlin, Germany
{fier,freytag}@informatik.hu-berlin.de
[2] Simon Fraser University, Burnaby, Canada
tzwang@sfu.ca
[3] Microsoft Research, Redmond, USA
ekzhu@microsoft.com

Abstract. Set similarity join (SSJ) is a well studied problem with many
algorithms proposed to speed up its performance. However, its scalability
and performance are rarely discussed in modern multicore environments.
Existing algorithms assume a single-threaded execution that wastes the
abundant parallelism provided by modern machines, or use distributed
setups that may not yield efficient runtimes and speedups that are proportional to the amount of hardware resources (e.g., CPU cores). In this
paper, we focus on a widely-used family of SSJ algorithms that are based
on the filter-verification paradigm, and study the potential of speeding
them up in the context of multicore machines. We adapt state-of-the-art
SSJ algorithms including PPJoin and AllPairs. Our experiments using
12 real-world data sets highlight important findings: (1) Using the exact
number of hardware-provided hyperthreads leads to optimal runtimes for
most experiments, (2) hand-crafted data structures do not always lead
to better performance, and (3) PPJoin's position filter is more effective
in the multithreaded case compared to the single-threaded execution.

1 Introduction

The set similarity join (SSJ) operation takes two collections (or a single collection) of records and finds all pairs of records with similarities greater than a
user-defined threshold. Many data management problems are modeled as SSJ,
such as fuzzy join of two tables on a pair of text columns, record deduplication to
remove highly similar near-duplicates, and plagiarism detection to find similar
sentences or paragraphs.

In particular, a fruitful line of research work has contributed to speeding
up filter-verification based SSJ algorithms [1,3,4,12]. The basic idea is to generate candidate pairs of records, which are a superset of the result set. Com-
putationally cheap *filters* are used to keep the number of candidate pairs far
below the size of the cross product of the input collection(s). Then, a *verification*
step computes the similarity of each candidate pair. However, filter-verification

S. Satoh et al. (Eds.): SISAP 2020, LNCS 12440, pp. 62–75, 2020.
https://doi.org/10.1007/978-3-030-60936-8_5

approaches either use single-threaded or shared-nothing, distributed computing paradigms (e.g., MapReduce [6,11]). Neither approach fully exploits the parallelization potential provided by modern multi-socket multicore machines. Single-threaded solutions waste the available parallelism on modern hardware. Distributed solutions can use local parallelism by running multiple executors on one node, but assume a shared-nothing architecture that replicates data structures such as inverted indexes multiple times on the same machine, wasting memory and introducing cache inefficiency [6]. Surprisingly, they often cannot compete with single-threaded algorithms in terms of runtime and data set sizes [8].

In this paper, we explore the potential of parallelizing such filter-verification based SSJ algorithms on modern multicore machines. These machines often feature high core counts over multiple processors and exhibit non-uniform memory access (NUMA) in which remote memory access is much slower than local accesses. In this context, we adapt existing single-core SSJ algorithms to become parallel algorithms and discuss the performance impact of various design decisions such as thread placement, filtering and record inlining to improve locality.[1] Experimentally, we show how local parallelization significantly speeds up existing single-threaded approaches, without data replication and the high complexity and cost of managing a cluster of machines. Furthermore, we find some optimizations such as position filters can work even better in parallel SSJ algorithms than in sequential algorithms.

We provide the necessary background in Sect. 2. We then describe our approach to parallelizing SSJ algorithms in Sect. 3, followed by experimental evaluation in Sect. 4. Section 5 concludes this paper.

2 Background

In this section, we first define the exact SSJ problem formally and survey state-of-the-art algorithms and related work that parallelizes SSJ on different hardware platforms. We then give background on the hardware platform we target at, i.e., multi-socket, multicore servers with large main memory.

2.1 Exact Set Similarity Join

There are two categories of SSJ problems: approximate SSJ and exact SSJ. For approximate SSJ problems, it is acceptable to output pairs that are below the similarity threshold, and miss pairs that are above the threshold. We focus on the exact SSJ problem: the output pairs must be correct and there should be no missing correct pairs.

Given two collections (sets), S and R, formed over the same universe U of tokens (set elements), and a similarity function between two sets, $sim : \mathscr{P}(U) \times \mathscr{P}(U) \to [0,1]$, the SSJ between S and R computes all pairs of sets $(s,r) \in S \times R$ whose similarity exceeds a user-defined threshold t, where $0 < t \leq 1$. That is,

---

[1] Our implementation is available at https://github.com/fabiyon/ssj-sisap.

Algorithm 1: Sequential `AllPairs` algorithm.

Data: R, *invertedIndex*, t
Result: $\{(r,s)|(r,s) \in R \times R, r \neq s, sim(r,s) \geq t\}$

1 **foreach** $r \in R$ **do**
2 $\quad$ *candidates* $\leftarrow \{\}$
3 $\quad$ **foreach** *token* $\in$ GETPREFIX(r,t) **do**
4 $\quad\quad$ **foreach** $s \in$ GETLIST(*invertedIndex*, *token*) **do**
5 $\quad\quad\quad$ *candidates* $\leftarrow$ *candidates* $\cup \{s\}$
6 $\quad$ **foreach** $s \in$ *candidates* **do**
7 $\quad\quad$ VERIFY(r,s,t)

---

the output is the set of all pairs (s,r) with $sim(s,r) \geq t$. Following previous work [1,8,12] on SSJ algorithms, we hereafter exemplarily focus on all-pairs self-joins using the inverse Jaccard distance as a similarity function. However, all subsequently described approaches can be adapted to the $R \times S$ join and are applicable to other set similarity measures such as Cosine or Dice. All our datasets are a single collection R of sets consisting of sorted tokens. In the following, we use the terms *set* and *record* interchangeably.

2.2 State-of-the-Art Approaches

The exact SSJ computation can be expensive: to compute the SSJ over R, $|R| \cdot |R|$ set comparisons need to be performed in the worst case. To speed up SSJ, a line of prior work focused on minimizing the number of candidates generated. Efficient techniques for SSJ use filters to avoid comparing hopeless record pairs, i.e., pairs that provably cannot pass the threshold [1,4,12]. We distinguish two classes of filters. *Filter-verification* techniques use set prefixes or signatures followed by an explicit verification of candidate pairs (e.g., [1,12]). *Metric-based* approaches regard each record as a point in space with each token as dimension. It partitions the space such that similar records fall into the same or nearby partitions (e.g. [7]). The study in [6] suggests that this approach is not efficient. Thus, we focus on the filter-verification approach.

To generate candidate pairs, for each set r in the input collection R, the filter-verification approach aims to find other sets s in R which contain tokens (set elements) from r. Inverted indexes are used to speed up the process. For a given similarity threshold t and a set R, we only need to probe the inverted index for a subset of tokens in R (i.e., the *prefix*) to discover all possible candidates. For a Jaccard similarity threshold t, the size of the prefix can be computed as $|R| - \lceil |R| \cdot t \rceil + 1$ (referred to as *prefix filter* [4]). Any prefix-sized subset of tokens in R can be used as prefix. Thus, choosing the subset of the least frequent tokens would be the most efficient and likely yield the least number of candidates. As a result, most exact SSJ algorithms sort tokens in every set using inverse global token frequency, so that the prefix can be obtained by reading the inverted lists of the tokens starting from the beginning.

Several SSJ algorithms use prefix filtering, such as AllPairs [1], PPJoin [12] and GroupJoin [3]. Algorithm 1 shows the major steps of AllPairs. In lines 2–5, the candidates of set R are found from the inverted lists of the prefix of R; at lines 6–7 the exact similarity of each pair (r, s) is computed. PPJoin extends AllPairs by using a *position filter* which helps remove candidates based on the position of the first intersecting tokens in R and S [12]. GroupJoin further extends PPJoin by merging identical prefixes over multiple sets; this avoids the re-computation of the same overlaps [3]. MPJoin [10] introduces a removal filter. It disregards entries in the inverted index that do not pass future applications of the position filter. It uses the observation that records are indexed and probed in ascending order in length such that the required overlap increases monotonically.

Besides the CPU-based algorithms described previously, some recent work focuses on speeding up SSJ using different hardware platforms, notably GPUs. Quirino et al. proposed a standalone GPU algorithm that runs both candidate generation and verification within GPU using a block-based probing approach [9]. Bellas et al. proposed a different framework that uses GPU for candidate verification, while keeping candidate generation a CPU task [2]: working under a much-limited GPU memory, candidate pairs are verified by GPU in chunks. The experimental result in [2] indicates that the CPU-GPU solution out-performs the standalone GPU algorithm in [9]. It batches candidate pairs for verification when the number of candidates are large, which is the case for low similarity thresholds. As noted by the authors of [2], the bottleneck in SSJ is often the candidate generation rather than candidate verification, thus the acceleration provided by GPU is limited. In comparison, our work exclusively focuses on the parallelization potential of multi-core hardware in combination with existing filtering approaches and implementation optimizations – it is orthogonal to the recent work on GPU-based SSJ.

2.3 Modern Multicore Systems

We target single-node shared memory systems with multiple processors and a high core count. All processors in such a system are connected through an interconnect (e.g., Intel QPI) that implements a coherence protocol. Each processor can access it's local memory through an integrated memory controller. Local memory access is fast, while accessing remote memory attached to other processors on the same machine, comes with additional latency. This is referred to as "NUMA effect."

Modern processors use caches for performance. There are usually three levels of caches with a total size of tens of MBs. The last level cache (LLC) is usually shared among all cores in a processor. The first-level cache L1 is typically small and core-local. Modern processors usually provide hyperthreading. The idea is to better utilize a processor by allowing two processes to concurrently access different resources of one core, i. e. the arithmetic logic unit (ALU) or the floating point unit. Another important technique for performance is prefetching. Processors probabilistically read data from main memory which is likely to be used subsequently by a running program. Prefetching can hide memory stalling, i.e., a core waiting for data to arrive from main memory. Prefetching is done automatically or explicitly as instructed by software.

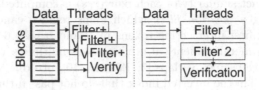

Fig. 1. Data-parallel (left) vs. pipelined (right) execution models.

3 Parallelizing Filter-Verification Based Set Similarity Join

We choose the `AllPairs` algorithm [1] as the basis filter-verification algorithm. First, we discuss the execution model (i.e., how to assign threads tasks to run SSJ algorithms), and then discuss the design considerations in the context of multicores. We show the impact of different design decisions experimentally in Sect. 4.

3.1 Execution Model

An SSJ algorithm can be parallelized using *data parallelization* or *pipelining*. Figure 1 shows the basic idea of each design choice. In data parallelization, the input data is partitioned into disjoint batches consist of a tunable number of records. Each thread then runs the `AllPairs` algorithm (or `PPJoin` when the position filter is activated) on a different batch. Multiple threads can proceed in parallel without conflicts. In practical implementations, a pool of threads can be created upon system start. After a batch is processed, the thread continues with the next batch, avoiding the cost of creating and destroying threads at runtime.

Another possible parallelization model is *pipelined* execution. The SSJ task can be subdivided into sub-tasks, each of which can be executed on a dedicated thread. The entire join algorithm is finished cooperatively by multiple threads which communicate through message passing.

Compared to data parallelization, pipelining requires frequent inter-thread communication and synchronization using message queues. We experimentally verified that such overheads were too high to make the parallel algorithm efficient. Therefore, in the rest of the paper we focus on data parallelization.

3.2 Design Considerations

Under the data-paralell execution mode, we identify four important issues in designing parallel SSJ algorithms.

Filters. It was not clear how filter techniques used in single-core algorithms may behave on multicores. Prior study [8] has shown that the prefix filter in AllPairs is the most effective filter technique. Besides the prefix filter, AllPairs also includes the important length filter. Furthermore, we explore the use of PPJoin's position filter. Our parallel SSJ algorithm ignores by default the entries in the inverted index which cannot be similar due to length differences. This is comparable to the deletion filter of MPJoin, which deletes such entries in the inverted index which cannot pass the position filter for following probe records. Since it has a significant positive impact in all our experimental cases, we decided to use this optimization by default.

Record Inlining. Records consist of a record ID (integer) and a variable number of integer tokens. A straightforward record implementation is to use a struct which contains the id and a pointer to an array of integer tokens. In order to access the tokens, the pointer has to be dereferenced first which often incurs expensive cache misses and CPU stalls as the processor waits for data to be fetched from memory to CPU caches. By inlining, we co-locate the tokens with the record ID without such extra layer of indirection. We expect this to be more efficient as it avoids pointer-chasing during record access. AllPairs reads the records including their tokens one-by-one in the filter phase, so we expect a positive effect on runtime. On the other hand, inlining introduces overhead when accessing records randomly due to the variable-length tokens. For random reads, we introduce a pointer array that maps token IDs to the location of the corresponding record in the record array. AllPairs accesses records randomly in the probe phase. As a result, in the probe phase, both variants (with and without inlining) do pointer chasing once per record.

Thread Affinity. Threads running SSJ tasks may get migrated among cores by the OS scheduler because of various events if they are not pinned to specified hardware threads or cores. This can degrade performance due to the NUMA effect in case a thread is migrated to a socket but the data it is accessing is on another socket.

Batching. In the data-parallel execution model, each thread runs the SSJ algorithm by batches. The batch size controls the number of records that are joined on one thread without synchronizing with other threads. Thus, we expect the batch size to influence the runtime.

4   Experiments

In this section, we empirically quantify the impact of the design considerations discussed in the previous section.

4.1   Setup

Our implementation uses C++11 and allows tuning of various parameters as described in the previous section. We run experiments on a server with two Intel(R) Xeon(R) CPU E5-2620 processors clocked at 2.0 GHz and 32 GB of DRAM. Each CPU has six cores (12 hyperthreads) and 32 KB/256 KB/15 MB of L1/L2/L3 caches. We use the machine exclusively for the experiments. Since system processes and hardware events (network etc.) can influence the measurements, we repeat each experiment three times and report the average to even out such effects.

Datasets. We use 10 real-world and two synthetic datasets from a prior non-distributed experimental survey [8]. Table 1 summarizes the characteristics of these datasets. We omit more detailed descriptions of the datasets available elsewhere [8]. Similar to prior work [8], we assume that records in the input collection R are sorted by ascending lengths. This is important for applying length filter and reducing index accesses. Tokens within each record are sorted by global token frequency. Using the least frequent tokens for the prefix reduces the number of candidates. There are no exact duplicates in the datasets; finding exact duplicates is an important but orthogonal problem, and does not impact our design. In order to analyse the runtime behavior on larger datasets we scale each dataset 5 and 10 times using the method from [11]. We copy each record n times, and in the copied records, each contained token is replaced with the next token in the global token frequency. Token lengths and distributions remain unchanged. Note that the approach does not introduce duplicates. Similar to prior work on sequential SSJ algorithms, we assume the input datasets fit in main memory: modern multicore servers often feature 100s of GBs or even TBs of main memory. Even if the dataset does not fully fit in memory, we expect for a majority of workloads, the working set will fit in memory such that during SSJ execution no disk I/O is involved on the critical path.

Metrics. We vary the Jaccard similarity threshold among 0.6, 0.75, and 0.9. We assume these are sensible values for many SSJ applications. We measure the runtime within our program from including the index build until the join computation is completed. We do not store the join result itself, only its size. We run our code with all combinations of variables described in the previous section and report on the results in the following. In each run, we also profile the execution using `perf` to gather metrics such as cache misses. This adds a small and constant amount of overhead, however, it does not affect relative runtimes, which are important in our discussion.

Table 1. Characteristics of the experimental datasets.

Dataset	# recs ·10^5	Record length		Universe ·10^3		Size (B)
		max	avg	size	maxFreq	
AOL	100	245	3	3900	420	396M
BPOS	3.2	164	9	1.7	240	17M
DBLP	1.0	869	83	6.9	84	41M
ENRO	2.5	3162	135	1100	200	254M
FLIC	12	102	10	810	550	92M
KOSA	6.1	2497	12	41	410	46M
LIVE	31	300	36	7500	1000	873M
NETF	4.8	18000	210	18	230	576M
ORKU	27	40000	120	8700	320	2.5G
SPOT	4.4	12000	13	760	9.7	41M
UNI	1.0	25	10	0.21	18	4.5M
ZIPF	4.4	84	50	100	98	33M

4.2 Speedups and Scalability

We first investigate how the number of threads affects runtime.

Speedup over Single-Core Execution. Using multithreading is benefitial for the SSJ runtime under all combinations of our input datasets and thresholds. We observed speedups of roughly 2–10 times on our hardware. We omit the detailed results for brevity. The absolute runtimes of the multithreaded version vary between rougly 0.2 and 262 s for all datasets and thresholds. For each combination of input and threshold, we evaluated the parameter combination of number of threads, core affinity, position filter, inlining, and batch size leading to the lowest runtime. Overall, the best runtimes were achieved by using 24 or 32 threads. 70% of the best runtimes were achieved for a batch size of 125 or 250. The position filter is effective for most (70%) of the cases. However, we did not find an optimal parameter configuration for all the combinations of dataset and threshold. In the following, we discuss the influence and interdependencies of and between the variables and draw conclusions under which conditions which variable values are favorable.

Scalability. Figure 2 shows the speedup of our experiments relative to the number of threads. Without loss of generality we only consider results with the following parameters: no inlining, batch size 500, no CPU affinity, and no position filter. Other parameter combinations show a similar behavior, so we omit them here. For the majority of results, the speedup increases linearly up to 12 threads (number of physical cores). Starting with 16 threads, we record decreasing speedups. The optimal runtime is achieved at 24 threads for all datasets except ORKU and LIVE (0.75 and 0.9 similarity thresholds), and ENRO

(0.9 similarity threshold). Since the machine has 12 physical cores, this result shows that SSJ algorithms generally benefit from hyperthreading which can hide memory access latency caused by cache misses. This is not a trivial result, as hyperthreading was shown to be only benefitial in a limited range of cases [5].

The results show that the scalability varies depending on the input dataset, the threshold, and the number of threads. Note that SPOT is an exception showing a hard limit at a speedup of 2, independent of the threshold and the number of threads. We found that the scability behavior is related to index lengths (the number of record IDs for each token). For SPOT, the average index length varies between 2.19 and 0.82 (for thresholds 0.6 and 0.9, respectively).

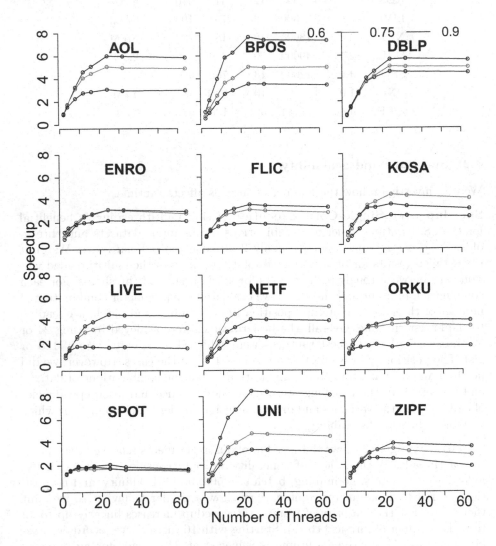

Fig. 2. Speedup under various datasets and similarity thresholds.

The index lengths of other datasets and thresholds varies between roughly 1500 (for UNI and threshold 0.6) and 2.6 (AOL, threshold 0.9). Only ENRO, FLIC, LIVE and ORKU reveal comparably short index lengths for a threshold of 0.9. The low speedup of SPOT can be explained by data access patterns which can improve or prevent prefetching. Our implementation probes the inverted index for each prefix token in each record. If there are sufficiently many entries in the postings list, the CPU can guess that they are needed subsequently. If there is only a small number contained, prefetching does not apply and wait can occur. Longer postings lists in the inverted index give better scaleups as we show in the following section.

4.3 Impact of Data Size

We enlarge our datasets synthetically as described in Sect. 4.1. With $5\times$ larger data, the runtime increases between $3.6\times$ and $44\times$; the numbers for $10\times$ larger data are $6.1\times$ and $182\times$. In most cases, the runtime does not increase linearly with respect to data size. This is expected because SSJ is a quadratic operation. The filter-verification framework only optimizes the operation depending on favorable data characteristics.

Only ENRO, FLIC, LIVE, ORKU, SPOT, and ZIPF show roughly a linear runtime increase for a threshold of 0.9; for SPOT we observe linear runtime increases under thresholds 0.75 and 0.6. As we have shown in the previous section with the original datasets, SPOT was not well parallelizable. The relative runtime increase for $5/10\times$ larger data is below $5/10\times$ for all thresholds, hence the scalability is better for the enlarged datasets. With larger datasets, the postings list lengths in the inverted index also increase. We attribute the reason to be that this makes it easier for the CPU to prefetch the larger SPOT datasets.

4.4 Impact of Inlining

Inlining only has a positive impact on runtime for a minority of our experiments. It has a generally positive impact on experiments with the AOL dataset. Figures 3–4 show the runtime gain of AOL compared to the non-inlined version relative to the parameters method (single-threaded [allp], multithreaded [allph], multithreaded with CPU affinity [allps]) and threshold. There are no clear interdependencies to the other parameters number of threads, batching and position filter. The figures show that the biggest runtime gain occurs for a threshold of 0.6. Furthermore, only the multicore implementations profit from inlining. BPOS shows a similar behavior like AOL. We omit the figures for brevity. DBLP shows only small positive effects using inlining. The biggest runtime gain occurs for a threshold of 0.9. For KOSA, there is only a positive effect at 0.6. SPOT only shows a positive effect for 0.9. Inlining has a generally neutral or negative effect on the runtimes of ENRO, FLIC, LIVE, NETF, ORKU, UNI, and ZIPF.

We expected inlining to have a positive effect on the filter phase, because it saves pointer chasing to obtain the prefix tokens. It helps the CPUs to perform prefetching. However, if prefixes are much shorter than the complete records, many tokens must be skipped to read the next record. As shown in Table 1, AOL has the smallest average record size of 3. For such short lengths, the prefix is usually not much shorter than the record. Thus, prefetching may increase the runtime if there are many short records in the input dataset.

4.5 Impact of Batching

We grouped the experimental results by all variables except the batch size and computed the percentaged difference between the lowest and the highest runtime. It varies between 0.7% and 1%. Thus, we consider the impact of batching on the runtime as rather low. Our runtime experiments suggest that the batch size of 125 is the best in most cases (23 times) and 250 is the second best (10 times). We could not find a pattern that tells us when which batch size is optimal. It seems to be a complex relation with other variables and with the data characteristics.

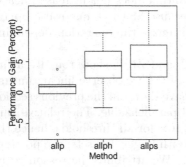
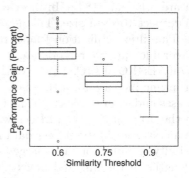

Fig. 3. AOL: Runtime gain of inlining relative to single-threaded (allp), multithreaded without (allph), and with CPU affinity (allps).

Fig. 4. AOL: Runtime gain of inlining relative to thresholds.

4.6 Position Filter

Using position filter is benefitial in most cases with thresholds of 0.6 and 0.75. Figure 5 shows the relative runtime gains using the position filter grouped by threshold. For a threshold of 0.6 the runtime gain varies between 20–50% except for SPOT, where the median is close above zero. The position filter only has a small impact on SPOT for all thresholds. This can be explained by the number of candidates. For SPOT 0.6 the position filter saves roughly 8% of candidates, or in absolute numbers 50 000. The verification of this number of additional candidates is cheaper than to filter them out before. On other datasets this filter saves 28% of candidates on average. Only for AOL, the savings with the position filter are equally low with 7%. However, the absolute number of saved candidates is orders of magnitude higher with 86 600 000, so the position filter pays off for

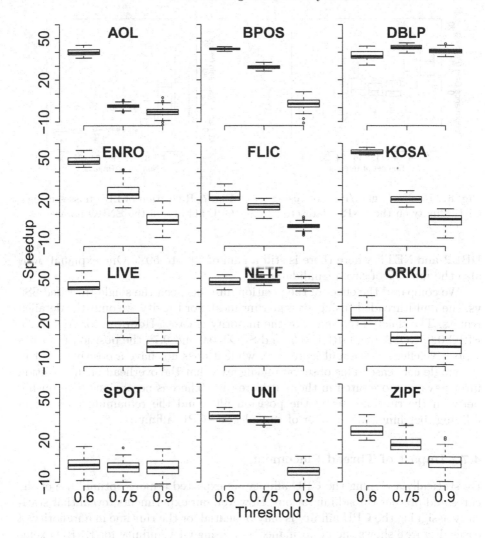

Fig. 5. Runtime gain/loss of using the position filter grouped by similarity threshold.

AOL. The `maxFreq` of tokens (cf. Table 1) gives a hint on the effectiveness of the prefix filter. The most frequent token in SPOT occurs roughly 9 700 times, which is very low compared to all other datasets. This implies that the prefix filter generates few candidates. The position filter only pays off if the prefix filter is less effective, which is the case for all other datasets and thresholds below 0.9. For a threshold of 0.75, the gain varies between 5-50% for all datasets except for SPOT (as discussed before) and AOL. For AOL the number of saved candidates relative to the number of candidates without position filter is 0.3% and thus comparably low. For a threshold of 0.9, all gains are close to zero except for

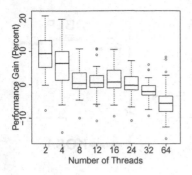
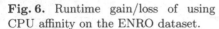
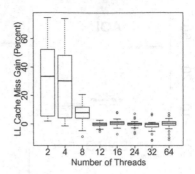

Fig. 6. Runtime gain/loss of using CPU affinity on the ENRO dataset.

Fig. 7. Reduction of LLC misses using CPU affinity on the ENRO dataset.

DBLP and NETF where there is still a gain of 40% to 50%. One explanation is also the number of saved candidates.

We compared the effect of the position filter between the single-threaded SSJ vs. the multithreaded (using average runtimes). For brevity, we omit the detailed results. The effect is the same for the majority of cases. However, for AOL 0.75, FLIC 0.9, KOSA 0.9, ORKU 0.9, and SPOT 0.6 and 0.75 the position filter has a positive effect in the multicore case, while it does not have a positive effect in the single-core case. This observation suggests that the overhead of the position filter pays off more often in the multicore case. There is no obvious relationship between the runtimes using the position filter and the remaining parameters inlining, batching, the number of threads, and CPU affinity.

4.7   Impact of Thread Placement

By statically assigning the CPU affinity we expected a more optimal use of the cores and prevent thread migrations. However, our experiments reveal that statically assigning the CPU affinity is only benefitial for the runtime in a minority of cases. Figure 6 shows the performance gain using CPU affinity for ENRO exemplarily. There is a performance gain for 2 to 4 threads. This gain decreases down to 12 threads, stays nearly the same up to 24 threads, and decreases for more threads. This can be explained by the saved cache misses. Figure 7 shows the percentage of saved LLC misses with CPU affinity. The runtime is generally the best from a number of threads starting from 24. Our results show that manually setting CPU affinity is not very helpful for optimizing SSJ algorithms.

5   Conclusion

Filter-verification based SSJ algorithms were either single-threaded or distributed, wasting much computing capability provided by multicore processors. In this paper, we fill the gap to explore the potential of parallelizing SSJ on multicores. We propose a data-parallelization execution model along with various design considerations, including the use of filters, CPU affinity, record inlining and batching. Experiments using real-world datasets revealed several important insights. Using multithreading improves SSJ runtime by 2–10× on a 12-core machine; the optimal number of threads is often the number of hardware threads (hyperthreads). Surprisingly, unlike in many other workloads, using hand-crafting data structures (e.g., using inlining) or CPU affinity do not always lead to significantly higher performance. We also find that the position filter is more effective than in the single-core scenario and should generally be used for parallel SSJ. One interesting direction of future work is to use a multithreaded CPU and GPU parallelization for the computation of the SSJ and find the optimal point (i.e., number of candidates) from where the usage of the GPU is benefitial.

Acknowledgements. This work was partially supported by a LexisNexis research grant. We thank Panagiotis Bouros for the sequential SSJ source code.

References

1. Bayardo, R.J., Ma, Y., Srikant, R.: Scaling up all pairs similarity search. In: Proceedings of the 16th International Conference on World Wide Web, pp. 131–140. ACM (2007)
2. Bellas, C., Gounaris, A.: Exact set similarity joins for large datasets in the GPGPU paradigm. In: Neumann, T., Salem, K. (eds.) Proceedings of the 15th International Workshop on Data Management on New Hardware, pp. 5:1–5:10. ACM (2019)
3. Bouros, P., Ge, S., Mamoulis, N.: Spatio-textual similarity joins. Proc. VLDB Endow. **6**(1), 1–12 (2012)
4. Chaudhuri, S., Ganti, V., Kaushik, R.: A primitive operator for similarity joins in data cleaning. In: Proceedings of the 22nd International Conference on Data Engineering (ICDE), pp. 5–5. IEEE (2006)
5. Drepper, U.: What Every Programmer Should Know About Memory, vol. 11. Red Hat Inc, Raleigh (2007)
6. Fier, F., Augsten, N., Bouros, P., Leser, U., Freytag, J.C.: Set similarity joins on MapReduce: an experimental survey. PVLDB **11**(10), 1110–1122 (2018)
7. Jacox, E.H., Samet, H.: Metric space similarity joins. ACM Trans. Database Syst. (TODS) **33**(2), 7 (2008)
8. Mann, W., Augsten, N., Bouros, P.: An empirical evaluation of set similarity join techniques. PVLDB **9**(9), 636–647 (2016)
9. Quirino, R.D., Ribeiro-Júnior, S., Ribeiro, L.A., Martins, W.S.: Efficient filter-based algorithms for exact set similarity join on GPUS. In: Hammoudi, S., Smialek, M., Camp, O., Filipe, J. (eds.) ICEIS 2017. LNCS, vol. 321, pp. 74–95. Springer, Cham (2017). https://doi.org/10.1007/978-3-319-93375-7_5
10. Ribeiro, L.A., Härder, T.: Generalizing prefix filtering to improve set similarity joins. Inf. Syst. **36**(1), 62–78 (2011)
11. Vernica, R., Carey, M.J., Li, C.: Efficient parallel set-similarity joins using MapReduce. In: Proceedings of the 2010 ACM SIGMOD International Conference on Management of data, pp. 495–506. ACM (2010)
12. Xiao, C., Wang, W., Lin, X., Yu, J.X., Wang, G.: Efficient similarity joins for near-duplicate detection. ACM Trans. Database Syst. (TODS) **36**(3), 15 (2011)

Similarity Search with Tensor Core Units

Thomas D. Ahle[1] and Francesco Silvestri[2]($\boxtimes$)

[1] IT University and BARC, Copenhagen, Denmark
thdy@itu.dk
[2] University of Padova, Padova, Italy
silvestri@dei.unipd.it

Abstract. Tensor Core Units (TCUs) are hardware accelerators developed for deep neural networks, which efficiently support the multiplication of two dense $\sqrt{m} \times \sqrt{m}$ matrices, where m is a given hardware parameter. In this paper, we show that TCUs can speed up similarity search problems as well. We propose algorithms for the Johnson-Lindenstrauss dimensionality reduction and for similarity join that, by leveraging TCUs, achieve a $\Omega(\sqrt{m})$ speedup up with respect to traditional approaches.

Keywords: Similarity search · Tensor core units · Dimensionality reduction · Similarity join · Locality sensitive hashing

1 Introduction

Several hardware accelerators have been introduced to speed up deep neural network computations, such as Google's Tensor Processing Units [13] and NVIDIA's Tensor Cores [16]. The most important feature of these accelerators is a hardware circuit to efficiently compute a small and dense matrix multiplication between two $\sqrt{m} \times \sqrt{m}$ matrices, where m is a given hardware parameter. On modern chips m can be larger than 256 [13]. Matrix multiplication is indeed one of the most frequent operations in machine learning, and specialized hardware for supporting this operation can significantly reduce running times and energy requirements [12]. We refer to these accelerators as *Tensor Core Units* (TCUs).

Recently, several studies have been investigating how to use TCUs in other domains. For instance, TCUs have been used for scanning and prefix computations [10], linear algebra primitives like matrix multiplication and FFT [9,15], and graph problems [9]. The key designing goal when developing TCU algorithms is to decompose the problem into several small matrix multiplications of size $\sqrt{m} \times \sqrt{m}$, which are then computed on the accelerator. Such algorithms also imply fast external memory algorithms, though not the other way around, since the matrix multiplication chip can be seen as a restricted cache [9].

This work was partially supported by UniPD SID18 grant, PRIN17 20174LF3T8, MIUR "Departments of Excellence".

S. Satoh et al. (Eds.): SISAP 2020, LNCS 12440, pp. 76–84, 2020.
https://doi.org/10.1007/978-3-030-60936-8_6

The goal of this paper is to show that TCUs can also speed up similarity search problems. As case studies, we propose TCU algorithms for the Johnson-Lindenstrauss dimensionality reduction and for similarity join. In both cases, our results improve the performance by a factor $\sqrt{m}$ with respect to state of the art approaches without hardware accelerators.

We analyze our algorithms on the (m, τ)-*TCU model*, which is a computational model introduced in [9] and capturing the main hardware features of TCU accelerators. In the (m, τ)-TCU model, it is possible to compute the matrix multiplication between two matrices of size $\sqrt{m} \times \sqrt{m}$ in time τ, where m and τ are given parameters. In a traditional machine, without accelerators, we have $\tau = \Theta(m^{3/2})$.[1] In contrast, with TCUs, we have $\tau = O(m)$ (i.e., input size complexity) or even sublinear time under some assumptions.

The *Johnson-Lindenstrauss (JL)* dimensionality transform reduces the dimension of a vector $x \in \mathbb{R}^d$ to roughly $k = \varepsilon^{-2} \log(1/\delta)$ while preserving its norm up to a factor $1 \pm \varepsilon$ with probability at least $1 - \delta$. It is an important primitive in many learning algorithms, since it dramatically reduces the number of trained variables, while preserving important characteristics of the feature vectors, such as their pairwise inner products. The JL transform can be represented as a multiplication of the input vector $x \in \mathbb{R}^d$ by a $k \times d$ matrix. This naively takes time $\Omega(dk)$. In this paper we use recent breakthroughs in dimensionality reduction techniques, combined with TCU's to reduce the time to $O(dk/\sqrt{m} + d + k^2 \log^3 \frac{d}{k})$. This is significant, since TCUs typically cut a factor $\sqrt{m}$ off matrix-matrix multiplication, but here we cut $\sqrt{m}$ off *matix-vector* multiplication! When $\sqrt{m} \geq k$ our dimensionality reduction takes time linear in the input dimension. This improves upon even the famous "Fast Johnson Lindenstrauss" transform [6], which takes time $\Omega(d \log d + k^{2+\gamma})$ for any $\gamma > 0$ [7], or $\Omega(d \frac{\log d}{\log m})$ with TCU optimized FFT [9].

The *Similarity Join* on two sets P and Q of n points each in $\mathbb{R}^d$, asks us to find all pairs $(x, y) \in P \times Q$ whose distance is below a given threshold r (i.e., all near pairs). Similarity join occurs in numerous applications, such as web deduplication and data cleaning. As such applications arise in large-scale datasets, the problem of scaling up similarity join for different metric distances is getting more important and more challenging. Exact similarity join cannot be faster than brute force [4], but by leveraging Locality Sensitive Hashing (LSH), we will develop a TCU approximate algorithm that, under some assumptions, finds all pairs in expected time $O((\frac{n}{\sqrt{m}})^\rho (\frac{|P \bowtie_r Q| d}{\sqrt{m}} + n))$, where $|P \bowtie_r Q|$ is the number of near pairs. When $\tau = O(m)$, the TCU algorithm exhibits a $\Omega(\sqrt{m})$ speedup with respect to traditional approaches (even those based on LSH).

---

[1] Fast matrix multiplication algorithms require $O(m^{\omega/2})$ time with $\omega \in [2, 3]$, [8], but they exhibit poor experimental performance than traditional $\Theta(m^{3/2})$ algorithms.

2   Preliminaries

2.1   The TCU Model

(m, τ)-TCU model is a RAM model with an instruction to multiply two dense matrices of size $\sqrt{m} \times \sqrt{m}$ in time τ, where m and τ are given parameters depending on the underline platform.[2] It is reasonable to assume that $\tau = O(m)$, that is matrix multiplication takes linear time: indeed, on TCUs, the cost of the operation is upper bounded by the time for reading/writing the $\sqrt{m} \times \sqrt{m}$ matrices, while the cost of the $m^{3/2}$ elementary products is negligible due to the high level of parallelism inside TCU accelerators (e.g., systolic array). Moreover, under some conditions on high bandwidth connections, we might have τ to be even sublinear (e.g., $O(\sqrt{m})$). We recall a result from [9] that will be used later:

Theorem 1. *Let A and B be two matrices of size $p \times r$ and $r \times q$ with $p, r, q \geq \sqrt{m}$, then there exists an algorithm for computing $A \cdot B$ on a (m, τ)-TCU model in time $O(prqm^{-3/2}\tau)$.*

2.2   Johnson-Lindenstrauss Dimensionality Reduction

We say a distribution over random matrices $M \in \mathbb{R}^{k \times d}$ is a (ε, δ)-Johnson-Lindenstrauss (JL) distribution, if we have $\Pr\left[|\|Mx\|_2 - 1| \leq \varepsilon\right] \geq 1 - \delta$ for all unit vectors $x \in \mathbb{R}^d$ In this section we will note some definitions and lemmas related to building and combining random matrices in ways related to JL distributions. The first property was introduced by Kane and Nelson [14]:

Definition 1 (JL-moment property). *We say a distribution over random matrices $M \in \mathbb{R}^{k \times d}$ has the (ε, δ, p)-JL-moment property, when $E[\|Mx\|_2^2] = 1$ and $\left(E\left[\left|\|Mx\|_2^2 - 1\right|^p\right]\right)^{1/p} \leq \varepsilon\delta^{1/p}$ for all $x \in \mathbb{R}^d$, $\|x\|_2 = 1$.*

A distribution with the (ε, δ, p)-JL-moment property is (ε, δ)-JL because of Markov's inequality: $\Pr\left[|\|Mx\|_2 - 1| > \varepsilon\right] \leq E\left[\left|\|Mx\|_2^2 - 1\right|^p\right]/\varepsilon \leq \delta$.

An interesting property of the JL Moment Property is related to the tensor product of matrices. The tensor (or Kronecker) product between two matrices $A \in \mathbb{R}^{m \times n}$ and $B \in \mathbb{R}^{k \times \ell}$ is defined as below. In particular, if we take the tensor product $I_k \otimes A$, where I_k is the $k \times k$ identity matrix, we get a $km \times kn$ block matrix with A on the diagonal:

$$A \otimes B = \begin{bmatrix} A_{1,1}B & \cdots & A_{1,n}B \\ \vdots & \ddots & \vdots \\ A_{m,1}B & \cdots & A_{m,n}B \end{bmatrix}, \quad I_k \otimes A = \begin{bmatrix} A & 0 & \cdots & 0 \\ 0 & A & \ddots & \vdots \\ \vdots & \ddots & \ddots & 0 \\ 0 & \cdots & 0 & A \end{bmatrix}.$$

---

[2] The model in [9] is slightly different, and we use here a simplified version for the clarity of exposition.

The tensor product relates to the JL-moment property by the following simple lemma from [1]:

Lemma 1 (JL Tensor lemma). *For any matrix, Q, with (ε, δ, p)-JL moment property, $I_k \otimes Q$ has (ε, δ, p)-JL moment property.*

By the simple property $A \otimes B = (I \otimes B)(A \otimes I)$ this lemma allows studying the JL properties of general tensor products, as long as we can also handle matrix products. The following generalization of the JL Moment Property will be key to doing exactly that:

Definition 2 ((ε, δ)-Strong JL Moment Property). *Let $\varepsilon, \delta \in [0, 1]$. We say a distribution over random matrices $M \in \mathbb{R}^{m \times d}$ has the (ε, δ)-Strong JL Moment Property, when $E\left[\|Mx\|_2^2\right] = 1$ and $\left(E\left[\left|\|Mx\|_2^2 - 1\right|^p\right]\right)^{1/p} \leq \frac{\varepsilon}{e}\sqrt{\frac{p}{\log 1/\delta}}$, for all $x \in \mathbb{R}^d$, $\|x\|_2 = 1$ and all p such that $2 \leq p \leq \log 1/\delta$.*

Note that the (ε, δ)-Strong JL Moment Property implies the $(\varepsilon, \delta, \log 1/\delta)$-JL Moment Property, since then $\varepsilon \delta^{1/p} = \varepsilon/e$. Similarly, having the $(\varepsilon\sqrt{2/e}, \delta, p)$-JL-moment property for all $p \in [2, \log 1/\delta]$ implies the Strong JL Moment Property, since $\delta^{1/p} \leq \frac{1}{\sqrt{2e}}\sqrt{\frac{p}{\log 1/\delta}}$.

The key workhorse is the following lemma by Ahle and Knudsen [2]. Note that the original lemma required the $(\varepsilon/(C_0\sqrt{k}), \delta)$-Strong JL Moment Property, but a quick scan of the proof shows that $(\varepsilon/(C_0\sqrt{i}), \delta)$-Strong suffices.

Lemma 2 (JL Product lemma). *There exists a universal constant C_0, such that, for any constants $\varepsilon, \delta \in [0, 1]$ and positive integer $k \in \mathbb{Z}_{>0}$. If $M^{(1)} \in \mathbb{R}^{d_2 \times d_1}, \ldots, M^{(k)} \in \mathbb{R}^{d_{k+1} \times d_k}$ are independent random matrices satisfying the $(\varepsilon/(C_0\sqrt{i}), \delta)$-Strong JL Moment Property, then the matrix $M = M^{(k)} \cdot \ldots \cdot M^{(1)}$ has the (ε, δ)-Strong JL Moment Property.*

Intuitively this says that combining k JL reductions, we don't get an error of εk, as we would expect from the triangle inequality, but only $\varepsilon\sqrt{k}$, as we would expect from a random walk.

2.3 Locality Sensitive Hashing

Much of recent work on similarity search and join has focused on Locality Sensitive Hashing: at a high level, similar points (i.e., with distance $\leq r$) are more likely to collide than far points (i.e., with distance $\geq cr$ for a given approximation factor c). Formally, an LSH is an (r, cr, p_1, p_2)-sensitive hashing scheme:

Definition 3. *Fix a distance function $D : \mathbb{U} \times \mathbb{U} \to \mathbb{R}$. For positive reals r, c, p_1, p_2, and $p_1 > p_2, c > 1$, a family of functions $\mathcal{H}$ is (r, cr, p_1, p_2)-sensitive if for uniformly chosen $h \in \mathcal{H}$ and all $x, y \in \mathbb{U}$:*

- *If $D(x, y) \leq r$ then $Pr[h(x) = h(y)] \geq p_1$;*
- *If $D(x, y) \geq cr$ then $Pr[h(x) = h(y)] \leq p_2$.*

We say that $\mathcal{H}$ is monotonic if $Pr[h(x) = h(y)]$ is a non-increasing function of the distance function $D(x, y)$.

LSH schemes are characterized by the $\rho = \log_{p_2} p_1$ value, with $\rho \in [0, 1]$: small values of ρ denote LSHs that well separate near points from far points. Term c is the approximation factor.

3   Dimensionality Reduction

We will describe a construction of a matrix $M \in \mathbb{R}^{k \times d}$ which is (ε, δ)-JL as described in the preliminaries, and for which there is an efficient algorithm for computing the matrix vector product Mx on a TCU. We first give a general lemma describing the construction, then show how it applies to TCUs:

Lemma 3. *Let $T(a, b, c)$ be the time for multiplying two matrices of size $(a \times b)$ and $(b \times c)$. For a constant $C > 0$ and for any $d, \varepsilon, \delta > 0$, there exists a matrix $M \in \mathbb{R}^{k \times d}$, with $k = \lceil C\varepsilon^{-2} \log 1/\delta \rceil$, such that $|\,\|Mx\|_2 - \|x\|_2| \leq \varepsilon \|x\|_2$ for any $x \in \mathbb{R}^d$ with probability $1 - \delta$ (i.e., M is (ε, δ)-JL). The multiplication Mx can be computed in time $\sum_{i=1}^{\ell} T(ik, \zeta ik, \zeta^{\ell-i})$ for any $\zeta > 1$ and ℓ such that $\zeta^\ell = d/k$.*

Note that, depending on the speed of the rectangular matrix multiplication, it might be beneficial to pick different values for ζ.

Proof. We define the JL transformation by the following matrix:

$$M = (I_{r_\ell} \otimes A_\ell) \cdots (I_{r_1} \otimes A_1) \in \mathbb{R}^{r_m k_\ell \times r_1 c_1},$$

where $r_1, \ldots, r_\ell$ is a sequence of positive integers, I_r is the $r \times r$ identity matrix, and $A_1, \ldots, A_{\ell-1}$ are independent $k_i \times c_i$ matrices, where A_i has the $(\varepsilon/(C_0\sqrt{i}), \delta)$-Strong JL Moment Property (SJLMP). By Lemmas 1 and 2 we get that the tail $(I_{r_{\ell-1}} \otimes A_{\ell-1}) \cdots (I_{r_1} \otimes A_1) \in \mathbb{R}^{r_m k_\ell \times r_1 c_1}$ has the $(\varepsilon/\sqrt{C_0}, \delta)$-SJLMP. We further assume A_ℓ has the $(\varepsilon/(\sqrt{2C_0}), \delta)$-SJLMP. Again by Lemmas 1 and 2 we get that M has the (ε, δ)-SJLMP, and thus M is a JL reduction as wanted.

Next we prove the running time of the matrix-vector multiplication. The key is to note that $I \otimes A$ is the "block identity matrix" with A copied along the diagonal. The following figure should give some some intuition:

$$(I_{r_i} \otimes A_i)x = \underset{\text{blocks}}{\overset{r_i}{}}\left\{ \begin{bmatrix} k_i \left\{\underbrace{A_i}_{c_i}\right. & & \\ & A_i & \\ & & A_i \end{bmatrix} x \simeq A_i \begin{bmatrix} x_1 \ldots x_{r_i} \end{bmatrix} \right\} c_i = \begin{bmatrix} y_1 \ldots y_{r_i} \end{bmatrix} \right\} k_i.$$

By splitting x into r_i blocks, the multiplication $(I_{r_i} \otimes A)x$ corresponds to reducing each block of x by identical JL matrices. Repeating this process for a logarithmic number of steps, we get the complete dimensionality reduction.

To make sure the matrix sizes match up, we have

$$d = r_1 c_1, \quad r_1 k_1 = r_2 c_2, \quad r_2 k_2 = r_3 c_3, \quad \ldots, \quad r_{\ell-1} k_{\ell-1} = r_\ell c_\ell, \quad r_\ell k_\ell = k.$$

We will define $k = \lceil C\varepsilon^{-2}\log 1/\delta \rceil$, $k_{i<\ell} = ik$, $k_\ell = k$, $c_1 = k\zeta$, $c_{i>1} = \zeta k_{i-1}$, $r_i = \zeta^{\ell-i}$ and $\ell = \frac{\log(d/k)}{\log \zeta}$ such that $c_1 r_1 = k\zeta^\ell = d$. The constant C depends on the constant of the JL lemma we use for the individual A_i, but in general $10C_0^2$ will suffice, where C_0 is the constant of Lemma 2.

Recall the assumption that rectangular multiplication takes time $T(a, b, c)$, and hence the ith step thus takes time $T(k_i, c_i, r_i)$. Adding it all up we get

$$\sum_{i=1}^{\ell} T(k_i, c_i, r_i) = T(k, \zeta k(\ell-1), 1) + \sum_{i=1}^{\ell-1} T(ik, \zeta k \max(1, i-1), \zeta^{\ell-i})$$

which is then upper bounded by $\sum_{i=1}^{\ell} T(ik, \zeta ik, \zeta^{\ell-i})$. The claim follows.

By the above theorem and by using the matrix multiplication algorithm of Theorem 1, we get the following theorem (see the full version [5] for the proof).

Theorem 2. *For any $d, \varepsilon, \delta > 0$, there exists a (ε, δ)-JL matrix $M \subset \mathbb{R}^{k \times d}$ such that the product Mx can be computed in time $O((dk + k^2\sqrt{m}\log^3 \frac{d}{k})\tau m^{-3/2})$, on the (m, τ)-TCU model, assuming $k \geq \sqrt{m}$.*

In particular for $\tau = O(m)$ it takes time $O(dk/\sqrt{m} + k^2 \log^3 \frac{d}{k})$. If $\sqrt{m} > k$ we can "pad" the construction by increasing k to $\sqrt{m}$ and simply throw away the unneeded rows. The running time is then $O(d + k^2 \log^3 \frac{d}{k})$. We observe that if $\tau = O(m)$ and d dominates k^2, then we get time $O(dk/\sqrt{m}))$, which improves a factor $\sqrt{m}$ over a standard application of the standard JL transform in the case of dense vectors, and for $m \approx k$ this even improves upon the so-called "Fast JL transform" [6].

Finally, we note the following extra properties of the construction:

1. In the case of sparse vectors, where many blocks of x are empty, we can skip them in the computation.
2. The computation can be easily parallelized, with different blocks of x being reduced on different machines. Our construction also implies a $O(dk/\sqrt{m})$ upper bound in the external memory model.
3. Our construction improves upon the standard matrix-vector multiplication for JL, even in the RAM model, by using the Coppersmith-Winograd method for fast matrix multiplication. In particular we can do JL in time $dk^\varepsilon + k^{2+\varepsilon}$ if matrix multiplication takes time $n^{2+\varepsilon}$.
4. The construction works with any distribution of matrices that have the Strong JL Moment Property. This means we can use random ± 1 matrices or even ε-Sparse JL matrices.

4 Similarity Join

We now study the similarity join problem: given two sets P and Q of n points each in $\mathbb{R}^d$ and a distance function $D : \mathbb{R}^d \to R_0^+$, compute the set $P \bowtie_r Q = \{(x, y) : x \in P, y \in Q, D(x, y) \leq r\}$. We consider distance functions that can be computed with an inner product on a suitable transformation of the two points: a distance function D is an *ip-distance* is there exist two functions $f, g : \mathbb{R}^d \to \mathbb{R}^{d'}$ such that $D(x, y) = f(x) \cdot g(y)$ for each pair $x, y \in \mathbb{R}^d$ For the sake of simplicity, we assume $d' = \Theta(d)$. Notable examples of ip-distances are Hamming, squared L_2 distance, and cosine similarity: for Hamming, $f(x) = (x_0, 1 - x_0, x_1, 1 - x_1, \ldots, x_{d-1}, 1 - x_{d-1})$ and $g(x) = (1 - y_0, y_0, 1 - y_1, y_1 \ldots, 1 - y_{d-1}, y_{d-1})$; for the squared L_2 distance, $f(x) = (x_0^2, 1, -2x_0, x_1^2, 1, -2x_1 \ldots, x_{d-1}^2, 1, -2x_{d-1})$ and $g(x) = (1, y_0^2, y_0, 1, y_1^2, y_1 \ldots, 1, -y_{d-1}^2, y_2)$; for cosine similarity, $f(x) = g(x) = x/\|x\|_2$.

The simplest way to exploit TCUs is a brute force approach, where all pair distances are computed. As ip-distance computations can be translated into inner products, we can reduce the similarity join problem to a simple matrix multiplication between two $n \times d'$ matrices F_P and G_Q: F_P and G_Q are the matrices representing, respectively, the sets $\{f(p), \forall p \in P\}$ and $\{g(q), \forall q \in Q\}$. By exploiting TCUs, we can compute $P \cdot Q^T$ in time $O(dn^2 m^{-3/2}\tau)$.

A more efficient approach uses LSH for reducing the number of candidate pairs for which we have to compute distances. The proposed algorithm finds all $P \bowtie_r Q$ pairs in expectation, but it can be easily modified to return all near pairs with high probability by running $O(\log n)$ instances of the algorithm and merging the results.

The standard LSH approach for similarity join (see e.g. [11,17]) partitions the points in $P \cup Q$ into buckets using an (r, cr, p_1, p_2)-sensitive monotone LSH. A brute force algorithm is then used for searching similar pairs within each bucket. The procedure is repeated L times with independent LSHs to guarantee that all near pairs are found. The LSH is usually set so that $p_2 = 1/n$, which implies that each point collides once (in expectation) with a point at distance larger than cr (i.e., a far point), while L is set to $\tilde{O}\left(p_1^{-1}\right) = \tilde{O}\left(p_2^{-\rho}\right) = \tilde{O}\left(n^\rho\right)$ to guarantee that each near pair is found once (in expectation).

As for similarity join in the external memory model [17], we can improve the performance in the TCU model by increasing the value of p_2 (i.e., by allowing for more collisions between far points), which implies that the number L of repetitions decreases since $L = p_1^{-1} = \tilde{O}\left(p_2^{-\rho}\right)$. We observe that a TCU unit can multiply two matrices of size $\sqrt{m'} \times \sqrt{m'}$ in a $\text{TCU}(m, \tau)$ in τ time for each $m' \leq m$, and we exploit this fact by increasing the number of collisions with far points. We set $p_2 = m^{3/2}/(\tau n)$: each point collides in expectation with at most $m^{3/2}/\tau$ far points, but the overhead due to the respective inner products do not dominate the running time.

As an LSH is usually given as a black box $\mathcal{H}'$ with fixed probability values p_1' and p_2', we can get the desired probability $p_2 = m^{3/2}/(\tau n)$ by concatenating $k = \log_{p_2'} p_2$ hash functions. However, if k is not an integer, the rounding gives $L = O(n^\rho p_1^{-1})$. A more efficient approach has been recently proposed in [3] that uses L_{high} hash tables by concatenating $\lceil k \rceil$ LSHs $\mathcal{H}'$, and L_{low} hash tables by concatenating $\lfloor k \rfloor$ LSHs $\mathcal{H}'$, and where $L = L_{low} + L_{high} = O(n^\rho p_1^{-(1-\rho)})$. The right values of L_{low} and L_{high} depend on the decimal part of k.

We have the following result (see the full version [5] for the proof).

Theorem 3. *Given two sets $P, Q \subset R^d$ of n points, with $n, d \geq \sqrt{m}$, a threshold value $r > 0$, and an (r, c, p_1, p_2)-sensitive monotone LSH, then the set $P \bowtie_r Q$ for an ip-distance can be computed on a $TCU(m, \tau)$ in expected time:*

$$O(p_1^{\rho-1}(n\tau m^{-3/2})^\rho \left(\frac{|P \bowtie_r Q|\tau}{m^{3/2}} + n \right) + \tau m^{-3/2}|P \bowtie_{cr} Q|).$$

When $\tau = O(m)$, there are at least $n\sqrt{m}$ near pairs, and the number of pairs with distance in $[r, cr]$ is at most linear with the number of near pairs (which happens in several datasets [17]), the cost is $O(p_1^{\rho-1}(n/\sqrt{m})^\rho |P \bowtie_r Q|/\sqrt{m})$, a factor at least $\sqrt{m}$ faster than an LSH solution without TCU (e.g., $O(p_1^{\rho-1}n^\rho|P \bowtie_r Q|)$).

5 Conclusion

In this paper, we have investigated from a theoretical point of view how to exploit TCU accelerators for similarity search problems, showing a $\Omega(\sqrt{m})$ improvement over algorithms for traditional architectures. As future work, we plan to experimentally evaluate our algorithms on common TCU accelerators, such as the GPU Nvidia Tesla.

References

1. Ahle, T.D., et al.: Oblivious sketching of high-degree polynomial kernels. In: Proceedings of the 40th Symposium on Discrete Algorithms (SODA), pp. 141–160 (2020)
2. Ahle, T.D., Knudsen, J.B.: Almost optimal tensor sketch. arXiv preprint arXiv:1909.01821 (2019)
3. Ahle, T.D.: On the problem of p_1^{-1} in locality-sensitive hashing. In: Proceedings of the 13th International Conference on Similarity Search and Applications (SISAP) (2020)
4. Ahle, T.D., Pagh, R., Razenshteyn, I., Silvestri, F.: On the complexity of inner product similarity join. In: Proceedings of the 35th Symposium on Principles of Database Systems (PODS), pp. 151–164 (2016)
5. Ahle, T.D., Silvestri, F.: Similarity search with tensor core units. arXiv preprint arXiv:2006.12608 (2020)
6. Ailon, N., Chazelle, B.: Approximate nearest neighbors and the fast johnson-lindenstrauss transform. In: Proceedings of the 38th Symposium on Theory of computing (STOC), pp. 557–563 (2006)
7. Ailon, N., Liberty, E.: Fast dimension reduction using rademacher series on dual BCH codes. Discr. Comput. Geom. **42**(4), 615 (2009)
8. Alman, J.: Limits on the universal method for matrix multiplication. In: Proceedings of the 34th Computational Complexity Conference (CCC), vol. 137, pp. 12:1–12:24 (2019)
9. Chowdhury, R., Silvestri, F., Vella, F.: Brief announcement: a computational model for tensor core units. In: Proceedings of the 32nd Symposium on Parallelism in Algorithms and Architectures (SPAA) (2020)

10. Dakkak, A., Li, C., Xiong, J., Gelado, I., Hwu, W.M.: Accelerating reduction and scan using tensor core units. In: Proceedings of the International Conference on Supercomputing (ICS) (2019)
11. Gionis, A., Indyk, P., Motwani, R.: Similarity search in high dimensions via hashing. In: Proceedings of VLDB'99, pp. 518–529 (1999)
12. Jouppi, N.P., Young, C., Patil, N., Patterson, D.A.: A domain-specific architecture for deep neural networks. Commun. ACM **61**(9), 50–59 (2018)
13. Jouppi, N.P., et al.: In-datacenter performance analysis of a tensor processing unit. In: Proceedings of the 44th International Symposium on Computer Architecture (ISCA), pp. 1–12 (2017)
14. Kane, D.M., Nelson, J.: Sparser Johnson-Lindenstrauss transforms. J. ACM (JACM) **61**(1), 1–23 (2014)
15. Lu, T., Chen, Y.F., Hechtman, B., Wang, T., Anderson, J.: Large-scale discrete fourier transform on TPUs (2020)
16. Nvidia Tesla V100 GPU architecture. http://images.nvidia.com/content/volta-architecture/pdf/volta-architecture-whitepaper.pdf
17. Pagh, R., Pham, N., Silvestri, F., Stöckel, M.: I/O-efficient similarity join. Algorithmica **78**(4), 1263–1283 (2017)

On the Problem of p_1^{-1} in Locality-Sensitive Hashing

Thomas Dybdahl Ahle[✉][iD]

IT University and BARC, Copenhagen, Denmark
thomas@ahle.dk
http://thomasahle.com

Abstract. A Locality-Sensitive Hash (LSH) function is called (r, cr, p_1, p_2)-sensitive, if two data-points with a distance less than r collide with probability at least p_1 while data points with a distance greater than cr collide with probability at most p_2. These functions form the basis of the successful Indyk-Motwani algorithm (STOC 1998) for nearest neighbour problems. In particular one may build a c-approximate nearest neighbour data structure with query time $\tilde{O}(n^\rho/p_1)$ where $\rho = \frac{\log 1/p_1}{\log 1/p_2} \in (0, 1)$. This is *sub-linear* as long as p_1 is not too small. Such an algorithm is significant, since most high dimensional nearest neighbour problems suffer from the curse of dimensionality, and can't be solved *exact*, faster than a brute force *linear-time* scan of the database.

Unfortunately many of the best LSH functions tend to have very low collision probabilities, including the best functions for Cosine and Jaccard Similarity. This means that the n^ρ/p_1 query time of *LSH is often not sub-linear* after all, even for *approximate* nearest neighbours!

In this paper, we improve the general Indyk-Motwani algorithm to reduce the query time of LSH to $\tilde{O}(n^\rho/p_1^{1-\rho})$ (and the space usage correspondingly.) Since $n^\rho/p_1^{1-\rho} < n \Leftrightarrow p_1 > n^{-1}$, our algorithm always obtains sublinear query time, for all collision probabilities at least $1/n$. For p_1 and p_2 small enough, our improvement over all previous methods can be *up to a factor n* in both query time and space.

The improvement comes from a simple change to the Indyk-Motwani algorithm, which we call "LSH with High-Low Tables". This technique can easily be implemented in existing software packages.

Keywords: Locality-sensitive hashing · Nearest neighbour · Similarity search

1 Introduction

Locality Sensitive-Hashing (LSH) [16] is one of the most efficient approaches to the nearest neighbour search problem in high dimensional spaces. It comes with theoretical guarantees, and it has the advantage of easy adaption to nearly any metric or similarity function one might want to use for search.

© Springer Nature Switzerland AG 2020
S. Satoh et al. (Eds.): SISAP 2020, LNCS 12440, pp. 85–93, 2020.
https://doi.org/10.1007/978-3-030-60936-8_7

The (r_1, r_2)-near neighbour problem is defined as follows: Given a set X of points, we build a data-structure, such that given a query, q we can quickly find a point $x \in X$ with distance $< r_2$ to q, or determine that X has no points with distance $\leq r_1$ to q. Given a solution to this "gap" problem, one can obtain a r_1/r_2-approximate nearest neighbour data structure, or even an exact[1] solution using known reductions [2,13,15].

For any measure of similarity, the gap problem can be solved by LSH: we find a distribution of functions H, such that $p_1 \geq \text{Pr}_{h \sim H}[h(x) = h(y)]$ when x and y are similar (distance $\leq r_1$), and $p_2 \leq \text{Pr}_{h \sim H}[h(x) = h(y)]$ when x and y are dissimilar (distance $\geq r_2$). Such a distribution is called (r_1, r_2, p_1, p_2)-sensitive. If $p_1 > p_2$ the LSH framework gives a data-structure with query time $\tilde{O}(n^\rho/p_1)$ for $\rho = \frac{\log 1/p_1}{\log 1/p_2}$, which is usually significantly faster than the alternatives. *At least when p_1 is not too small.*

The two most common families of LSH is Cross-Polytope (or Spherical) LSH [6] for Cosine similarity and MinHash [10] for Jaccard Similarity.

Cross-Polytope is the basis of the Falconn software package [19], and solves the (r, cr)-near neighbour problem on the sphere in time $\tilde{O}(n^{1/c^2}/p_1)$. Here $p_1 = \exp(-\frac{\tau^2}{4-\tau^2}(1 - o(1))) \log d)$, where $\tau = \|p - q\|_2 \in [0, 2]$ is the distance between two close points. We see that already at $\tau \approx \sqrt{2}$ (which corresponds to near orthogonal vectors) the $1/p_1$ factor results in a factor d slow-down. For larger $\tau \in (\sqrt{2}, 2]$ the slow-down can grow arbitrary large. Using dimensionality reduction techniques, like the Johnson Lindenstrauss transform, one may assume $d = \varepsilon^{-2} \log n$ at the cost of a factor $1 + \varepsilon$ distortion of the distances. However if ε is just $1/100$, the slow-down factor of d is still worse than, say, $n^{1/2}$ for datasets of size up to 10^8, and so if $c \leq \sqrt{2}$ we get that n^ρ/p_1 is larger than n. So *worse than a brute force scan of the database!*

The MinHash algorithm was introduced by Broder et al. for the Alta Vista search engine, but is used today for similarity search on sets in everything from natural language processing to gene sequencing. MinHash solves the (j_1, j_2) gap similarity search problem, where $j_1 \in (0, 1)$ is the Jaccard Similarity of similar sets, and j_2 is that of dissimilar sets, in time $\tilde{O}(n^\rho/j_1)$ where $\rho = \frac{\log 1/j_1}{\log 1/j_2}$. (In particular MinHash is (j_1, j_2, j_1, j_2)-sensitive in the sense defined above.) Now consider the case $j_1 = n^{-1/4}$ and $j_2 = n^{-3/10}$. This is fairly common as illustrated in Fig. 1a. In this case $\rho = \frac{\log 1/j_1}{\log 1/j_2} = 5/6$, so we end up with $n^\rho/j_1 = n^{13/12}$. Again *worse than a brute force scan of the database!*

In this paper we reduce the query time of LSH to $n^\rho/p_1^{1-\rho}$, which is less than n for all $p_1 > 1/n$. In the MinHash example above, we get $n^\rho/p_1^{1-\rho} = n^{5/6+1/4(1-5/6)} = n^{7/8}$. More than a factor $n^{0.208}$ improvement(!) In general the improvement of $p_1^{-\rho}$ may be as large as a factor of n when p_1 and p_2 are both close to $1/n$. This is illustrated in Fig. 1b.

The improvements to LSH comes from a simple observation: During the algorithm of Indyk and a certain "amplification" procedure has to be applied

---

[1] In general we expect the exact problem to be impossible to solve in sub-linear time, given the hardness results of [1,5]. However for practical datasets it is often possible.

$\kappa = \frac{\log n}{\log 1/p_2}$ times. When $\log 1/p_2$ does not divide n, which is extremely likely, the amount of amplification has to be approximated by the next integer. We propose instead an ensemble of two kinds of LSH tables with respectively $\lceil \kappa \rceil$ and $\lfloor \kappa \rfloor$ concatenations of the hash function. We call those respectively "High" and "Low" tables. When analysed sufficiently precisely yields the improvements described above.

1.1   Related Work

We will review various directions in which LSH has been improved and generalized, and how those results related to what is presented in the present article.

In many cases, the time required to sample and evaluate the hash functions dominate the time required by LSH. Recent papers [11] have reduced the number of distinct calls to the hash functions which is needed. The most recent paper in the line of work is [11], which reduces the number of calls to $(\frac{\log n}{\log 1/p_2})^2/p_1$. On top of that, however, they still require n^ρ/p_1 work, so the issue with small p_1 isn't touched upon. In fact, some of the algorithms in [11] *increase* the dependency from n^ρ/p_1 to $n^\rho/(p_1 - p_2)$.

Other work has sought to generalize the concept of Locality Sensitive Hashing to so-called Locality Sensitive Filtering, LSF [9]. However, the best work for set similarity search based on LSF [4,12] still have factors similar in spirit to p_1^{-1}. E.g., the Chosen Path algorithm in [12] uses query time $\tilde{O}(n^\rho/b_1)$, where b_1 is the similarity between close sets.

A third line of work has sought to derandomize LSH. The result is so-called Las Vegas LSH [3,20]. Here the families H are built combinatorially, rather than at random, to guarantee the data structure always return a near neighbour, when one exists. While these methods don't have probabilities, they still end up with similar factors for similar reasons.

As mentioned, the reason p_1^{-1} shows up in all these different approaches, is that they all rely on the same amplification procedure, which has to be applied an integer number of times. One might wonder if tree based methods, which do an adaptive amount of amplification, could get rid of the $1/p_1$ dependency. However as evidenced by the classical and current work [7,8,13,14] these methods still have a factor $1/p_1$. We leave it open whether this might be avoidable with better analysis, perhaps inspired by our results for "High-Low" tables.

2   Preliminaries

Before we give the new LSH algorithm, we will recap the traditional analysis. For a more comprehensive introduction to LSH, see the Mining of Massive Datasets book [17], Chap. 3. In the remainder of the article we will use the notation $[n] = \{1, \ldots, n\}$.

Assume we are given a (r_1, r_2, p_1, p_2)-sensitive LSH family, H, as defined in the introduction. Let k and L be some integers defined later, and let $[m]$ be the range of the hash functions, $h \in H$. Let n be an upper bound on the number

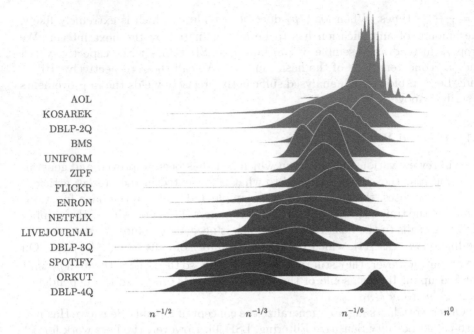

(a) Density plots of all pairwise Jaccard similarities in the datasets studied by Mann et al. [18] written in terms of the size of their corresponding datasets. Curiously the largest datasets, like Orkut and Spotify, have the smallest median Jaccard similarities, even when expressed in terms of the dataset size.

We see that reasonable values for $j_1 = p_1$ range between $n^{-1/3}$ and $n^{-1/6}$.

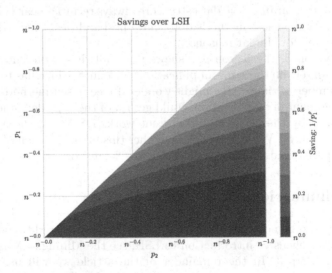

(b) $p_1^{-\rho}$: The possible improvements in query time and space, over classical LSH, as a function of p_1 and p_2. With $p_1 = n^{-1/4}$ and $p_2 = n^{-1/3}$ we save a factor of $n^{3/16} = n^{0.1875}$.

Fig. 1. Overview over available savings

of points to be inserted.[2] The Indyk-Motwani data-structure consists of L hash tables, each with m^k hash buckets.

To insert a point, x, we draw $L \cdot k$ functions from H, denoted by $(h_{i,j})_{i \in [L], j \in [k]}$. In each table $i \in [L]$ we insert x into the bucket keyed by $(h_{i,1}(x), h_{i,2}(x), \ldots, h_{i,k}(x))$. Given a query point q, the algorithm iterates over the L tables and retrieves the data points hashed into the same buckets as q. The process stops as soon as a point is found within distance r_1 from q.

The algorithm as described has the performance characteristics listed below. Here we assume the hash functions can be sampled and evaluated in constant time. If this is not the case, one can use the improvements discussed in the related work.

- Query time: $O(L(k + np_2^k)) = O(n^\rho p_1^{-1} \log n)$.
- Space: $O(nL) = O(n^{1+\rho} p_1^{-1})$ plus the space to store the data points.
- Success probability 99%.

To get these bounds, we have defined $k = \lceil \frac{\log n}{\log 1/p_2} \rceil$ and

$$L = \lceil p_1^{-k} \rceil \leq \exp\left(\log 1/p_1 \cdot \lceil \frac{\log n}{\log 1/p_2} \rceil\right) + 1 \leq n^\rho/p_1 + 1.$$

It's clear from this analysis that the p_1^{-1} factor is only necessary when $\frac{\log n}{\log 1/p_2}$ is not an integer. However in those cases it is clearly necessary, since there is no obvious way to make a non-integer number of function evaluations. We also cannot round k down instead of up, since the number of false positives would explode: rounding down would result in a factor of p_2^{-1} instead of p_1^{-1} — much worse.

3   LSH with High-Low Tables

The idea of the main algorithm is to create some LSH tables with k rounded down, and some with k rounded up. We call those respectively "high probability" tables and "low probability" tables. In short "LSH with High-Low Tables".

The main theorem is the following:

Theorem 1. *Let H be a (r_1, r_2, p_1, p_2)-sensitive LSH family, and let $\rho = \frac{\log 1/p_1}{\log 1/p_2}$. Assume $p_1 > 1/n$ and $p_2 > 1/n$. Then High-Low tables give a solution to the (r_1, r_2)-near neighbour problem with the following properties:*

- *Query time: $O(n^\rho/p_1^{1-\rho} \log n)$.*
- *Space: $O(nL) = O(n^{1+\rho}/p_1^{1-\rho})$ plus the space to store the data points.*
- *Success probability 99%.*

---

[2] If we don't know how many points will be inserted, several black box reductions allow transforming LSH into a dynamic data structure.

Proof. Assume r_1, r_2, p_1, p_2 are given. Define $\rho = \frac{\log 1/p_1}{\log 1/p_2}$, $\kappa = \frac{\log n}{\log 1/p_2}$, and $\alpha = \lceil \kappa \rceil - \kappa \in [0, 1)$. We build $\lfloor a \rfloor + \lceil b \rceil$ tables (for $a, b \geq 0$ to be defined), where the $\lfloor a \rfloor$ tables are "Low" tables, which use the hash function concatenated $\lfloor \kappa \rfloor$ times as keys, and the remaining $\lceil b \rceil$ are "High" tables, which use it concatenated $\lceil \kappa \rceil$ times.

The total number of High and Low tables to query is then $\lceil b \rceil + \lfloor a \rfloor$. The expected total number of far points we have to retrieve is

$$
\begin{aligned}
n(\lfloor a \rfloor p_2^{\lfloor \kappa \rfloor} + \lceil b \rceil p_2^{\lceil \kappa \rceil}) &= n(\lfloor a \rfloor p_2^{\kappa - 1 + \alpha} + \lceil b \rceil p_2^{\kappa + \alpha}) \\
&= \lfloor a \rfloor p_2^{-1+\alpha} + \lceil b \rceil p_2^{\alpha} \\
&\leq a p_2^{-1+\alpha} + (b+1) p_2^{\alpha} \\
&\leq a p_2^{-1+\alpha} + b p_2^{\alpha} + 1.
\end{aligned}
$$

For the second equality, we used the definition of κ: $p_2^{\kappa} = 1/n$. We only count the expected number of points seen that are at least r_2 away from the query. This is because the algorithm, like classical LSH, terminates as soon as it sees a point with distance less than r_2.

Given any point in the database within distance r_1 we must be able to find it with high enough probability. This requires that the query and the point shares a hash-bucket in one of the tables. The probability that this is doesn't happen in any of the $\lfloor a \rfloor$ low tables, and not any of the $\lceil b \rceil$ high tables is

$$
\begin{aligned}
(1 - p_1^{\lfloor \kappa \rfloor})^{\lfloor a \rfloor} (1 - p_1^{\lceil \kappa \rceil})^{\lceil b \rceil} &\leq (1 - p_1^{\lfloor \kappa \rfloor})^{a-1} (1 - p_1^{\lceil \kappa \rceil})^{b} \\
&\leq \exp(-a p_1^{\lfloor \kappa \rfloor} - b p_1^{\lceil \kappa \rceil})(1 - p_1^{\lfloor \kappa \rfloor})^{-1} \\
&= \exp(-(a p_1^{-1+\alpha} + b p_1^{\alpha}) n^{-\rho})(1 - p_1^{\lfloor \kappa \rfloor})^{-1} \\
&\leq \exp(-(a p_1^{-1+\alpha} + b p_1^{\alpha}) n^{-\rho}) \cdot 2.
\end{aligned}
$$

For the first inequality we used that $1 - x \leq \exp(-x)$. For the equality, we used the definition of κ and ρ: $p_1^{\kappa} = p_2^{\rho \kappa} = n^{-\rho}$. For the last inequality we have assumed $p_2 > 1/n$ so $\lfloor \kappa \rfloor \geq 1$, and that $p_1 < 1/2$, since otherwise we could just get the theorem from the classical LSH algorithm.

We now define a and b, both ≥ 0, such that

$$
a p_2^{-1+\alpha} + b p_2^{\alpha} = a + b \quad \text{and} \tag{1}
$$
$$
a p_1^{-1+\alpha} + b p_1^{\alpha} = n^{\rho}. \tag{2}
$$

By the previous calculations this will guarantee the number of false positives is not more than the number of tables, and a constant success probability.

We can achieve this by taking

$$
\begin{bmatrix} a \\ b \end{bmatrix} = \begin{bmatrix} p_2^{-1+\alpha} - 1 & p_2^{\alpha} - 1 \\ p_1^{-1+\alpha} & p_1^{\alpha} \end{bmatrix}^{-1} \begin{bmatrix} 0 \\ n^{\rho} \end{bmatrix} = \frac{n^{\rho}}{(p_2^{-1+\alpha} - 1)p_1^{\alpha} + (1 - p_2^{\alpha})p_1^{-1+\alpha}} \begin{bmatrix} 1 - p_2^{\alpha} \\ p_2^{-1+\alpha} - 1 \end{bmatrix}.
$$

We can check that both values are non-negative, since $\alpha \in [0, 1]$, so the definition is meaningful.

When actually implementing "LSH with High-Low Tables", these are the values you should use for the number of respectively the high and low probability tables. That will ensure you take full advantage of when α is not worst case, and you may do even better than the theorem assumes.

To complete the theorem we need to prove $a + b \leq n^\rho p_1^{\rho-1}$. For this we bound

$$\frac{a+b}{n^\rho} = \frac{p_2^{-1+\alpha} - p_2^\alpha}{(p_2^{-1+\alpha} - 1)p_1^\alpha + (1 - p_2^\alpha)p_1^{-1+\alpha}}$$

$$\leq \left(\frac{(p_1 - p_2)\log 1/p_1}{(1 - p_1)\log p_1/p_2}\right)^\rho \left(\frac{(1 - p_2)\log p_1/p_2}{(p_1 - p_2)\log 1/p_2}\right)$$

$$= \exp\left(D\left(\rho \,\middle\|\, \frac{1/p_1 - 1}{1/p_2 - 1}\right)\right)$$

$$\leq p_1^{\rho-1}.$$

Here $D(r\|x) = r\log\frac{r}{x} + (1 - r)\log\frac{1-r}{1-x}$ is the Kullback-Leibler divergence. The two inequalities are proven in the Appendix in the full version of the paper. The first bound comes from maximizing over $\alpha \in [0, 1]$, so in principle we might be able to do better if $\kappa = \frac{\log n}{\log 1/p_2}$ is close to an integer. The second bound is harder, but the realization that the left hand side can be written on the form of a divergence helps a lot, since those have known properties we can exploit. The bound is tight up to a factor 2, so no significant improvement is possible.

Bringing it all together the expected query time is equal to the number of tables we have to query plus the number of far points to inspect. Using Eq. (1) we have:

$$\lfloor a\rfloor + \lceil b\rceil + n(\lfloor a\rfloor p_2^{\lfloor \kappa\rfloor} + \lceil b\rceil p_2^{\lceil \kappa\rceil}) \leq (a + b + 1) + (ap_2^{1-\alpha} + bp_2^\alpha + 1)$$

$$= 2(a + b) + 2$$

$$\leq 2n^\rho/p_1^{\rho-1} + 2.$$

Similarly, recall that to succeed the data structure must be able to find a near point when one exists. Recalling the previous computations and Eq. (2) we have:

$$1 - (1 - p_1^{\lfloor \kappa\rfloor})^{\lfloor a\rfloor}(1 - p_1^{\lceil \kappa\rceil})^{\lceil b\rceil} \geq 1 - \exp(-ap_1^{\lfloor \kappa\rfloor} - bp_1^{\lceil \kappa\rceil})(1 - p_1^{\lfloor \kappa\rfloor})^{-1}$$

$$= 1 - \exp(-1) \cdot 2$$

$$\geq 0.26.$$

Finally we can boost the success probability from 26% to 99% by repeating the entire data-structure 16 times.

References

1. Abboud, A., Rubinstein, A., Williams, R.: Distributed PCP theorems for hardness of approximation in P. In: 2017 IEEE 58th Annual Symposium on Foundations of Computer Science (FOCS), pp. 25–36. IEEE (2017)

2. Ahle, T.D., Aumüller, M., Pagh, R.: Parameter-free locality sensitive hashing for spherical range reporting. In: Proceedings of the Twenty-Eighth Annual ACM-SIAM Symposium on Discrete Algorithms, pp. 239–256. SIAM (2017)
3. Ahle, T.D.: Optimal Las Vegas locality sensitive data structures. In: 2017 IEEE 58th Annual Symposium on Foundations of Computer Science (FOCS), pp. 938–949. IEEE (2017)
4. Ahle, T.D., Knudsen, J.B.T.: Subsets and supermajorities: optimal hashing-based set similarity search. arXiv preprint arXiv:1904.04045 (2020)
5. Ahle, T.D., Pagh, R., Razenshteyn, I., Silvestri, F.: On the complexity of inner product similarity join. In: Proceedings of the 35th ACM SIGMOD-SIGACT-SIGAI Symposium on Principles of Database Systems, pp. 151–164. ACM (2016)
6. Andoni, A., Indyk, P., Laarhoven, T., Razenshteyn, I., Schmidt, L.: Practical and optimal LSH for angular distance. In: Advances in Neural Information Processing Systems, pp. 1225–1233 (2015)
7. Andoni, A., Razenshteyn, I., Nosatzki, N.S.: LSH forest: practical algorithms made theoretical. In: Proceedings of the Twenty-Eighth Annual ACM-SIAM Symposium on Discrete Algorithms, pp. 67–78. SIAM (2017)
8. Bawa, M., Condie, T., Ganesan, P.: LSH forest: self-tuning indexes for similarity search. In: Proceedings of the 14th International Conference on World Wide Web, pp. 651–660 (2005)
9. Becker, A., Ducas, L., Gama, N., Laarhoven, T.: New directions in nearest neighbor searching with applications to lattice sieving. In: Proceedings of the Twenty-Seventh Annual ACM-SIAM Symposium on Discrete Algorithms, pp. 10–24. SIAM (2016)
10. Broder, A.Z., Charikar, M., Frieze, A.M., Mitzenmacher, M.: Min-wise independent permutations. In: Proceedings of the Thirtieth Annual ACM Symposium on Theory of Computing, pp. 327–336. ACM (1998)
11. Christiani, T.: Fast locality-sensitive hashing frameworks for approximate near neighbor search. In: Amato, G., Gennaro, C., Oria, V., Radovanović, M. (eds.) SISAP 2019. LNCS, vol. 11807, pp. 3–17. Springer, Cham (2019). https://doi.org/10.1007/978-3-030-32047-8_1
12. Christiani, T., Pagh, R.: Set similarity search beyond MinHash. In: Proceedings of the 49th Annual ACM SIGACT Symposium on Theory of Computing, STOC 2017, Montreal, QC, Canada, 19–23 June 2017, pp. 1094–1107 (2017)
13. Christiani, T., Pagh, R., Thorup, M.: Confirmation sampling for exact nearest neighbor search. arXiv preprint arXiv:1812.02603 (2018)
14. Christiani, T.L., Pagh, R., Aumüller, M., Vesterli, M.E.: PUFFINN: parameterless and universally fast finding of nearest neighbors. In: European Symposium on Algorithms, pp. 1–16 (2019)
15. Datar, M., Immorlica, N., Indyk, P., Mirrokni, V.S.: Locality-sensitive hashing scheme based on p-stable distributions. In: Proceedings of the Twentieth Annual Symposium on Computational Geometry, pp. 253–262. ACM (2004)
16. Indyk, P., Motwani, R.: Approximate nearest neighbors: towards removing the curse of dimensionality. In: Proceedings of the Thirtieth Annual ACM Symposium on Theory of Computing, pp. 604–613. ACM (1998)
17. Leskovec, J., Rajaraman, A., Ullman, J.D.: Mining of Massive Data Sets. Cambridge University Press, Cambridge (2020)
18. Mann, W., Augsten, N., Bouros, P.: An empirical evaluation of set similarity join techniques. Proc. VLDB Endow. 9(9), 636–647 (2016)

19. Razenshteyn, I., Schmidt, L.: FALCONN-fast lookups of cosine and other nearest neighbors (2018)
20. Wei, A.: Optimal Las Vegas approximate near neighbors in lp. In: Proceedings of the Thirtieth Annual ACM-SIAM Symposium on Discrete Algorithms, pp. 1794–1813. SIAM (2019)

Similarity Measures, Search,
and Indexing

Confirmation Sampling for Exact Nearest Neighbor Search

Tobias Christiani[1] , Rasmus Pagh[2,4](✉) , and Mikkel Thorup[3,4]

[1] Norwegian University of Science and Technology, Trondheim, Norway
`tobias.christiani@ntnu.no`
[2] IT University of Copenhagen, Copenhagen, Denmark
`pagh@itu.dk`
[3] University of Copenhagen, Copenhagen, Denmark
`mikkel2thorup@gmail.com`
[4] Basic Algorithms Research Copenhagen, Copenhagen, Denmark

Abstract. Locality-sensitive hashing (LSH), introduced by Indyk and Motwani in STOC '98, has been an extremely influential framework for nearest neighbor search in high-dimensional data sets. While theoretical work has focused on the *approximate* nearest neighbor problem, in practice LSH data structures with suitably chosen parameters are used to solve the *exact* nearest neighbor problem (with some error probability). Sublinear query time is often possible in practice even for exact nearest neighbor search, intuitively because the nearest neighbor tends to be significantly closer than other data points. However, theory offers little advice on how to choose LSH parameters outside of pre-specified worst-case settings.

We introduce the technique of *confirmation sampling* for solving the exact nearest neighbor problem using LSH. First, we give a general reduction that transforms a sequence of data structures that each find the nearest neighbor with a small, unknown probability, into a data structure that returns the nearest neighbor with probability $1 - \delta$, using as few queries as possible. Second, we present a new query algorithm for the *LSH Forest* data structure with L trees that is able to return the exact nearest neighbor of a query point within the same time bound as an LSH Forest of $\Omega(L)$ trees with internal parameters specifically tuned to the query and data.

Keywords: Nearest neighbor search · Locality-sensitive hashing

T. Christiani—The research leading to these results has received funding from the European Research Council under the European Union's 7th Framework Programme (FP7/2007-2013)/ERC grant agreement no. 614331.
R. Pagh—Supported by Villum Foundation grant 16582 to Basic Algorithms Research Copenhagen (BARC). Part of this work was done while visiting Simons Institute for the Theory of Computing.
M. Thorup—Supported by an Investigator Grant from the Villum Foundation, Grant No. 16582.

© Springer Nature Switzerland AG 2020
S. Satoh et al. (Eds.): SISAP 2020, LNCS 12440, pp. 97–110, 2020.
https://doi.org/10.1007/978-3-030-60936-8_8

1   Introduction

Locality-sensitive hashing [13] (LSH) is the leading theoretical approach to nearest neighbor problems in high dimensions. In nearest neighbor search we seek to preprocess a point set P such that given a query point q, we can quickly return the point in P that is closest to q according to some distance measure $\text{dist}(\cdot, \cdot)$. Theoretical results are typically formulated as *approximation* algorithms that allow a point at distance cr to be returned if the nearest neighbor has distance r from the query point, where $c > 1$ is a user-specified approximation factor. In practice the quality parameter of interest is the *recall*, i.e., the empirical probability of retrieving the nearest neighbor (see e.g. [1]). As we will see below is not hard to show that LSH methods can obtain recall arbitrarily close to 1 if parameters are suitably chosen according to the given query and data set. However, choosing parameters well, in an efficient way, is a challenge [15].

Background on Locality-Sensitive Hashing. A locality-sensitive family of hash functions $\mathcal{H}$ (an "LSH family") has the property that hash collision probability decreases as distance increases. Specifically, for $h \sim \mathcal{H}$ the "hash bucket" $S_h(q) = \{x \in P \mid h(x) = h(q)\}$ is more likely to contain the nearest neighbor of q than any other element of P. For a given data set P one would typically use a family $\mathcal{H}$ such that the expected size of $S_h(q)$ is constant (for every q or on average for a certain query distribution) [17]. Given such a family $\mathcal{H}$, suppose the nearest neighbor is $x_1 \in P$, and define $p_1 = \Pr[x_1 \in S_h(q)]$ to be the probability of a hash collision with the nearest neighbor. Then inspecting $S_{h_i}(q)$ for a sequence of hash functions $h_1, \ldots, h_L$ independently sampled from $\mathcal{H}$ we will fail to find x_1 with probability $(1 - p_1)^L \approx \exp(-p_1 L)$. To make this as efficient as possible we can use a hash table that given q allows us to retrieve $S_{h_i}(q)$ in time $O(1 + |S_{h_i}(q)|)$. If we assume that the distance between q and $x \in P$ can be computed in constant time, the expected time for this procedure is $O(L \, \mathbb{E}[1 + |S_h(q)|])$. There are several issues with the above construction:

- If p_1 is large then the query algorithm still goes through L hash buckets, even though we expect to see x_1 within the first $O(1/p_1)$ buckets.
- If $p_1 L$ is small, the recall $1 - \exp(-p_1 L)$ is close to zero.

Notice that p_1 depends on the nearest neighbor that we are searching for, resulting in a chicken-and-egg situation: we would like to conduct the search with knowledge of p_1, but we only know p_1 if the search finds x_1 (and we know how the collision probability depends on $\text{dist}(q, x_1)$). We will introduce a technique called *confirmation sampling* for dealing with the former problem of when to terminate the search when we have no knowledge of p_1. The latter problem requires us to take a new look at how to query the so-called LSH forest data structure, described below.

Approximation Versus Recall. Early theoretical work on high-dimensional nearest neighbor search dealt with the simpler case of *near neighbor* search where it is assumed that a maximum distance r to the nearest neighbor is known and a point within distance cr must be returned. A reduction with logarithmic overhead in time and space extends this to solve the approximate nearest neighbor problem with unknown distance r [12,13]. These reductions increase the approximation factor by $1+\gamma$, with space usage proportional to $1/\gamma$, and do not seem to provide any guarantee on recall even if used with a near-neighbor data structure with approximation factor $c = 1$.

A data structure known as *LSH forest*, first described by Charikar [7] and later generalized and baptized by Bawa et al. [6], removes the logarithmic overhead in space but the query algorithm still only provides c-approximate results and does not guarantee a specific recall. Indeed, it is not hard to construct examples where there are many c-approximate nearest neighbors and the probability of returning the exact nearest neighbor is negligible.

LSH Forest. Since we will describe a new query algorithm for the LSH Forest data structure we review the data structure here. We will again make use of an LSH family $\mathcal{H}$, but this family can be "weak" in the sense that collision probabilities are large, say, $Pr[h(q) = h(x)] = \Omega(1)$ for $x \in P$. Assume for simplicity that we can sample $h \sim \mathcal{H}$ and evaluate $h(x)$ in constant time. For parameters K and L and $(i,j) \in \{1,\ldots,K\} \times \{1,\ldots,L\}$, independently sample hash functions $h_{i,j} \sim \mathcal{H}$. Associate each point $x \in P$ with a string $h_j(x) = h_{1,j}(x)h_{2,j}(x)\ldots h_{K,j}(x)$. For $j = 1,\ldots,L$ the jth part of the LSH Forest is a trie that stores prefixes of the set of strings $h_j(P) = \{h_j(x) \mid x \in P\}$. Specifically, for each $x \in P$ it stores the shortest prefix of $h_j(x)$ that is unique among strings in $h_j(P)$ (if such a prefix exists, otherwise the whole string $h_j(x)$). A pointer to x is placed in the leaf corresponding to a prefix of $h_j(x)$. The space for the data structure, not counting space for storing the n points in P, is $O(nKL)$ words naïvely, and can be improved to $O(nL)$ words using path compression [6].

Querying LSH Forest. For a given query q and a parameter $i \in \{1,\ldots,K\}$, LSH Forest allows us to retrieve the hash bucket $S_{i,j}(q)$ of points in P matching a length-i prefix of $h_j(q)$ in time $O(i + |S_{i,j}(q)|)$. We will use $p(q,x) = Pr_{h\sim\mathcal{H}}[h(q) = h(x)]$ as shorthand for the collision probability between q and x. We have

$$\mathbb{E}[|S_{i,j}(q)|] = \sum_{x \in P} p(q,x)^i . \tag{1}$$

The larger the "level" i, the smaller $S_{i,j}(q)$ is in expectation. Conversely the probability of finding x_1 in the hash bucket is $Pr[x_1 \in S_{i,j}(q)] = p(q,x_1)^i$ which decreases exponentially with i. The query algorithm described in [6] chooses the level i_0 to inspect as the smallest level where the number of collisions is linear, $i_0 = \min\{i \mid \sum_{j=1}^{L} |S_{i_0,j}(q)| \leq cL\}$, for some constant c. The probability of failing to find the nearest neighbor by inspecting all buckets $S_{i,1},\ldots,S_{i,L}$ at level i is $(1 - p_1^i)^L \approx \exp(-p_1^i L)$, so to bound the failure probability we need to choose L large enough. For example, if the nearest neighbor of q is in a dense cluster of $2cL$ points whose points almost surely reside in the same LSH bucket, the algorithm fails to find the nearest neighbor almost surely. So LSH Forest is only "self-tuning" to a limited extent if high recall is desired: choosing a suitable parameter L requires at least approximate knowledge of the distance distribution from q to points of P. Instead, we would like L to be simply a parameter that determines the space usage, and use a query algorithm that adapts to the data.

Our Results. LSH methods work by performing many iterations, each inspecting a hash table $\mathcal{D}_i$ with a small (and unknown) probability p_1 of finding the nearest neighbor. To stress that p_1 depends on the query we will sometimes denote it $p_1(q)$. It is easy to see that after $\ln(1/\delta)/p_1(q)$ iterations the nearest neighbor will be retrieved with probability at least $1 - \delta$. We show that this number of iterations can be matched in expectation without knowledge of $p_1(q)$, and in fact even without estimating any collision probabilities. Using a technique we call *confirmation sampling* we obtain the following result on LSH-like methods:

Theorem 1. *Suppose there is a sequence of independent, randomized data structures $\mathcal{D}_1, \mathcal{D}_2, \ldots$, such that on query q, $\mathcal{D}_i$ returns the nearest neighbor of q in P with probability at least $p_1(q)$ and each other point in P with probability at most $p_1(q)$. Let $\delta > 0$ be given. There is an algorithm that depends on δ but not on $p_1(q)$ that on input q queries data structures $\mathcal{D}_1, \ldots, \mathcal{D}_{j_q}$, performs j_q distance computations, where $\mathbb{E}[j_q] = O(\ln(1/\delta)/p_1(q))$, and returns the nearest neighbor of q with probability at least $1 - \delta$.*

Theorem 1 shows that if we use quadratic space to store a sufficiently long sequence of data structures $\mathcal{D}_i$, it suffices to focus on minimizing the product of the expected time for $\mathcal{D}_i$ and the number $1/p_1(q)$ of iterations.

In practice one would of course not have access to an unbounded sequence of data structures, but rather to a fixed number L of data structures. If these data structures offer a trade-off between query time and probability of returning the nearest neighbor it is still possible to apply Theorem 1: For $i = 1, 2, \ldots, \log n$ run confirmation sampling in rounds of L steps with time budget 2^i for each data structure $\mathcal{D}_i$. Terminate as soon as confirmation sampling returns a result—by a union bound over the $\log n$ rounds the error probability is at most $\delta \log n$.

Our second result addresses how to adapt not only to the collision probability of the nearest neighbor, but to the whole distance distribution from q to points in P. In particular, we design and analyze a new adaptive query algorithm for the LSH Forest data structure [6,7] discussed above. LSH Forest is known to be able to adapt to the distance distribution to some extent, but previous work has required the query algorithm to depend on the distance to the nearest neighbor in P. In contrast our query algorithm is independent of properties of the data. The only requirement is that the LSH family used is *monotone* in the sense that collision probability is non-increasing with distance. We compare our adaptive algorithm to an optimal algorithm in a class of *natural* algorithms that choose a level i^* and a number of tries j^* (which may depend on the distance distribution between q and P) and inspect the first j^* buckets at level i^*.

Theorem 2. *Let $OPT(L, K)$ denote the optimal cost of a natural algorithm that queries an LSH Forest data structure with L trees and K levels and returns the nearest neighbor with probability at least $1 - 1/n$. Further assume that the LSH family is monotone. Then there is an adaptive algorithm that queries an LSH Forest data structure with $O(L)$ trees and K levels that returns the nearest neighbor using time $O(OPT(L, K))$ with probability $1 - 1/n$.*

LSH Forest is not an asymptotically optimal data structure for approximate nearest neighbor search in general. For example, it is known that data-dependent methods can be asymptotically faster in several important spaces, and data structures obtaining better space-time trade-offs are known [2, 3]. Generalizing our results for exact nearest neighbors to a data-dependent setting, say, in Euclidean space, is an interesting open direction. Note that the data structures $\mathcal{D}_i$ in Theorem 1 could be data dependent, though present data-dependent LSH techniques rely on knowing the (approximate) distance to the nearest neighbor.

1.1 Related Work

There is a large literature on using LSH for nearest neighbors search in practice, often generalized to the k-nearest neighbor problem where the k closest points in P must be returned. For simplicity we concentrate on the case $k = 1$, but most results extend to arbitrary k. Many heuristics that work well in practice come without guarantees on either result quality or query time in high dimensions, or provides guarantees only under certain assumptions on the data set.

Guarantees on Recall. In practice, the performance of locality-sensitive hashing techniques is usually measured by their recall: the fraction of the true k-nearest neighbors found on average, see e.g. [1, 4]. From a theoretical point of view it is natural to bound the *expected recall*, i.e., the probability that the nearest neighbor is found. We are only aware of very few works that provide theoretical guarantees on expected recall in conjunction with sublinear query time in high dimensions and without assumptions on data.

Dong et al. [11] outline an "adaptive" method for achieving a given expected recall in the context of multiprobe LSH (with no formal statement of guarantees). The idea is to determine, after inspecting i buckets, whether to terminate or to inspect bucket $i + 1$ based on the collision probability $p(q, \hat{x}_1)$ between q and the nearest neighbor $\hat{x}_1$ found in the first i buckets. This requires an efficient method for computing $p(q, \hat{x}_1)$, which might not be known, especially for small collision probabilities. This is not just a theoretical problem: Prominent LSH methods such as p-stable LSH [10] and cross-polytope LSH [1] do not have closed-form expressions for collision probabilities. Our adaptive algorithm is similar in spirit, but entirely avoids having to compute collision probabilities.

Recently, Aumüller et al. [5] introduced a practical implementation of the adaptive method of Dong et al. with several additional optimizations from the literature such as b-bit hashing for faster distance estimation [14] and LSH-pooling [8] for reducing the number of independent LSH evaluations by recycling.

They circumvent the problem of computing collision probabilities by instead relying on estimated collision probabilities from synthetic data, using the LSH property that collision probabilities only depend on the distance between points.

A downside of previous adaptive approaches is that they proceed by searching all L tries of the LSH Forest in a bottom-up fashion starting at level K. This approach results in an overhead of $O(LK)$ even if the desired recall can be achieved at a lower expected cost by searching $i^* < L$ tries at a level $j^* < K$. Our adaptive algorithm avoids this overhead and is able to match the running time of an algorithm that chooses a number of tries to search i^* and the depth j^* in order to minimize the expected cost of finding the nearest neighbor with high probability, even when that cost is dominated asymptotically by $O(LK)$.

Parameter Tuning. Since the performance of LSH data structures depends on parameter choices, a lot of work has gone into devising ways of choosing good parameters for a given data set, both during data structure construction and adaptively for the query algorithm. Slaney et al. [17] propose to select parameters based on the "distance profile" of a data set, but needs a bound on the distance to the nearest neighbor to function.

The state-of-the-art FALCONN library [1] uses grid search over parameters to empirically estimate the best parameters, assuming that the data and query distributions are identical.

We note that the adaptive method of Dong et al. [11] does not adapt search depth to the distance distribution from the query point q. Choosing good parameters for LSH and especially multi-probe LSH was mentioned by Lv et al. [15] as a challenge in the paper celebrating their VLDB 10-year Best Paper Award.

2   Confirmation Sampling

Let Q denote a probability distribution with finite support S. Further assume that elements of S are equipped with a total ordering relation $\prec$, and define $x_1 = \min(S)$ as the smallest element in the support with respect to the ordering $\prec$. Consider the problem of identifying x_1 given that we only have access

---

Algorithm 1: CONFIRMATIONSAMPLING($Q, t, \prec$)

1 $\beta \leftarrow \infty$, count $\leftarrow 0$
2 **while** count $< t$ **do**
3 | sample $X \sim Q$
4 | **if** $X = \beta$ **then**
5 | | count $\leftarrow$ count $+ 1$
6 | **else if** $X \prec \beta$ **then**
7 | | $\beta \leftarrow X$
8 | | count $\leftarrow 0$

9 **return** β

---

to samples from the distribution Q and to the ordering, i.e., given elements $x, y \in S$ we can determine whether $x \prec y$, $x = y$, or $y \prec x$. We propose a simple randomized algorithm for solving this problem that we call confirmation sampling. The algorithm works by drawing samples from Q while keeping track of the smallest element seen so far together with the number of times it has been sampled in addition to the first sample—the number of *confirmations*. Once the smallest element has been confirmed t times, the algorithm reports that element and terminates. We use ∞ to denote an element that is larger than all elements of S.

Theorem 3. *Let Q denote a probability distribution with finite support S. For $x_1 = \min(S)$ and $X \sim Q$ let $p_1 = \Pr[X = x_1]$ and let $p_2 = \max\{\Pr[X = x] \mid x \in S\backslash\{x_1\}\}$ be the largest sampling probability among elements of $S\backslash\{x_1\}$. Then:*

$$\Pr[\text{CONFIRMATIONSAMPLING}(Q, t) \neq x_1] \leq (1 - p_1)\left(\frac{p_2}{p_1 + p_2}\right)^t$$

The expected number of samples made by CONFIRMATIONSAMPLING *is bounded by* $(t + 1)/p_1$.

Before we show Theorem 3 we observe that it implies Theorem 1: Define an ordering on P by $x \preceq_q y \iff \text{dist}(q, x) \leq \text{dist}(q, y)$. It can be turned into a total ordering $\prec_q$ by an arbitrary but fixed tie-breaking rule. Choose $t = \lceil \log_2(1/\delta) \rceil$ and run CONFIRMATIONSAMPLING$(Q, t, \prec_q)$ with the ith sample from Q being produced by querying $\mathcal{D}_i$ for the nearest neighbor of q. Since $p_1 \geq p_2$ we have that the error probability is bounded by $2^{-t} \leq \delta$.

Proof. If the algorithm fails to report x_1 it must have happened at least t times that the confirmation counter was incremented (line 5) due to a sample X satisfying the condition $X = \beta$ for $\beta \neq x_1$. We will refer to such events as *false confirmations* and proceed by upper bounding the probability that the algorithm performs t false confirmations. Prior to each sample the probability of performing a false confirmation is maximized if $\beta = x_2$ for some $x_2 \neq x_1$ maximizing the sampling probability, i.e., $\Pr[X = x_2] = p_2$. Note also that the first sample can never result in a false confirmation. The probability of the algorithm performing t false confirmations before sampling x_1 can therefore be upper bounded by the probability that the first sample is not equal to x_1 and that we in the following samples observe t samples of x_2 before sampling x_1. The probability that we sample x_2 conditioned on sampling either x_1 or x_2 is exactly $\frac{p_2}{p_1 + p_2}$, and the probability of this happening t times in a row is $\left(\frac{p_2}{p_1 + p_2}\right)^t$.

To analyze the number of samples, consider an infinite sequence of independent samples $X_1, X_2, \cdots \sim Q$, and suppose that in the ith iteration the algorithm uses sample X_i. Observe that the algorithm terminates no later than iteration i if x_1 is sampled $t + 1$ times in $X_1, \ldots, X_i$. The expected number of iterations needed to sample x_1 $t + 1$ times is exactly $(t + 1)/p_1$.

Theorem 3 is tight in the case where Q only assigns non-zero probability to two elements. The exact distribution of the output of CONFIRMATIONSAMPLING is derived in Appendix A of the arXiv version of this paper [9]. We observe that for the proof to work, the distribution from which samples are drawn does not need to be the same in each iteration of CONFIRMATIONSAMPLING, as long as p_1 is a lower bound on sampling x_1 and p_2 is an upper bound on sampling each element other than x_1. If for some $\gamma \in [0,1]$ we have that every distribution satisfies $p_2/(p_1 + p_2) \leq \gamma$ then we can upper bound the error probability by γ^t.

2.1   Application to Locality-Sensitive Hashing

Assume that we have an LSH family that is tuned to give few collisions between query and non-neighbor points for a given query and data distribution. Such a "tuned" LSH family may be obtained if the query distribution is known as discussed in Sect. 1.1. We can use confirmation sampling to adjust query time according to the distance to the nearest neighbor.

Let (V, dist) denote a distance space. That is, V is equipped with a distance function $\text{dist}\colon V \times V \to \mathbb{R}$. We define locality-sensitive hashing [13] as follows:

Definition 1. *Let $\mathcal{H}$ denote a distribution over functions $h\colon V \to R$. We say that $\mathcal{H}$ is locality-sensitive over (V, dist) if there exists a non-increasing $f\colon \mathbb{R} \to [0,1]$ such that for all $x, y \in V$ we have that*

$$\Pr_{h \sim \mathcal{H}} [h(x) = h(y)] = f(\text{dist}(x,y)).$$

We use the ordering $\prec_q$ defined above and define a distribution $\mathcal{Q}_q$ that is most easily described as a sampling procedure. For now we will not care about the efficiency of implementing the sampling. To create a sample $X \sim \mathcal{Q}_q$, sample $h \sim \mathcal{H}$, compute the "bucket"

$$S(q) = \{x \in P \mid h(x) = h(q)\} \ .$$

Now define X as the element of $S(q)$ closest to q, if such an element exists, and otherwise a random element in P.[1] More precisely: If $S(q) \neq \emptyset$ we pick X as the unique minimum element in $S(q)$ according to the total order $\prec_q$, and if $S(q) = \emptyset$ we pick X uniformly at random from P.

Lemma 1. *For $X \sim \mathcal{Q}_q$ and any $x_2 \in P$, $\Pr[X = x_1] \geq \Pr[X = x_2]$.*

Proof. Since $\mathcal{H}$ is locality-sensitive we have that $\Pr[h(q) = h(x_1)] \geq \Pr[h(q) = h(x_2)]$. Thus

$$\Pr[X = x_1] = \Pr[h(q) = h(x_1)] + \frac{\Pr[S(q) = \emptyset]}{n}$$

$$\geq \Pr[h(q) = h(x_2)] + \frac{\Pr[S(q) = \emptyset]}{n} = \Pr[X = x_2] \ .$$

---

[1] The sampling of a random element ensures compatibility with CONFIRMATIONSAMPLING, which requires a sample to be returned even if there is no hash collision. It is not really necessary from an algorithmic viewpoint, but also does not hurt the asymptotic performance.

Theorem 3 implies that confirmation sampling succeeds with good probability:

Lemma 2. *Let x_1 be the nearest neighbor of q in P (breaking ties according to $\prec_q$).* CONFIRMATIONSAMPLING$(\mathcal{Q}_q, t, \prec_q)$ *returns x_1 with probability $\geq 1 - 2^{-t}$. The expected number of samples from $\mathcal{Q}_q$ is bounded by $(t + 1)/p_1$, where $p_1 \geq \Pr[h(q) = h(x_1)]$.*

To efficiently sample from $\mathcal{Q}_q$ we independently sample $h_1, h_2, \cdots \sim \mathcal{H}$, and construct a sequence of hash tables $\mathcal{D}_1, \mathcal{D}_2, \ldots$ that allow us to find $S_i(q) = \{x \in P \mid h_i(x) = h_i(q)\}$ in time $O(1 + |S_i(q)|)$. Random samples from P can be realized using an array of pointers to elements of P.

We note that the above is not an entirely satisfactory solution, since the number of data structures needed cannot be bounded ahead of time (or rather, $\Omega(n)$ data structures may be needed to succeed, resulting in quadratic space usage). A possible remedy if the algorithm does not terminate after inspecting L hash tables is *multi-probing* [15,16] where more than one bucket is inspected in each hash table. Multiprobing increases the probability p_1 of finding the nearest neighbor in each hash table. In the next section we consider another approach to dealing with a space-bounded data structure.

3   Fully Adaptive Nearest Neighbor Search

We present an adaptive algorithm for nearest neighbor search in an LSH Forest that succeeds with high probability[2] (w.h.p.) and matches the minimum expected running time that can be obtained by a natural algorithm that has full knowledge of the LSH collision probabilities between the query point and all the data points, provided we are are allowed a constant factor increase in the number of trees used by the algorithm. We define $OPT(L, K)$ as the minimum expected search time that can be achieved by an algorithm with access to an LSH Forest of L trees of depth K where the algorithm can choose to search $j \leq L$ trees at level $i \leq K$ with the requirement that the nearest neighbor should be reported with probability at least $1 - 1/n$.

$$OPT(L, K) = \min \left\{ \frac{i + \sum_{x \in P} p(q, x)^i}{p(q, x_1)^i} \ln n \;\middle|\; 0 \leq i \leq K, \, p(q, x_1)^i L \geq \ln n \right\}$$

We note that $OPT(L, K)$ only reflects the optimal running time under the assumption that p_1 is bounded away from 1. If for example we had $p_1 = 1$ the multiplicative overhead of $\ln n$ in the running time would not be needed.

---

[2] For every choice of constant $c \geq 1$ there exists a constant n_0 such that for $n \geq n_0$ we can obtain success probability $1 - 1/n^c$ where $n = |P|$ denotes the size of the set of data points.

Overview of Our Approach. The algorithm works by measuring the number of collisions at different levels in the LSH Forest and w.h.p. adapting to search at a level that will result in $O(OPT(L,K))$ running time. Ideally, given sufficiently many trees, we would like to search the level i that balances the number of hash function evaluations and the expected number of collisions with the query point. However such a level might not exist as the expected number of collisions can decrease by more than a constant factor as we increase the level. We begin by introducing some notation. Let $p_1 = p(q, x_1)$, where x_1 denotes the nearest neighbor to q in P and define:

$$C(i) = \sum_{x \in P} p(q, x)^i, \qquad T(i) = (i + C(i))/p_1^i .$$

Observe that $C(i)$ is the expected number of collisions with the query point at level i, and $T(i)$ is the expected running time of an algorithm that searches at level i and guarantees reporting the nearest neighbor of q with some constant probability. If we let i^* denote the choice of level resulting in the minimum value of $OPT(L,K)$ then $OPT(L,K) = T(i^*) \ln n$. Finally, define i' to be the smallest integer i such that $C(i) \le i$.

Given that the number of trees L is sufficiently large we can show that searching either the first $1/p_1^{i'}$ trees at level i' or the first $1/p_1^{i'-1}$ trees at level $i' - 1$ results in an expected running time that is bounded by $O(T(i^*))$ while we report the nearest neighbor with constant probability at least $1 - 1/e$. That is, one of the two levels right around where the number of hash function evaluations and the number of collisions balance out (we have $C(i') \le i'$ and $C(i' - 1) > i' - 1$) result in optimal running time for constant failure probability. Since we don't know p_1 we can search *both* of these levels using confirmation sampling, in parallel, until one of them terminates. This gives us an algorithm that with constant probability terminates in time $O(T(i^*))$ and reports the nearest neighbor. In order to reduce the failure probability to $1/n$ while obtaining optimal running time in the high probability regime we can perform $O(\log n)$ independent repetitions, so that conceptually there are $O(\log n)$ independent forests, and stop the search once a constant fraction terminates.

Query Algorithm and Parameters. There are two circumstances that prevent us from being able to use the approach outlined above. The primary problem is that we don't know the value of i' and estimating it appears to be difficult. The solution proposed by our algorithm is to instead search the "empirical" i' and $i' - 1$: we measure the number of collisions at different levels and search level i and $i - 1$ where i is set to the minimum level where the average number of collisions is smaller than i. This procedure is described in pseudocode in the for-loop section of Algorithm 2.

The second problem is that restrictions on L and K can make it necessary for us to search a level $i < i' - 1$, either because $K < i' - 1$ or because L is too small to ensure that we find the nearest neighbor by searching at level $i' - 1$. The second part of Algorithm 2 that runs when $j = L'$ deals with this problem by searching through the LSH forests bottom-up until a level that results in optimal running time is encountered.

Algorithm 2: ADAPTIVENEARESTNEIGHBOR(q)

---

1 **for** $j \leftarrow 1, 2, 4, \ldots, L'$ **do**
2   find the smallest level i such that the first j trees in at least half of the forests have at most $10ij$ collisions. If such a level does not exist set $i = K$.
3   in each forest run confirmation sampling at level i and $i - 1$ with a time budget of $10ij$ (looking at no more than j buckets and at no more than $10ij$ collisions).
4   **if** *confirmation sampling terminated in $1/4$ of the forests at level i or level $i - 1$* **then**
5   report the closest point seen so far and terminate.
6   **if** $j = L'$ **then**
7   run confirmation sampling in lock step across the forests starting at level $i - 1$, decreasing the level and starting over once $1/2$ of the searches have explored tree number L'. Do this until confirmation sampling terminates in $1/4$ of the forests.

We aim for matching the running time of $OPT(L, K)$ up to constant factors when we are allowed to use $O(L)$ trees. Algorithm 2 operates on $\Theta(\log n)$ LSH Forests that each has L' trees where $L' = O(L/\log n)$ is a sufficiently large power of two. The confirmation sampling used to search in these forests has a parameter setting of $t = 3$ since we only need each search to terminate and correctly report the nearest neighbor with a sufficiently large constant probability.

The proof of Theorem 2 is based on two arguments. First we will show that the stopping condition that $1/4$ of the forests at a given level terminates within the time budget ensures that w.h.p. the nearest neighbor is reported. Second, we show that w.h.p. the algorithm terminates in time $O(OPT(L, K))$.

Correctness. The choice of i made by the algorithm always satisfies $i \leq n$ since there can be no more than n collisions at any level. If we show correctness w.h.p. at a fixed level then we can use a simple union bound over the first n levels to show that w.h.p. at every level where $1/4$ of the searches terminate we have found the nearest neighbor of the query point. The instances of confirmation sampling used by Algorithm 2 use $t = 3$ confirmations before terminating. According to Lemma 2 the probability of terminating and reporting a point different from the nearest neighbor is at most $1/8$. By applying a standard Chernoff bound we can show that over $O(\log n)$ independent runs of confirmation sampling w.h.p. less than $1/4$ the instances will fail to report the nearest neighbor.

Bounding the Running Time. We remind the reader that we use i^* to denote the underlying choice of level that minimizes $OPT(L, K)$, that i' denotes the minimum level such that $C(i') \leq i'$, and that i is the choice of level made by the query algorithm.

Consider line 2 of Algorithm 2 where the level i is set to the smallest level where the first j trees in at least half the forest have at most $10ij$ collisions. This operation can be completed in $O(ij)$ time per forest by proceeding top-down across all the forests and for each forest summing up the number of collisions across all its tries at the current level until level i is reached. We make use of constant-time access to the size of buckets/subtrees as we search down in an LSH Forest trie (either by explicitly storing the size of subtrees when we construct the trie, or by inspecting the pointers to the bucket associated with a given prefix).

We will now argue that w.h.p. Algorithm 2 terminates in time $O(OPT(L, K))$ in each of the two following cases:

Case 1: $C(i^*) \leq i^*$. We will show that there exists a value of $j \leq L'$ such that w.h.p. the algorithm terminates at this value (or earlier) and in $O(OPT(K, L))$ time. Consider the first iteration of the for-loop where $100/p_1^{i^*} \leq j \leq L'$. Such a j exists by the restrictions underlying the choice of level that minimizes $OPT(L, K)$ and by our freedom to set $L' = O(L/\log n)$. By Markov's inequality the probability that the number of collisions in the first j trees of a forest at level i' is greater than $10i'j$ is at most $1/10$. Therefore it happens w.h.p. that the algorithm sets $i \leq i' \leq i^*$ where the last inequality follows from the definition of i' and the assumption that $C(i^*) \leq i^*$. By our choice of j we know that confirmation sampling at level i will terminate in each forest with a large constant probability, say, $9/10$. w.h.p. we therefore have that in at least $1/4$ of the forests confirmation sampling at level i terminates within the budget of $10ij$. To bound the total running time we use that w.h.p. $i \leq i'$ for every value of j and since j is doubled at every step of the for loop we can bound the running time in all $O(\log n)$ LSH forests by $O(i'j \log n) = O(T(i^*) \log n) = O(OPT(L, K))$.

Case 2: $C(i^*) > i^*$. Consider first the sub-case where $i^* = i' - 1$. Suppose there exists a minimum $j \leq L'$ such that $i'j \geq 100T(i' - 1)$, j is an integer power of 2, and $j \geq 100/p_1^{i'-1}$ (the latter condition holds by the assumption $i^* = i' - 1$). We previously argued that w.h.p. the algorithm sets $i \leq i'$. In the first iteration of the for-loop where j takes on this value the following holds: If $i = i'$ then level $i' - 1$ is searched with a sufficiently large budget to w.h.p. ensure termination. If $i < i'$ then level $i' - 1$ is searched up until tree number j, again w.h.p. ensuring termination. In both of these cases the running time is bounded by $O(OPT(L, K))$. Otherwise, if $L'i' < T(i'-1)/100$ then w.h.p. the time spent in the for-loop part of the algorithm is upper bounded by $O(T(i' - 1) \log n) = O(OPT(L, K))$, and if level $i' - 1$ was not searched in the for-loop then it will be searched in the first step of the bottom-up part of the algorithm (because $i \leq i'$ w.h.p.) and w.h.p. we are guaranteed to terminate in optimal time.

Consider now the sub-case where $i^* < i' - 1$. Let $\hat{i}$ denote the largest level satisfying $\hat{i} < i' - 1$ and $100/p_1^{\hat{i}} \leq L'$. The query algorithm will terminate w.h.p. when having searched sufficiently many trees at level $\hat{i} \geq i^*$. We will proceed by bounding the cost up to the point where $100/p_1^{\hat{i}}$ trees have been searched in half of the forests at level $\hat{i}$. The cost of running the for-loop part of the algorithm is w.h.p. bounded by $O(L'i' \log n)$. The number of collisions encountered through the bottom-up search when having searched level $\hat{i} + 1$ is w.h.p. bounded by $O(C(\hat{i}+1)L' \log n) = O((C(\hat{i}+1)/p_1^{\hat{i}+1}) \log n)$ since $100/p_1^{\hat{i}+1} > L'$ by our choice of $\hat{i}$. Finally, the cost of searching at level $\hat{i}$ until $1/4$ of the forests terminate is w.h.p. bounded by $O(T(\hat{i}) \log n)$.

Next we show that the sum of all these costs is bounded by $O(OPT(L, K))$. For every $x \in P$ it holds by monotonicity that $p_1 = p(x_1, q) \geq p(x, q)$ and it follows that for every i we have $C(i+1) \leq p_1 C(i)$. Applying this inequality we get the bound $C(\hat{i}+1)/p_1^{\hat{i}+1} \leq C(i^*)/p_1^{i^*} \leq T(i^*)$ that is used to bound the number of collisions from the bottom-up search. The same approach also gives a bound on the number of collisions at level $\hat{i}$. In order to bound the contribution from the for-loop note that $C(\hat{i}+1) \geq C(i'-1) > i'-1$ where the last inequality holds by the definition of i'. It also holds that $L' < 100/p_1^{\hat{i}+1}$ by the choice of $\hat{i}$. Combining these two inequalities $L'i' \leq 100\left(C(\hat{i}+1)+1\right)/p_1^{\hat{i}+1} = O(T(i^*))$. The bound on the total running time is then given by $O(T(i^*) \log n)) = O(OPT(L, K))$.

4   Conclusion and Open Problems

We have introduced confirmation sampling as a technique for identifying the minimum element from a discrete distribution. Confirmation sampling works particularly well when the minimum element is at least as likely to be sampled as other elements. Combining confirmation sampling with locality-sensitive hashing we obtain a randomized solution to the exact nearest neighbor search problem that works without knowledge of the probability of collision between pairs of points. We use these techniques to design a new adaptive query algorithm for the LSH Forest data structure with L trees that returns the nearest neighbor of a query point with the same time bound that is achieved if the query algorithm has access to an LSH forest of $\Omega(L)$ trees with internal parameters specifically tuned to the query and data.

We can use confirmation sampling with LSH to solve the k-nearest neighbor problem w.h.p. in k by keeping track of the top-k closest points and requiring each to be confirmed $O(\log k)$ times. If we are able to compute the collision probabilities we can use the adaptive stopping rule of Dong et al. [11] to stop the search once we have sampled $j \geq \ln(1/\delta)/\hat{p}_k$ buckets, where $\hat{p}_k$ is the collision probability between the query point and the kth nearest neighbor candidate found by the query algorithm. This stopping rule guarantees that if x is a k-nearest neighbor to the query point, and the LSH family is monotone, then x is reported with probability at least $1 - \delta$. It would be interesting to find a similarly efficient stopping rule for $\delta = \Theta(1)$ that works without knowledge of the collision probabilities.

Our adaptive query algorithm for the LSH Forest data structure makes use of union bounds over the K levels of the data structure when showing correctness and uses that w.h.p. it does not search too far (which could potentially cost time $O(n)$). When we compare our performance against an optimally tuned algorithm that must succeed w.h.p. we can afford to pay for this extra overhead. It remains an open problem to find an adaptive query algorithm that matches an optimally tuned algorithm when we only require constant success probability, even if we can compute collision probabilities.

References

1. Andoni, A., Indyk, P., Laarhoven, T., Razenshteyn, I., Schmidt, L.: Practical and optimal LSH for angular distance. In: Proceedings of the NIPS 2015, pp. 1225–1233 (2015)
2. Andoni, A., Laarhoven, T., Razenshteyn, I.P., Waingarten, E.: Optimal hashing-based time-space trade-offs for approximate near neighbors. In: Proceedings of the SODA 2017, pp. 47–66 (2017)
3. Andoni, A., Razenshteyn, I.: Optimal data-dependent hashing for approximate near neighbors. In: Proceedings of the STOC 2015, pp. 793–801 (2015)
4. Aumüller, M., Bernhardsson, E., Faithfull, A.: ANN-benchmarks: a benchmarking tool for approximate nearest neighbor algorithms. In: Beecks, C., Borutta, F., Kröger, P., Seidl, T. (eds.) SISAP 2017. LNCS, vol. 10609, pp. 34–49. Springer, Cham (2017). https://doi.org/10.1007/978-3-319-68474-1_3
5. Aumüller, M., Christiani, T., Pagh, R., Vesterli, M.: PUFFINN: parameterless and universally fast finding of nearest neighbors. In: Proceedings of the ESA 2019. LIPIcs, vol. 144, pp. 10:1–10:16 (2019)
6. Bawa, M., Condie, T., Ganesan, P.: LSH forest: self-tuning indexes for similarity search. In: Proceedings of the WWW 2005, pp. 651–660 (2005)
7. Charikar, M.: Similarity estimation techniques from rounding algorithms. In: Proceedings of the STOC 2002, pp. 380–388 (2002)
8. Christiani, T.: Fast locality-sensitive hashing frameworks for approximate near neighbor search. In: Amato, G., Gennaro, C., Oria, V., Radovanović, M. (eds.) SISAP 2019. LNCS, vol. 11807, pp. 3–17. Springer, Cham (2019). https://doi.org/10.1007/978-3-030-32047-8_1
9. Christiani, T., Pagh, R., Thorup, M.: Confirmation sampling for exact nearest neighbor search. CoRR abs/1812.02603 (2018). http://arxiv.org/abs/1812.02603
10. Datar, M., Immorlica, N., Indyk, P., Mirrokni, V.S.: Locality-sensitive hashing scheme based on p-stable distributions. In: Proceedings of the SOCG 2004, pp. 253–262 (2004)
11. Dong, W., Wang, Z., Josephson, W., Charikar, M., Li, K.: Modeling LSH for performance tuning. In: Proceedings of the CIKM 2008, pp. 669–678 (2008)
12. Har-Peled, S., Indyk, P., Motwani, R.: Approximate nearest neighbor: towards removing the curse of dimensionality. Theor. Comput. 8(1), 321–350 (2012)
13. Indyk, P., Motwani, R.: Approximate nearest neighbors: towards removing the curse of dimensionality. In: Proceedings of the STOC 1998, pp. 604–613 (1998)
14. Li, P., König, A.C.: Theory and applications of b-bit minwise hashing. Commun. ACM 54(8), 101–109 (2011)
15. Lv, Q., Josephson, W., Wang, Z., Charikar, M., Li, K.: Intelligent probing for locality sensitive hashing: multi-probe LSH and beyond. PVLDB 10(12), 2021–2024 (2017)
16. Panigrahy, R.: Entropy based nearest neighbor search in high dimensions. In: Proceedings of the SODA 2006, pp. 1186–1195 (2006)
17. Slaney, M., Lifshits, Y., He, J.: Optimal parameters for locality-sensitive hashing. Proc. IEEE 100(9), 2604–2623 (2012)

Optimal Metric Search Is Equivalent to the Minimum Dominating Set Problem

Magnus Lie Hetland[✉]🆔

Norwegian University of Science and Technology, Trondheim, Norway
mlh@ntnu.no

Abstract. In metric search, worst-case analysis is of little value, as the search invariably degenerates to a linear scan for ill-behaved data. Consequently, much effort has been expended on more nuanced descriptions of what performance might in fact be attainable, including heuristic baselines like the AESA family, as well as statistical proxies such as intrinsic dimensionality. This paper gets to the heart of the matter with an exact characterization of the best performance actually achievable for any given data set and query. Specifically, linear-time objective-preserving reductions are established in both directions between optimal metric search and the minimum dominating set problem, whose greedy approximation becomes the equivalent of an oracle-based AESA, repeatedly selecting the pivot that eliminates the most of the remaining points. As an illustration, the AESA heuristic is adapted to downplay the role of previously eliminated points, yielding some modest performance improvements over the original, as well as its younger relative iAESA2.

Keywords: Metric indexing · Baselines · Hardness · Dominating set

1 Introduction

Mapping out the complexity of a computational problem is generally a two-pronged affair. On the one hand, there will be algorithms solving the problem, whose performance is evaluated theoretically or empirically, providing ever-tightening pessimistic bounds on what is possible. On the other hand, there may be *lower* bounds, based on reasonable complexity-theoretical assumptions, as in the case of edit distance, for example [1], or on reasoning about the fundamentals of the computational model, as in the case of sorting [11]. The endgame is when these bounds meet, showing some algorithm to be optimal.

Such bounds generally apply to the worst case, as the best-case performance tends to be trivial. For metric search, however, both the best case and the worst are quite uninformative. For a range query, one could always construct an input where examining a single object is enough—or one where there is no escaping a full linear scan. The main thrust of research attempting to describe what performance is possible has thus been directed toward empirical baselines like the AESA family [10,25] and statistical hardness measures such as intrinsic

© Springer Nature Switzerland AG 2020
S. Satoh et al. (Eds.): SISAP 2020, LNCS 12440, pp. 111–125, 2020.
https://doi.org/10.1007/978-3-030-60936-8_9

dimensionality [5],[1] or in some cases restricting the type of structure studied, to permit a more nuanced analysis [20].

It is, however, possible to describe *exactly* what performance is attainable for a given data set and query, as I show in what follows. The main equivalence result, between metric search and dominating sets, provides just such a description, i.e., the lowest number of distance computations that can resolve the query. This performance will not, in general, be attainable without some lucky guesses, but it *is* attainable. In addition, it is possible to give a bound on how close to this performance a polytime algorithm may come in the worst case, under reasonable complexity assumptions. The bound is tight for a sufficiently precise pivot selection heuristic, i.e., one that is able to predict which point will eliminate the most of the remainder, if used as a pivot.

In the AESA method, the index is a distance matrix, and search alternates between heuristically selecting points close to the query and eliminating remaining objects that are shown to be irrelevant. The results in this paper are based on an idea developed by Ole Edsberg,[2] which involves computing an *elimination* matrix for a given query, with which one may implement an "oracle AESA," selecting pivots greedily based on elimination power, rather than on similarity to the query object. I build on this idea, establishing equivalence to the minimum dominating set problem.[3] The main results and contributions of the paper are summarized in the following.

Reduction to Domination. Sections 2 and 3 establish a linear-time objective-preserving reduction from the problem of resolving metric range queries (and certain kNN queries) with as few distance computations as possible to that of finding minimum dominating sets in directed graphs. This reduction applies to an *offline* variant of metric search, where all query–object distances are already known. It does, however, make it possible to compute the exact optimum attainable for the online version as well. Some experimental results are provided as an illustration.

Reduction from Domination. Section 4 describes a reduction in the other direction, from the dominating set problem in *undirected* graphs to minimizing distance computations, establishing the hardness of metric search. While it may in many cases still be feasible to determine the optimum using efficient solvers of various kinds, this does mean that under reasonable complexity-theoretical assumptions, no search method can, in general, *guarantee* attaining this optimum.

The reduction preserves the objective value, and for range search, the number of data objects equals the number of vertices, which means that inapproximability results for the dominating set problem carry over to metric search, with

---

[1] Other measures include the distance exponent [24] and the ball-overlap factor [21].

[2] Personal communication, July 2012.

[3] Note that the reductions are to and from two different versions of the dominating set problem (the directed and undirected version, respectively). At the price of slightly looser bounds, one could stick with just one of these.

approximation bounds for the former applying to the performance of the latter, i.e., the number of distance computations. Thus, for range search, one cannot even expect to get closer than within a log-factor of the optimum.

AESA and Greedy Approximation. Because the objective is preserved also in reducing *to* domination, and the number of objects equals the number of vertices, *approximability* results also translate, meaning that in principle the standard greedy selection strategy would yield the best feasible metric range search algorithm (or very close to it), in terms of distance computations in the worst case.[4] As discussed in Sect. 5, the greedy approach corresponds to the AESA family of algorithms, given the right selection heuristic, i.e., one that accurately estimates the elimination power of a potential pivot, among the remaining objects. An exact estimate here is, of course, not possible without knowing the query–pivot distances, but this correspondence does demonstrate that, in the limit, AESA is, indeed, as good as it gets. As an illustration, inspired by the greedy approximation, *greedy* AESA (gAESA) is proposed, taking into account which points remain to be eliminated.

2   Pivoting Is, of Course, Optimal

A *range search* using a metric δ over a set X means finding all points $x \in X$ within some search radius r of a given query point q, i.e., all points x for which $\delta(q, x) \leqslant r$. Given the distances between a query q and a set P of *pivots*, the distance $\delta(q, x)$ for any point x is bounded as follows:

$$\max_{p \in P} |\delta(q, p) - \delta(p, x)| \leqslant \delta(q, x) \leqslant \min_{p \in P} \delta(q, p) + \delta(p, x) \tag{1}$$

Leaving q and P implicit, we may refer to the lower and upper bounds as $\ell(x)$ and $u(x)$, respectively. If our search radius falls outside this range, there is no need to compute $\delta(q, x)$; either the radius is small enough that we simply eliminate x $(r < \ell(x))$, or it is great enough that x is "eliminated" by adding it to the search result, sight unseen $(r \geqslant u(x))$.

This very direct approach of using exact, stored distances $\delta(p, x)$, *pivoting*, is the gold standard for minimizing the number of distance computations needed. Other approaches, which all involve coarsening the stored information in some way, may reduce the computational resources needed to eliminate candidate objects, but it should be obvious that they cannot require fewer distance computations. As the following lemma shows, the lower and upper bounds are necessarily valid values for $\delta(q, x)$, so if $\ell(x) \leqslant r \leqslant u(x)$, x cannot safely be eliminated.

Lemma 1. *Let* (X, δ) *be a metric space, with* $X = \{p_1, \ldots, p_m, q, z\}$, *and let the distances* $\delta_1, \delta_2 : X \times X \to \mathbb{R}_{\geqslant 0}$ *be defined as follows:*

$$\delta_1(x, y) = \begin{cases} \max_i |\delta(q, p_i) - \delta(p_i, z)| & \text{if } \{x, y\} = \{q, z\}; \\ \delta(x, y) & \text{otherwise.} \end{cases}$$

---

[4] This is the worst case *given* that the optimal number of distance computations is some value γ, not the more general, non-informative worst-case of $\Omega(n)$.

$$\delta_2(x, y) = \begin{cases} \min_i \ \delta(q, p_i) + \delta(p_i, z) & if \ \{x, y\} = \{q, z\}; \\ \delta(x, y) & otherwise. \end{cases}$$

Then δ_1 is a pseudometric and δ_2 is a metric. If $\delta(q, p_i) \neq \delta(p_i, z)$ for some i, or if $q = z$, then δ_1 is a metric.

Proof. We have $\delta_j(x, y) = \delta_j(y, x)$ and $\delta_j(x, x) = 0$, for $x, y \in X, j \in \{1, 2\}$. We also have $\delta_2(x, y) = 0 \implies x = y$, and if $\delta(q, p_i) \neq \delta(p_i, z)$ for some i, or if $q = 0$, then $\delta_1(x, y) = 0 \implies x = y$. We have $\delta_1(q, z) \leqslant \delta(q, z)$, so triangularity can only be broken for δ_1 in the cases $\delta_1(q, p_k) \leqslant \delta_1(q, z) + \delta_1(z, p_k)$ or $\delta_1(z, p_k) \leqslant \delta_1(z, q) + \delta_1(q, p_k)$, for some k. Consider the first of these. We maximize over i, so we need only show the following for *some* choice of i:

$$\delta(q, p_k) \leqslant |\delta(q, p_i) - \delta(p_i, z)| + \delta(z, p_k) \tag{2}$$

This is satisfied for $k = i$. The other case is handled symmetrically. For δ_2, we have $\delta(q, z) \leqslant \delta_2(q, z)$, so triangularity can only be broken in $\delta_2(q, z) \leqslant \delta_2(q, p_k) + \delta_2(p_k, z)$. We minimize over i, so this need only hold for *some* choice of i, and again we may choose $i = k$, producing an equation. □

Corollary 1. *No search method can resolve a metric range query with fewer distance computations than pivoting.*

Proof. From Lemma 1, we know that after a set of distance computations making pivots $p_1, \ldots, p_m$ available, the pivoting bounds are *tight*; if pivoting cannot eliminate an object, no method can safely do so. (Note that an adversary would be free to let $q = z$ in the case where $\delta(q, z) = \ell(z) = 0$, ensuring that we are indeed dealing with a metric space.) And given that no method can eliminate more objects than pivoting for any distance count, no method can eliminate all the objects with a lower distance count than pivoting. □

In other words, any method using fewer distance computations than pivoting could be made to fail by an adversary in charge of the data set. This argument covers range queries, and it is not hard to translate it to the kNN case, where the k nearest neighbors of q are sought, as long as the result set is uniquely determined. A radius must then exist, separating the k nearest neighbors from the others, and the tightest possible upper bound on this radius is the maximum of the k lowest pivoting bounds we have. Pivoting must then be able to eliminate all points outside this radius, or our adversary might strike again. The following corollary covers the more general case.

Corollary 2. *No search method can resolve a metric kNN query with fewer distance computations than pivoting.*

Proof. We need to establish $\delta(q, x) \leqslant \delta(q, y)$ for every x in the result and every y outside it. Assume that, given some pivot set P, there is one such inequality that cannot be established by pivoting, i.e., $u(x) > \ell(y)$. An adversary could then ensure $\delta(q, x) > \delta(q, y)$, as follows. First, let $\delta(q, x) = u(x)$. The only effect on the valid range for $\delta(q, y)$ is found in the lower bound $\delta(q, y) \geqslant |u(x) - \delta(x, y)|$. If $\delta(x, y) \leqslant u(x)$, then the relevant lower bound is $u(x) - \delta(x, y)$, which is *strictly* less than $\delta(q, x) = u(x)$ (because $\delta(x, y) > 0$, as $x \neq y$), and so it is still possible to have $\delta(q, y) < \delta(q, x)$.

If, however, $\delta(x, y) > u(x)$, the relevant lower bound is $\delta(x, y) - u(x)$. Let p be the pivot that produced the pivoting bound $u(x)$. We then have:

$$\delta(x, y) - u(x) = \delta(x, y) - \big(\delta(q, p) + \delta(p, x)\big)$$
$$= \big(\delta(x, y) - \delta(p, x)\big) - \delta(q, p) \leqslant \delta(p, y) - \delta(q, p) \leqslant \ell(y)$$

In other words, $\delta(q, y) = \ell(y)$ is still a valid choice for the adversary, yielding the desired $\delta(q, y) < \delta(q, x)$.   □

The upshot is that the optimal distance count (for range and kNN queries) can be found by considering only elimination *using individual pivots*.

The range and kNN search modes are closely related, and yet there are cases where they behave quite differently, as shown in Fig. 1.

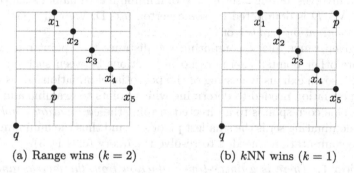

(a) Range wins ($k = 2$)   (b) kNN wins ($k = 1$)

Fig. 1. Differences between range search and kNN in the presence of ties for the kth position, using $(\mathbb{R}^2, \mathrm{L}_1)$. In both configurations, we have $r = 8$. In (a), a range search need only compute $\delta(q, p)$, while kNN much also compute $\delta(q, x_i)$ for all but one of the x_i. In (b), the kNN search need only compute $\delta(q, p)$ and $\delta(q, x_i)$ for one of the x_i, while a range search must compute all distances $\delta(q, -)$

It is, however, possible to establish some correspondence between the two, when the kNN result set is uniquely determined.

Lemma 2. *If the kNN result is uniquely determined, the optimum number of distance computations for kNN is no worse than for a range search with the smallest possible kNN radius, even if the radius is unknown initially. Furthermore, there is a radius for which range queries and kNN will produce the same search result using the same number of distance computations.*

Proof. When the kNN result is unique, there is a radius r corresponding to the k resulting objects. Resolving a range query with radius r must necessarily yield upper bounds of at most r for returned objects and lower bounds greater than r for the remainder. These same bounds can also be used to separate the k nearest objects from the remainder, *without a specified radius*, so kNN cannot require more distance computations.

Conversely, consider a kNN query. By Corollary 2, no method requires fewer distance computations than pivoting, so in the optimal case we will have actual distance bounds available, strictly separating the k nearest from the remainder. Any range query with a search radius falling between the upper and lower bounds can then be resolved with the same number of distance computations. $\square$

It *is* possible to increase the radius such that a range query would require additional distance computations, while still just returning k objects (cf. Fig. 2).

3   Elimination as Domination

Given a (directed) graph $G = (V, E)$, a vertex u is said to *dominate* another vertex v if the graph has an edge from u to v. The *(directed) minimum dominating set problem* involves finding a set $D \subseteq V$ of minimum cardinality, such that every vertex $v \in V \setminus D$ is dominated by some vertex $u \in D$. We call $\gamma(G) = |D|$ the *(directed) domination number* of G.

For a given range query, computing the distance to a point may eliminate one or more other points. There are no interactions between such eliminations (see Sect. 2), so an exhaustive listing of the potential eliminations gives us all the relevant information needed to determine which points to examine and which to eliminate. This corresponds to a directed graph—the *elimination graph*—whose minimum dominating set is the smallest pivot set, and thus the minimum number of distance computations, needed to resolve the query (cf. Fig. 3).

Proposition 1. *There is a linear-time reduction from the metric range search problem to the directed minimum dominating set problem, which preserves the objective values of the solutions exactly.* $\square$

If the result of a kNN query is uniquely determined, and we ignore elimination based on upper bounds (usually done in practice), the number of distance computations correspond to a range query with the smallest kNN radius.[5] Of course, finding a minimum dominating set is NP-hard,[6] and given the rather unusual clash between large-scale information retrieval and combinatorial optimization, we may quickly end up with overwhelming instance sizes. Still, with a suitable mixed-integer programming solver, for example, the optimization may very well be feasible in many practical cases. As an example, Fig. 4 shows some computations made using the Gurobi solver [12]. Many of these optima were found rather quickly, as presumably the structure of the elimination graph was amenable to the solution methods of the solver. Others, such as those for the

---

[5] Optimal kNN *with* upper bounds does not map as cleanly to dominating sets.

[6] The undirected version is most commonly discussed, with a reduction, e.g., from set covering [13, Th. A.1]. A similar reduction to the directed version is straightforward.

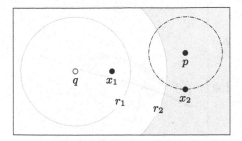

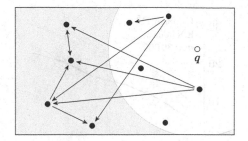

Fig. 2. The nearest neighbor can be determined by examining x_1 and p, as we then have $u(x_1) < \ell(x_2)$. Range search with r_1 can be resolved similarly, but using r_2 requires three distance computations, while still returning the single nearest neighbor

Fig. 3. The directed elimination graph G resulting from a specific range query, with the domination number $\gamma(G) = 5$ corresponding to the minimum number of distance computations needed to separate relevant objects from irrelevant ones

DNA data set, took several days to compute. And even for some of the easier cases, there were outliers. For example, for the 2NN radius in 15-dimensional Euclidean space, all of the 10 randomly selected queries led to computations lasting 10–200 s, except for one, which took almost twenty hours. As with many such cases, however, being satisfied with a solution that is a couple of percentage points shy of perfect could drastically cut down on the computation time (i.e., by setting the absolute or relative MIP gap), as illustrated in Fig. 5.

Figure 4 also includes results for several other methods, beyond the optimum. These are all versions of the AESA approach [25], as discussed in more depth in Sect. 5. At the opposite end of the spectrum of the optimum, there's the incremental random selection of pivots. Separating the feasible from the infeasible, is an *oracle* AESA, which has access to the elimination power of each potential pivot, i.e., how many of the remaining objects will be eliminated if a given pivot is selected. In the feasible region we find AESA, iAESA2 [10], and the new gAESA, which is explained in Sect. 5.

It is worth noting that $\gamma(G)$ is a more precise lower bound than an ordinary *best-case* analysis, which only takes input size into account, and which is therefore always 1. Rather, this is the lowest possible number of distance computations needed *for a given dataset and query*. In order to *guarantee* using at most $\gamma(G)$ distance computations, you would need to somehow determine G, which is quite unrealistic. And, as the next section shows, it is also far from enough.

4   Metric Search Is Hard, Even If You're Omniscient

Obviously, a major challenge in choosing the right pivots is that you don't know what the elimination graph looks like—you can only make heuristic guesses. But what if you *did* know? As it turns out, that wouldn't be the end of your worries.

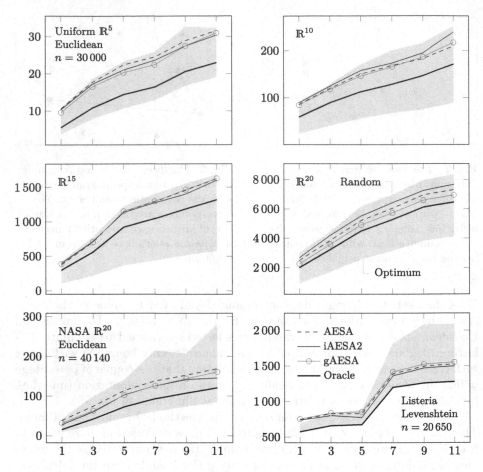

Fig. 4. Number of distance computations as a function of k, the number of nearest neighbors covered by the chosen radius used for a range query. The first four datasets are uniformly random vectors, while the last two are taken from the SISAP dataset collection [9], with queries withheld. The listeria string lengths vary from 39 to 6579. The results are the average over 10 randomly selected queries. The oracle AESA uses elimination power among remaining points as its heuristic

Section 3 showed that it is possible to find the optimum by framing the problem as that of looking for a minimum directed dominating set. Of course, this is an NP-hard problem, so there's no real surprise in that we can reduce *to* it. But what about reducing in the other direction? That is, unless P = NP, is there any hope of finding some feasible way of determining the optimum? Alas, no: reducing from the general minimum undirected dominating set problem to finding the optimum for metric search is quite straightforward, and the reduction

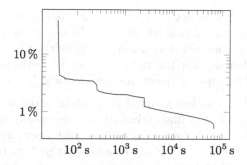

Fig. 5. Bound on relative error (MIP gap) as a function of time, when computing the optimal number of distance computations in a particularly difficult instance with $k = 2$ over uniformly random vectors in 15-dimensional Euclidean space. Finding the optimum took over nineteen hours. After 41 s, the gap was 39.1 %, but already at 44 s, it was down to 4.34 %. Getting to 1 % took 2.21 h.

preserves the both the objective value and the problem size exactly,[7] meaning that approximation hardness results apply as well.

Theorem 1. *There is a linear-time reduction from the undirected minimum dominating set problem on n vertices to the metric range search problem on n objects, which preserves the objective values of the solutions exactly.*

Proof. We first consider range search. To encode any instance $G = (V, E)$ of the minimum dominating set problem, we construct a metric space (X, δ), where $X = V \cup \{q\}$, with $q \notin V$, and design the metric so that the elimination graph corresponds to G. We define the metric as follows:

$$\delta(x, y) = \begin{cases} 0 & \text{if } x = y; \\ 1 & \text{if } \{x, y\} \in E; \\ 2 & \text{otherwise.} \end{cases}$$

In particular, $\delta(q, x) = 2$ for all $x \in V$. This definition of δ satisfies all the metric properties. Specifically, note that triangularity holds, because for any objects $x, y, z \in X$, we have $\delta(x, z) \leqslant 2 \leqslant \delta(x, y) + \delta(y, z)$ (assuming $x \neq y \neq z$; otherwise triangularity is trivial).

It should be clear that the elimination graph for q with $r < 1$ corresponds exactly to the original graph G.[8] The closed neighborhood $N[x]$ of x (that is, x and the set of objects dominated or eliminated by x) is $\{y : \delta(q, x) - \delta(x, y) \geqslant 1\}$, which corresponds exactly to the cases where $\delta(x, y) = 0$ (that is, $x = y$) and where $\delta(x, y) = 1$ (that is, $\{x, y\} \in E$). In other words, any set of pivots that eliminate the remaining objects corresponds to a dominating set in G, and vice

---

[7] In terms of vertices, not edges.

[8] Note that only the lower bound is relevant, as the upper bound is always greater than the search radius.

versa. If we find such a pivot set of minimum cardinality, we will have solved the undirected minimum dominating set problem. In other words, we have a valid reduction from the undirected dominating set problem to metric range search. It should also be obvious that the reduction can be performed in linear time, and that the size of the optimal solutions are identical.[9]   □

The previous reduction can be extended to a polytime reduction to kNN search quite easily, showing NP-hardness (though not necessarily preserving approximation results). We simply set $k = 1$ and add another object $\bar{x}$ so that $\delta(q, \bar{x}) = r < 1$ and $\delta(\bar{x}, y) = 2$ for any other object y. Now the minimum kNN radius will automatically be r, which gives us the same reduction as before.

The reduction in the proof of Theorem 1 constructs a metric range search problem on n objects from an undirected dominating set problem on n nodes, so that *if* (and only if) we can solve the search problem (that is, find a minimum pivot set), we have also solved the minimum dominating set problem. Approximation bounds thus carry over from the dominating set problem, so for any $\epsilon > 0$, finding solutions that are within a factor of $(1 - \epsilon) \ln n$ is unfeasible, unless $\mathrm{NP} \subseteq \mathrm{DTIME}(n^{O(\lg \lg n)})$ [6].

Corollary 3. *For instances of the metric range search problem over n objects where the optimal number of distance computations is γ, the worst-case running time of any algorithm is $\Omega(\gamma \log n)$, unless $\mathrm{NP} \subseteq \mathrm{DTIME}(n^{O(\lg \lg n)})$.*

Proof. An algorithm with a (polynomial) running time of $o(\gamma \log n)$ would necessarily use $o(\gamma \log n)$ distance computations, yielding an approximation algorithm for the dominating set problem with an approximation ratio $o(\log n)$.   □

Note that the worst-case running time *in general* is still $\Omega(n)$, as we may very well have $\gamma = n$, in degenerate workloads.

5   Omniscience Is Overrated

In the discussion so far, what has been described is a scenario where all potential eliminations are known. Even then, as we have seen, it is only realistically feasible to get to within a log-factor of the optimum. And as it turns out, achieving this log-factor is possible, even *without* knowing all the potential eliminations. What is assumed instead is a more limited oracle that can tell us which of the remaining points has the highest *elimination power*, that is, the highest out-degree among the remaining vertices.

The thing is, a minimum dominating set may be approximated to within a log-factor using a simple greedy strategy—a strategy that most likely cannot be significantly improved upon; it gets within a factor of $\ln n + 1$, and as discussed in the previous section, we have a lower bound of $(1 - \epsilon) \ln n$ for any $\epsilon > 0$.[10]

---

[9] If the new distance is allowed to use the original graph as part of its definition, the reduction can be performed in *constant* time—it is merely a reinterpretation.

[10] The *upper* bound is easily shown by reinterpreting the minimum dominating set problem for a directed graph $\mathrm{G} = (\mathrm{V}, \mathrm{E})$ as the problem of covering V with the closed out-neighborhoods of G, translating the standard set covering approximation [26].

What is more, this is exactly the approach taken by the AESA family of indexing methods: they greedily pick one point at a time, based on estimated elimination power, eliminating others as they go (cf. Fig. 6). In other words, full omniscience wrt. the elimination graph is not needed; if we can formulate a heuristic returning the most useful next pivot at each step, the algorithm is already as good as it realistically can be, or at least very nearly so.

Proposition 2. *Greedily selecting pivots based on high elimination power is an asymptotically optimal polytime strategy for minimizing distance computations in metric range search, unless* $\mathrm{NP} \subseteq \mathrm{DTIME}(n^{O(\lg \lg n)})$.   □

To say that AESA picks pivots based on elimination power may be overstating it, however. Rather, Vidal Ruiz talks about "successive approximation to nearest points" [25], while Figueroa et al. state that their goal is "to define an order such that the first element is very close to the query," because "[t]he closer the pivot to the query q, the more effective the pruning is" [10]. Of course, all manner of regression and learning methods might be used with the specific goal of estimating which points are close to the query [8,16], or which are likely to be part of the search result [17].

$\mathrm{AESA}(q, r; V, \delta)$	$\mathrm{GREEDY\text{-}DOM\text{-}SET}(V, E)$
1 $U \leftarrow V$	1 $U \leftarrow V$
2 $P = \emptyset;\ R = \emptyset$	2 $D \leftarrow \emptyset$
3 **while** $U \neq \emptyset$	3 **while** $U \neq \emptyset$
4 $p \leftarrow \arg\min_{x \in U} h_P(x)$	4 $p \leftarrow \arg\max_{x \in U} |N^+(x) \cap U|$
5 **if** $\delta(q,p) \leqslant r$: $R \leftarrow R \cup \{p\}$	5
6 $P \leftarrow P \cup \{p\}$	6 $D \leftarrow D \cup \{p\}$
7 $U \leftarrow U \setminus (\{p\} \cup \{x : \ell_P(x) > r\})$	7 $U \leftarrow U \setminus N^+[p]$
8 **return** R	8 **return** D

Fig. 6. Side-by-side comparison of the AESA metric search algorithm and the greedy approximation for the directed minimum dominating set problem

There has been work on pivot selection focusing directly on elimination power [4], but this does not seem to have been central in AESA-like methods, using a full distance matrix. One selection method, which maximizes the lower bound used for elimination, and skips over pivots that don't contribute, has been explored in the fixed, initial pivot list of PiAESA [22], but the second phase, where pivots are selected dynamically, still follows the heuristic of selecting those that seem close to the query.

Following the analogy with the greedy approximation for the directed dominating set problem, there are two modifications one might make. The first is to look for high elimination power in the data set overall, rather than closeness to the query. For example, it is quite possible that a pivot that is far away might be able to eliminate an entire nearby cluster. The second modification, which I will briefly explore, is to modify the selection based on redundancy, i.e., how much of a point's elimination power actually applies to *remaining* points. If one selects pivots that are as similar to the query as possible, they are bound to be similar to each other as well; and even if a pivot is able to eliminate many other points, that is of little use if those points have already been discarded.

A simple version of this second modification is the following: rather than merely minimizing the sum of lower bounds, as in AESA, we divide this by the sum of distances to remaining points. This will not only prefer pivots that seem to be close to q, but those that seem close to q *relative to* how far they are from the remaining points, meaning they ought to be able to eliminate more of them. Some preliminary results on the performance of this *greedy AESA* (gAESA) are shown in Fig. 4. As can be seen, it does seem to perform on par with AESA and iAESA2, at times outperforming both. Given the rather arbitrary nature of the heuristic, better variants might very well exist.

6   Concluding Remarks and Future Work

The previous sections have established an equivalence between the minimal number of distance computations needed to resolve an exact metric range query, on the one hand, and the size of a minimum dominating set in a directed graph on the other.[11] The result also applies to uniquely determined kNN queries, if upper bounds are ignored. One might object that the scenario is too limited— that in practice, one would be contented with an *approximate* or *probabilistic* search. In fact, the results do also apply for certain approximations, such as those that merely modify the query, resulting in a new, simpler exact search [18]. But beyond this, the main uses of these results are precisely in establishing the limits of exact search for given workloads; if one can show that any exact algorithm must examine an excessively large portion of the data set, that is a forceful argument in favor of approximation or randomization. What is presented here only scratches the surface, however. What follows is a sketch of possible directions for future research based on the established equivalence.

Heuristic Development. The gAESA heuristic is somewhat arbitrary. While it picks pivots that seem close to the query, relative to the remaining points, the *goal* is to pick the pivot with the highest elimination power. There may be many ways of estimating this more directly, either using hand-crafted heuristics (e.g., including pivots that are far away from the query compared to remaining points) or machine learning (which has so far been focused on distance or relevance).

Algorithm Development. In the interest of constructing better baselines, one might take the development further. Rather than going with the AESA approach, one might attempt to solve the dominating set problem without actually knowing the graph. This would be different from the more common forms of

---

[11] That is, for any range search instance, there is a directed graph with the objects as its nodes for which the equivalence holds. Reducing in the other direction preserves the objective value, but not necessarily the number of nodes/objects.

online dominating set problems [3], where vertices are provided in some arbitrary order. Rather, this would presumably involve link prediction [15], at each step selecting a pivot deemed likely to be included in the optimal solution or to provide good support for future predictions.

Problem Variants. The dominating set problem provides a new perspective on the problem of metric search, and variants of the former might find analogies for the latter. For example, the *weighted* dominating set problem can also be approximated greedily, and the analogous metric search method would be a weighted AESA, where selection is based on the ratio of weight to elimination power. The weight could, for example, represent the actual cost of computing the query–pivot distance, which is the effort that is being minimized, after all. For many distances, this cost is identical for all points, but for, e.g., the signature quadratic form distance [2], it may vary wildly.

One might also look for analogies in the other direction. For example, probabilistic methods (such as probabilistic iAESA [10]) do not aim to eliminate all vertices; in these cases, one could instead consider *partial* domination [7].

Probabilistic Analysis. There is a substantial literature on the topic of random graphs. For example, it is known that for random digraphs whose edges are independent Bernoulli variables with probability p,[12] the domination number is logarithmic, with base $1/(1-p)$ [14]. In fact, it is not hard to modify the results of Telelis and Zissimopoulos [23] to show that in this scenario, even AESA selecting pivots *arbitrarily* would yield a logarithmic number of pivots, staying within a doubly logarithmic additive term of the optimum, results that match those of Navarro [19].

Workload Descriptions. Beyond finding γ, the dominating set perspective may inspire other hardness measures and workload descriptions. For example, the greedy approximation is, more precisely, logarithmic in the *maximum degree* $\Delta(G)$, a value that could be used as an indicator of *how hard it is to get close to the optimum.* And although the independence assumption on elimination may be too strong, one could still use the elimination probability p, perhaps estimated by averaging over several queries, as an indication of general workload hardness.

Acknowledgements. The author would like to thank Ole Edsberg, both for discussions providing the initial idea for this paper, and for substantial later input. He would also like to thank Jon Marius Venstad and Bilegsaikhan Naidan for reading early drafts of the paper and providing feedback.

---

[12] Chávez et al. say that such independence is a "reasonable approximation" [5].

References

1. Backurs, A., Indyk, P.: Edit distance cannot be computed in strongly subquadratic time (unless SETH is false). In: Proceedings of the 47th Annual ACM Symposium on Theory of Computing (2015). https://doi.org/10.1145/2746539.2746612
2. Beecks, C., Uysal, M.S., Seidl, T.: Signature quadratic form distance. In: Proceedings of the ACM International Conference on Image and Video Retrieval. ACM, New York, NY, USA (2010). https://doi.org/10.1145/1816041.1816105
3. Boyar, J., Eidenbenz, S.J., Favrholdt, L.M., Kotrbčík, M., Larsen, K.S.: Online dominating set. Algorithmica **81**(5), 1938–1964 (2018). https://doi.org/10.1007/s00453-018-0519-1
4. Bustos, B., Navarro, G., Chávez, E.: Pivot selection techniques for proximity searching in metric spaces. Pattern Recogn. Lett. **24**(14), 2357–2366 (2003). https://doi.org/10.1016/S0167-8655(03)00065-5
5. Chávez, E., Navarro, G., Baeza-Yates, R., Marroquín, J.L.: Searching in metric spaces. ACM Comput. Surv. **33**(3), 273–321 (2001). https://doi.org/10.1145/502807.502808
6. Chlebík, M., Chlebíková, J.: Approximation hardness of dominating set problems. In: Albers, S., Radzik, T. (eds.) ESA 2004. LNCS, vol. 3221, pp. 192–203. Springer, Heidelberg (2004). https://doi.org/10.1007/978-3-540-30140-0_19
7. Das, A.: Partial domination in graphs. Iran. J. Sci. Technol. Trans. A Sci. **43**(4), 1713–1718 (2018). https://doi.org/10.1007/s40995-018-0618-5
8. Edsberg, O., Hetland, M.L.: Indexing inexact proximity search with distance regression in pivot space. In: Proceedings of the 3rd International Conference on Similarity Search and Applications (2010). https://doi.org/10.1145/1862344.1862353
9. Figueroa, K., Navarro, G., Chávez, E.: Metric spaces library (2007). http://www.sisap.org/Metric_Space_Library.html
10. Figueroa, K., Chávez, E., Navarro, G., Paredes, R.: Speeding up spatial approximation search in metric spaces. J. Exp. Algorithmics **14**, 3–6 (2010). https://doi.org/10.1145/1498698.1564506
11. Ford Jr., L.R., Johnson, S.M.: A tournament problem. Am. Math. Mon. **66**(5), 37–40 (1959). https://doi.org/10.1080/00029890.1959.11989306
12. Gurobi Optimization, LLC.: Gurobi optimizer reference manual (2020). http://gurobi.com
13. Kann, V.: On the Approximability of NP-complete Optimization Problems. Ph.D. thesis, Department of Numerical Analysis and Computing Science, Royal Institute of Technology, Stockholm (1992)
14. Lee, C.: Domination in digraphs. J. Korean Math. Soc. **35**(4), 843–853 (1998)
15. Lü, L., Zhou, T.: Link prediction in complex networks: a survey. Physica A stat. Mech. Appl. **390**(6), 1150–1170 (2011). https://doi.org/10.1016/j.physa.2010.11.027
16. Mao, R., Liu, X., Tang, H., Luo, Q., Chen, J., Wu, W.: Multivariate regression for pivot selection: a preliminary study. In: 2011 3rd Symposium on Web Society. IEEE (2011). https://doi.org/10.1109/SWS.2011.6101281
17. Murakami, T., Takahashi, K., Serita, S., Fujii, Y.: Probabilistic enhancement of approximate indexing in metric spaces. Inf. Syst. **38**(7), 1007–1018 (2013). https://doi.org/10.1016/j.is.2012.05.012
18. Naidan, B., Hetland, M.L.: Shrinking data balls in metric indexes. In: DBKDA (2013)

19. Navarro, G.: Analyzing metric space indexes: what for? In: Proceedings of the 2009 2nd International Workshop on Similarity Search and Applications, SISAP 2009. IEEE Computer Society (2009). https://doi.org/10.1109/SISAP.2009.17

20. Pestov, V.: Lower bounds on performance of metric tree indexing schemes for exact similarity search in high dimensions. Algorithmica (2013). https://doi.org/10.1007/s00453-012-9638-2

21. Skopal, T.: Unified framework for exact and approximate search in dissimilarity spaces. ACM Trans. Database Syst. (TODS) **32**(4), 1–45 (2007). https://doi.org/10.1145/1292609.1292619

22. Socorro, R., Micó, L., Oncina, J.: A fast pivot-based indexing algorithm for metric spaces. Pattern Recogn. Lett. **32**(11), 1511–1516 (2011). https://doi.org/10.1016/j.patrec.2011.04.016

23. Telelis, O.A., Zissimopoulos, V.: Absolute $o(\log m)$ error in approximating random set covering: an average case analysis. Inf. Process. Lett. **94**(4), 171–177 (2005). https://doi.org/10.1016/j.ipl.2005.02.009

24. Traina Jr., C.: Distance exponent: a new concept for selectivity estimation in metric trees. In: Proceedings of the 16th International Conference on Data Engineering (2000). https://doi.org/10.1109/ICDE.2000.839409

25. Vidal Ruiz, E.: An algorithm for finding nearest neighbours in (approximately) constant average time. Pattern Recogn. Lett. **4**(3), 145–157 (1986). https://doi.org/10.1016/0167-8655(86)90013-9

26. Williamson, D.P., Shmoys, D.B.: The Design of Approximation Algorithms. Cambridge University Press, Cambridge (2011)

Metrics and Ambits and Sprawls, Oh My

Another Tutorial on Metric Indexing

Magnus Lie Hetland$^{(\boxtimes)}$ (ID)

Norwegian University of Science and Technology, Trondheim, Norway
mlh@ntnu.no

Abstract. A follow-up to my previous tutorial on metric indexing, this paper walks through the classic structures, placing them all in the context of the recently proposed *sprawl of ambits* framework. The indexes are presented as configurations of a single, more general structure, all queried using the same search procedure.

Keywords: Metric indexing · Tutorial · Sprawls · Ambits

1 Introduction

About ten years ago, I wrote a tutorial on metric indexing [12], and last year, I finally finished a unifying framework for metric indexing and other comparison-based indexing [13]. That paper, however, is perhaps not the most inviting, containing quite a bit of detail and formalism, so in this paper, I'll revisit my earlier tutorial, in light of this new framework. This approach has two main benefits. First, the result should ideally be a streamlined, unified tutorial, rather than a smorgasbord of disjoint techniques; and, second, it provides an example-based introduction to the framework of sprawls and ambits, which might be useful to researchers who are already familiar with metric indexing. I focus primarily on "the classics"; for an overview of many variants, see, e.g., the recent paper by Chen et al. [6].

In contrast to the full paper introducing sprawls and ambits [13], I will try to keep this tutorial brief and to the point—more so, even, than my previous tutorial. In keeping with that, let's get going!

2 Framework

This section presents a thumbnail sketch of the framework used throughout. It may be easier to understand *after* you've read some of the example applications, so feel free to skim it, then skip ahead to Sect. 3, returning here later.

We have a data set V drawn from some universe U with an associated *metric*, i.e., a symmetric function $\delta : U \times U \to \mathbb{R}_{\geqslant 0}$, where $\delta(u,v) = 0$ iff $u = v$, and

$$\delta(u,v) \leqslant \delta(u,w) + \delta(w,v)\,, \tag{2.1}$$

© Springer Nature Switzerland AG 2020
S. Satoh et al. (Eds.): SISAP 2020, LNCS 12440, pp. 126–139, 2020.
https://doi.org/10.1007/978-3-030-60936-8_10

for all $u, v, w \in U$. The problem we are trying to solve is storing V in some data structure, which we may later traverse to efficiently extract those points relevant to some query. Intuitively, we view this data structure as a bipartite digraph of points and *regions*, i.e., *sets* of points. This is referred to as a *sprawl* of regions.[1]

A region R with parents $p_1, \ldots, p_m$ is then *defined* in terms of these. That is, whether $u \in R$ depends on the distances $x = [\delta(u, p_1), \ldots, \delta(u, p_m)]$, a vector in the so-called *pivot space* of $p_1, \ldots, p_m$. Specifically, we use a linear function $f(x)$ and a threshold, or *radius r*, so $u \in R$ iff $f(x) \leqslant r$. Such a region is called a linear *ambit*.

The regions partition the space, representing a coarsening of the data. For a query in the form of a ball $Q = \{u : \delta(q, u) \leqslant s\}$ of relevant points, we are only interested in the contents of a region R *if it intersects* Q. The idea, then, is to have the children of R point the way to smaller subsets of the data set. Search becomes a traversal of our graph, where each region is checked for overlap with the query before possibly traversing its children.

What is more, because a region is defined by its parents, we require *all* the parents to be traversed before traversing the region, and possibly its children. When we traverse a point u, we compute $\delta(q, u)$, so that when we traverse a region, we have all of $\delta(q, p_1), \ldots, \delta(q, p_m)$ available, giving us a distance vector z representing the query. If we assume, for now, that f is nondecreasing, Q and R intersect only if:[2]

$$f(z) \leqslant r + f(s) \tag{2.2}$$

For more advanced queries (kNN), and when we permit elimination, the order of traversal is significant. In these cases, we'd use a priority queue of nodes to traverse, updating their priorities each time we encounter them. In the basic scenario sketched out here, though, we might as well use a depth-first approach, as in the following mutually recursive procedures:

Simplified sprawl search algorithm

Visit-Point(u, q, s)	Visit-Region(R, q, s)
1 **if** $\delta(q, u) \leqslant s$	1 get z from R.*parents*
2 **print** u	2 get f and r from R
3 **for** $R \in u.children$	3 **if** $f(z) > r + f(s)$
4 R.*count* = R.*count* + 1	4 **return**
5 **if** R.*count* == |R.*parents*|	5 **for** $v \in R.children$
6 Visit-Region(R, s)	6 **if** $v.color$ == WHITE
7 $u.color$ = BLACK	7 Visit-Point(v, q, s)

In general, the idea is that δ is memorized in some way, so once $\delta(q, u)$ is computed on line 1 of Visit-Point, it is subsequently available when we gather

---

[1] Equivalently, a hyperdigraph on V, with one region per hyperedge [13, Remark 2.3.12].

[2] Here $f(s)$ is shorthand for $f(s, \ldots, s)$.

up z in VISIT-REGION. Normally, one would have one or more designated root nodes, and call VISIT-POINT on them in turn to initiate the search.

The way this is set up, one would need to run a reinitialization in-between queries, resetting the memo, coloring nodes white and setting counts to zero. There are many ways of handling this, of course. One could have actual attributes in the nodes, and maint a list of those that need resetting, requiring constant amortized time. An even simpler approach might be to simply use hash tables that are reset between searches. With some additional memory, one could even do the reset in actual constant time, using the standard trick for constant-time array initialization. In this case, one could keep a stack of nodes whose attributes are valid, and let each node keep its index in the stack. Then the reset would simply require setting the stack length to zero.

3   Ball Trees

A metric ball tree is a form of search tree where subtrees and their points are enclosed in balls. A subtree is then only explored if its ball intersects the query. For example, the simple BS-tree is a tree where each node is associated with a single point and a radius that covers the points below it in the tree [14]. The idea of a sprawl is for the graph (in this case, a tree) to express dependencies, where we have edges from points to the regions they tell us about, and from regions to the points they tell us to explore (*if* we intersect them). In the case of the BS-tree, then, each BS-tree node would be split into two sprawl nodes: one for the point, and one for the radius (i.e., region). For example:

Handling a BS-node then means first computing $\delta(q, p)$ and considering p for inclusion in the result, and then determining whether $\delta(q, p)$ is greater than $r + s$. If so, no further action is taken, as the query ball Q does not overlap the region (i.e., ball) R. Otherwise, the two child pointers are followed recursively.

In the sprawl version, we've split out the point p as a parent node of the region. Initially, we visit this node, compute $z = \delta(q, p)$, and increment a counter associated with the child node. In general, we'll need to hang on to the z value as well as the counter; we could keep those in some separate memo, or perhaps store them in the nodes themselves. The counter is only useful if a region has multiple parents, so we know when we've visited them all; in this case, as soon as the counter goes from 0 to 1, we're done. Also, storing z is mostly useful if we're not going to use it immediately, and so it may be a bit wasted in this case.

Be that as it may, once the counter hits the threshold m (the number of parents of the region), we visit the region node. Here we store the radius r, but also one or more coefficients in a vector a. Note that m here is the number of entries in a, stored as part of the vector (or implicit, if the length is fixed). In this case, $m = 1$, $a = [1]$ and $f(x) = ax = x$. That means the overlap check reduces to that of the BS-tree:

$$f(z) \leqslant r + f(s) \iff az \leqslant r + as \iff z \leqslant r + s$$

There is no magic in the use of two children here; we may very well increase this number, as in the M-tree, for example [7]. (The M-tree adds another twist, which we'll return to in Sect. 4.)

There's also the VP-tree [22, 24] and its relatives such as LC [5], where there's a single ball that separates the inside from the outside. In that case, we get a different transformation:

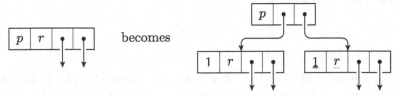

The idea here is that the center point p is shared between the ball (left subtree) and its complement, the outside (right subtree). The only difference between the two region nodes is that the outside one has its coefficient and radius negated.[3] At this point, a slight revision is in order. We have previously assumed that f is nondecreasing, i.e., that $a \geqslant 0$. That is no longer the case! The more general version of the overlap check then uses $|a|s$, rather than as. What happens, then, is that the overlap criterion for the left subtree is still $z \leqslant r + s$, but for the right one, we get:

$$az \leqslant -r + |a|s \iff -z \leqslant -r + s \iff z \geqslant r - s$$

This is exactly as in the VP-tree, except that the surface of the ball is included both for the inside *and* the outside; we'd really like $z > r - s$. This is a detail not handled by the framework (though it easily could be amended to); however, it could only (presumably in rare cases) lead to false positives, i.e., exploring subtrees unnecessarily, which won't produce any wrong results. However, except for the goal of emulating the VP-tree, there is no need to use the same radius in both regions. One could use r_1 and $-r_2$, for example, and adapt each to cover only the points in each subtree.

4   Intersections

In Sect. 3, our regions were individual balls and their complements.[4] We can combine these two kinds of regions to create *shell* regions, by turning a and r into column vectors:

$$a = \begin{bmatrix} -1 \\ 1 \end{bmatrix} \qquad r = \begin{bmatrix} -r' \\ r'' \end{bmatrix}$$

---

[3] Here $\bar{x}$ is a space-saving shorthand for $-x$.

[4] Strictly speaking, the closure of their complements, as we don't use strict inequalities.

This gives a shell region around the single parent point p, as follows:

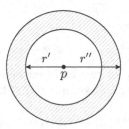

The membership check for a point u with distance $x = \delta(p, u)$ is still $ax \leqslant r$, but in this case, that means:

$$-x \leqslant -r'$$
$$x \leqslant r''$$

This is, of course, equivalent to $r' \leqslant x \leqslant r''$. For the overlap check, we take the absolute value for each row separately, so we still have $az \leqslant r + s$, which becomes (with some simplification):

$$z + s \geqslant r'$$
$$z - s \leqslant r''$$

That is, q must be so far away (z) that the s-ball around it reaches the inside radius (r') but not so far away that it ends up beyond the outside radius (r'').

A classic metric index—the Burkhard–Keller tree—branches out using multiple shells around a single center [3]. In this case, we'd simply use multiple shell regions, all with the same parent point.

There's not much point in using more than two rows when we have a single focus, i.e., a center, as we'll only end up with a single ball, inverted ball or shell, anyway. However, if we have more than one focus, we can add multiple *columns* to represent the intersection of multiple shells with *different* centers, yielding a coefficient matrix A. For simplicity, let's say we wish to represent the intersection of two balls, with respective centers p_1 and p_2. We use those points as the region's parents, and region membership becomes $Ax \leqslant r$, with coefficients and radii as follows:

$$A = \begin{bmatrix} 1 & 0 \\ 0 & 1 \end{bmatrix} \qquad r = \begin{bmatrix} r_1 \\ r_2 \end{bmatrix}$$

The intersection of multiple shells has been used in, e.g., Brin's GNAT [2] and its descendants, as well as the PM-tree family of structures [20] (see also Sect. 5), and was later dubbed a *cut region* by Lokoč et al. [15].

The M-tree combines balls and shells in an interesting way. Before even computing $\delta(q, u)$ to perform the overlap check $\delta(q, u) \leqslant r + s$, it executes a preliminary filtering step, with the check

$$|\delta(q, p) - \delta(p, u)| \leqslant r + s,$$

or, with our established notation, $|z - x| \leqslant r + s$. The intuition here is that $|z - x|$ is a lower bound for $d(q, u)$, a fact used in the standard pivot filtering check $|z - x| \leqslant s$ (see Sect. 5). Here, however, it's plugged in as a lower bound in our *ball* overlap check (with u as our ball center), creating a weakened, preliminary version. This might seem like it requires introducing some new concept or indirection, but that is not so. The check is still linear and is equivalent to a standard shell region. This is easily seen by rewriting the check as follows:

$$-z + x \leqslant r + s$$
$$z - x \leqslant r + s$$

We can rewrite this to match our previous shell overlap check:

$$z + s \geqslant x - r$$
$$z - s \leqslant x + r$$

In other words, we here simply have a shell region with inner radius $x - r$ and outer radius $x + r$, corresponding to our knowledge of the r-ball around u before computing $\delta(q, u)$:

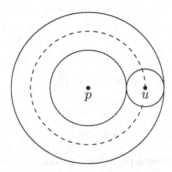

It would seem like we now have to store additional distances. Rather than just keeping x and r, we need to store r, $x - r$ and $x + r$. But is that really so? Given our M-tree to sprawl translation, each point node is now the center of multiple (quite possibly overlapping) shell regions, as well as a single ball region enclosing them all. The only reason to keep this ball region is if its radius is lower than the greatest radius of the shells. If we stuck rigidly to our translation, this could happen—but if we simply kept our shells as tight as possible around the subtrees, it could not. We then end up storing just two distances per subtree, once more, and have a structure with a behavior quite similar to, and no worse than, the M-tree.

5   Elimination

The most common purpose of a pointer in an index structure is to lead you toward further data to explore. There is a certain genre of structures, however, that do the exact opposite—where instead of discovering data, you *eliminate* it. Take, for example, the LAESA structure [17]: a table of distances between so-called *pivots* and the other points in the data set. The query is compared to each of these pivots, and the computed query–pivot distances, along with the stored pivot–data-point distances, are used to determine whether any given data point may possibly be relevant. In sprawl terms, each pivot–data-point distance represents a region:

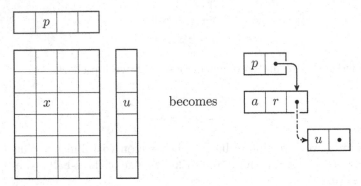

In this case, the region is a *sphere*, a shell of width zero (i.e., with identical inner and outer radii):

$$a = \begin{bmatrix} 1 \\ -1 \end{bmatrix} \qquad r = \begin{bmatrix} x \\ -x \end{bmatrix}$$

As before, our overlap check is $az \leqslant r + s$, or:

$$z \leqslant x + s$$
$$-z \leqslant -x + s$$

Combined, this is the standard pivoting bound, $s \geqslant |x - z|$. Now, however, we get to the more interesting point. The dotted pointer indicates that we've got a potential *elimination* on our hands. That is, rather than saying "if there's overlap, let's look at u, otherwise, let's ignore it" we turn it around, and say "if there's overlap, let's ignore it, otherwise, let's eliminate it." In the terminology of our earlier pseudocode, that essentially means setting $u.color$ to BLACK. The exact implementation here could be done in several ways. One could have different region node types, for positive and negative regions (leading to discovery and elimination, respectively), or have separate lists of positive and negative child-edges, so the same region could both discover and eliminate points. These are possible optimizations, but they don't substantially change the behavior of the search.

It's possible to combine discovery and elimination, such as in the PM-tree. A simplified version would consist of a ball tree, such as in Sect. 3, along with a set of pivots with eliminating regions around the subtrees. Specifically, the PM-tree uses shells around each subtree, with global pivots, yielding something like the following:

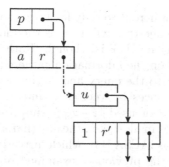

An important thing to note about elimination is that it may be performed *lazily*. That is, we need not check for overlap with the various shell regions associated with the shared, global points in the PM-tree until we've established that we intersect the ball region in the tree itself. This kind of laziness could be implemented by having pointers in the reverse direction, without a need for counter updating. When considering a point, we would simply look at the region parents and see if they had been examined yet (i.e., if they were colored black).

It's possible to implement such things in different ways, of course; one could, for example, have some *parents* of a region be lazy, explored on demand, or the like. Having such a mechanism, one could simply use the global pivots of a PM-tree as lazy parents of every region in the tree, turning them from balls into cut regions, removing the need for elimination altogether.

The elimination perspective in LAESA could similarly be turned on its head, if instead of multiple regions, we use a single region for each point, with all pivots as its parents. This region would then be the intersection of all the spheres, and a point would simply not be discovered if there were no intersection.

This does not mean that we can do entirely without elimination, however. In any scenario where we at one time are able to traverse a point, and at a later time are not, this is the result of elimination. To my knowledge, the only current structure where it is truly needed, even if one were to introduce various forms of laziness optimization, is the AESA family of indices [23], where all points are available initially, and the set of candidates is gradually whittled down. The order of traversal then becomes crucial, as discussed in the next section.

6   Priority

The AESA family of indexing methods are all based on the same simple data structure: a complete distance matrix between the data points.[5] The points are explored one by one, and at every step we eliminate any of the remaining points we're able to. The elimination works just as in LAESA; the difference is that the pivots aren't kept separate from the objects. Rather than simply examining all the pivots in an arbitrary order, we now need to be quite careful about which object to examine next, to minimize the number of objects explored overall.

---

[5] Because of symmetry, one need only store half of it, of course.

A ubiquitous simplification here is to only focus on the elimination power of the next object and select the one that will give us the most bang for our buck.

We don't know which one that is, though. Rather, we must perform this choice *heuristically*, based on the information gathered so far, i.e., the distances from each examined point to the query and to the candidates for examination. If we represent these distances by vectors z and x, the original AESA used $\|z - x\|_1$ while a revised version used $\|z - x\|_\infty$, simply choosing the object whose filtering lower bound is the smallest, i.e., the one that's furthest away from being eliminated. Later, there was iAESA [11], which instead used Spearman's footrule between permutations of the previously examined objects, sorted by distance to the query and the candidate. Even more recently, Socorro et al. introduced the two-phase PiAESA method [21], which initially uses a set of preselected pivots (like LAESA), chosen for their general filtering power; once enough objects have been explored, it switches to the classic AESA behavior. Many variations are possible here, of course; for example, one might use regression or machine learning to estimate distances or filtering power or the like [10,18].

From a sprawl-of-ambits point of view, these methods are essentially the same: A complete directed graph of elimination edges, where each edge has a single sphere region. The priority or heuristic used to select the next available point is left unspecified. What *is* relevant, however, is when and how to compute or update the heuristic. In the simplest, most naive implementation, on might merely iterate over all available objects in each step, computing an arbitrary black-box priority for each, based on the knowledge gathered so far. It's possible, however, to let priority updates piggyback on other traversal operations.

For example, if the heuristic is based on how hard a point is to get rid of, one might update the priority every time the point is rediscovered and every time one fails to eliminate it. In each of these cases, a lower bound on the distance is computed, and one may then simply keep the sum or maximum, as in AESA.

For structures without elimination, such as the majority of search trees, priority is not relevant to the number of distance computations needed to resolve a range query; the behavior will be the same, regardless of the ordering. For kNN queries, however, priority can be crucial, as the covering radius of the result set tends to shrink as good candidates are found, and this will improve the chances of eliminating subtrees.

7   Non-trees

Index structures tend to be tree-shaped, more or less, especially if we ignore the eliminating parts. One early exception is the *excluded middle vantage point forest* introduced by Yianilos [25]. This structure is still *mostly* tree-shaped—or, as the name implies, forest-shaped. That is, it primarily consists of a collection of trees. However, these trees are connected to each other, making the whole thing a directed, acyclic graph.

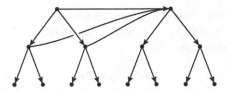

The trees are essentially VP-trees with three regions rather than two: an inner ball, a middle shell, and an outer inverted ball. For queries whose radius is less than half the width of the middle shell, the search will never traverse more than one of the inner and outer subtrees—a major selling-point of the structure. There may still be points located in the separating shell, though, and these must also be indexed!

The idea is to gather up all the points that end up in any separating shell throughout the tree, and build a *new* tree from those, in the same manner (possibly leading to a third tree, and so on). We then simply make the root of this new tree the single child of every shell region in the first tree, as in the following, where $a = [-1\ 1]^t$ and $r = [-r'\ r'']^t$:

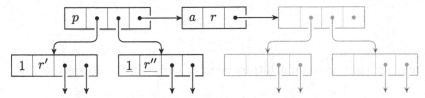

An essentially equivalent structure, at least from a bird's-eye view, is the D-index [9]. There, too, we have a multitude of shell regions separating inner and outer subsets, with the shells leading to a secondary structure, and so on. The main difference is that where the excluded middle vantage point forest uses tree traversal to determine which intersection of inner and outer regions a given point falls into, with the centers found along the path from the root, the D-index provides a fixed set of shared centers from the beginning, in a manner similar to the so-called fixed-queries tree [1]. Several levels are, in essence, collapsed, and the correct subtrees or leaves, representing the intersection of multiple shell or ball regions, are found directly, using hashing. This is an optimization that does not affect the high-level behavior (i.e., which points are examined).

8   Hyperplanes

In Sect. 5, we created sphere and shell regions by having two radii, and thus two rows in our coefficient matrix, ending up with a column vector $[1\ -1]^t$. But we could also just use a row vector $a = [1\ -1]$, along with a single radius. This means we need two parents, or *foci*, p_1 and p_2, and we finally get a pivot vector $z = [z_1\ z_2]^t$. The overlap check becomes:

$$az \leqslant r + \|a\|_1 s$$

Here $\|a\|_1$ is the sum of absolute values. If we use $r = 0$, this corresponds to a metric half-space, separated by a midset or hyperplane. The overlap check then simplifies to the standard one [22]:

$$z_1 - z_2 \leqslant 2s$$

We are here defining the region of points closer to p_1 than p_2. If we wish to have *multiple* contrasting objects, modeling general Voronoi cells or Dirichlet domains [19], we can just add parent points, as well as some rows and columns. Let's say, for example, we wish to describe the region of points that are closer to p_1 than both p_2 and p_3. We'd then use all three as parents of our region, and use the following coefficients and radii:

$$A = \begin{bmatrix} 1 & -1 & 0 \\ 1 & 0 & -1 \end{bmatrix} \qquad r = \begin{bmatrix} 0 \\ 0 \end{bmatrix}$$

This corresponds to the following overlap check, where both inequalities must hold for there to be overlap:

$$z_1 - z_2 \leqslant 2s$$
$$z_1 - z_3 \leqslant 2s$$

One may extend this to an arbitrary number of foci in the obvious manner.

9   Other Conics

The hyperplane case is easy enough to extend to (generalized) ellipses [8,22], by using coefficients $a = [1\ 1]$ and the appropriate radius, yielding the following overlap check:

$$az \leqslant r + \|a\|_1 s \iff z_1 + z_2 \leqslant r + 2s$$

Or we can get shifted half-spaces, what amounts to metric hyperbolas [8,16], by adjusting the radius away from 0. That is, we still have $a = [1\ -1]$ but we have $r \neq 0$, yielding the following slightly more general check:

$$z_1 - z_2 \leqslant r + 2s$$

This, then, represents not the points that are closer to z_1 than to z_2, but where the distances differ by a given value (i.e., the radius). That is, membership for a point with distance vector x is $ax \leqslant r$, i.e., $x_1 - x_2 \leqslant r$.

10   Other Queries

Nearest neighbor queries (kNN) have been mentioned briefly already. A general approach is to maintain the (up to) k points closest to q found so far, letting the search radius s be an upper bound on the distance to the kth nearest neighbor. Beyond updating s during the search, the procedure is the same.

A generalization that does not seem to have been explored is using other regions than balls as queries. After all, if a ball query works well in a tree built from hyperplanes, there's nothing stopping us from using a hyperplane query in a tree built from balls. That is, we might have a prototypical example object q, and a prototypical *counter*-example q', and we then search our index for objects closer to q than q'. (Such queries were briefly mentioned by Uhlmann [22].) Or maybe we have two prototypes, and wish to find the k objects with the lowest average distance to q and q', resulting in an ellipsoid query.

More generally, our query might consist of a *weighted combination* of query objects, looking for points with a low weighted sum of query distances. In other words, we may use an arbitrary linear ambit as our query [13, Subsect. 3.2.1]. As long as the ambit coefficients of the query, *or* those in the tree, are all non-negative, determining query–region overlap is straightforward, as we shall see.

11   ...and Beyond

It ought to be quite clear that the sprawls and ambits used so far have been quite limited. The sprawls have mostly been tree-like, and the coefficients of the ambits have been 1 or -1, with at most two nonzero coefficients to a row. Countless variations are possible, both in how the sprawls are put together and in how the ambits are parameterized.

Determining whether an arbitrarily constructed sprawl is correct is a hard problem [13, Theorem 2.3.2]. However, extrapolating from existing index structures, we may quite easily ensure that the sprawls we construct are *responsible*, in which case they are guaranteed to be correct. Roughly, responsibility means that for every point p, there is a set of edges we can traverse that will lead us to it, and that the regions of those edges contain p, as do the regions of any negative edges that might disrupt that traversal. For the case where the positive edges of our structure are acyclic, this can be dealt with locally, where the responsibilities of a node's incoming edges depend only on those of the outgoing ones [13, Observation 2.3.10]. Thus it ought to be possible to mix and match quite freely, perhaps even using heuristic search to look for efficient structures automatically.

As for regions, any coefficient matrix and radius vector yields a valid linear ambit, usable as a region or a query. For a query ambit Q with coefficient vector c and radius s, and a region ambit R with coefficient vector a and radius r, with a or c non-negative and $\|a\|_1, \|c\|_1 = 1$, if R and Q intersect, then

$$r + s \geqslant aZc^t, \tag{11.1}$$

where z_{ij} is the distance between focus p_i of R and focus q_j of Q [13, Theorem 3.1.2]. With this overlap check, one can use ambit queries with existing index structures, and one could extend existing indexes with additional regions, without adding any distance computations. In an tree structure where several points are explored when deciding which subtrees to visit, arbitrary subsets of these could be used to construct additional filtering predicates for any subtrees, merely by adding radii and possibly coefficients.[6]

Finally, one may go beyond the limits of linearity. For example, using any (multi-parameter) non-decreasing *metric-preserving* function f to calculate remoteness, we may still use the original overlap check (2.2) [13, Subsect. 3.5]. This opens the door to a wide range of learning and optimization methods for adapting regions to points in ways that improve search performance.

References

1. Baeza-Yates, R., Cunto, W., Manber, U., Wu, S.: Proximity matching using fixed-queries trees. In: Crochemore, M., Gusfield, D. (eds.) CPM 1994. LNCS, vol. 807. Springer, Heidelberg (1994). https://doi.org/10.1007/3-540-58094-8_18
2. Brin, S.: Near neighbor search in large metric spaces. In: Proceedings of the 21th International Conference on Very Large Data Bases, pp. 574–584 (1995)
3. Burkhard, W.A., Keller, R.M.: Some approaches to best-match file searching. Commun. ACM **16**(4), 230–236 (1973). https://doi.org/10.1145/362003.362025
4. Carlsen, S.M.Ø., Moe, H.H.: Similarity Search in Metric Spaces with Weighted Multi-focal Regions: Using the Ambit Region Type to Improve the Performance of the SSS-Tree. Master's thesis, Norwegian University of Science and Technology (2020)
5. Chávez, E., Navarro, G.: A compact space decomposition for effective metric indexing. Pattern Recogn. Lett. **26**(9), 1363–1376 (2005). https://doi.org/10.1016/j.patrec.2004.11.014
6. Chen, L., Gao, Y., Song, X., Li, Z., Miao, X., Jensen, C.S.: Indexing metric spaces for exact similarity search. arXiv preprint arXiv:2005.03468 (2020)
7. Ciaccia, P., Patella, M., Zezula, P.: M-tree: an effcient access method for similarity search in metric spaces. In: Proceedings of the 23rd VLDB Conference, Athens, Greece, pp. 426–435 (1997)
8. Dohnal, V., Gennaro, C., Savino, P., Zezula, P.: Separable splits of metric data sets. In: Proceedings of the Nono Convegno Nazionale Sistemi Evoluti per Basi di Dati (2001)
9. Dohnal, V., Gennaro, C., Savino, P., Zezula, P.: D-index: distance searching index for metric data sets. Multimed. Tools Appl. **21**(1), 9–33 (2003). https://doi.org/10.1023/A:1025026030880
10. Edsberg, O., Hetland, M.L.: Indexing inexact proximity search with distance regression in pivot space. In: Proceedings of the 3rd International Conference on SImilarity Search and APplications, pp. 51–58 (2010). https://doi.org/10.1145/1862344.1862353
11. Figueroa, K., Chávez, E., Navarro, G., Paredes, R.: Speeding up spatial approximation search in metric spaces. J. Exp. Algorithmics (JEA) **14**, 3–6 (2010). https://doi.org/10.1145/1498698.1564506
12. Hetland, M.L.: Swarm Intelligence for Multi-objective Problems in Data Mining. In: Coello, C.A.C., Dehuri, S., Ghosh, S. (eds.) The basic principles of metric indexing, vol. 242. Springer, Heidelberg (2009). https://doi.org/10.1007/978-3-642-03625-5_9

---

[6] This approach has been tentatively explored by my students Carlsen and Moe [4].

13. Hetland, M.L.: Comparison-based indexing from first principles. arXiv preprint arXiv:1908.06318 (2019)
14. Kalantari, I., McDonald, G.: A data structure and an algorithm for the nearest point problem. IEEE Trans. Softw. Eng. **5**, 631–634 (1983). https://doi.org/10.1109/TSE.1983.235263
15. Lokoč, J., Moško, J., Čech, P., Skopal, T.: On indexing metric spaces using cut-regions. Inf. Syst. **43**, 1–19 (2014). https://doi.org/10.1016/j.is.2014.01.007
16. Lokoč, J., Skopal, T.: On applications of parameterized hyperplane partitioning. In: Proceedings of the 3rd International Conference on Similarity Search and Applications, pp. 131–132. ACM (2010). https://doi.org/10.1145/1862344.1862370
17. Micó, M.L., Oncina, J.: A new version of the nearest-neighbour approximating and eliminating search algorithm (AESA) with linear preprocessing time and memory requirements. Pattern Recogn. Lett. **15**(1), 9–17 (1994). https://doi.org/10.1016/0167-8655(94)90095-7
18. Murakami, T., Takahashi, K., Serita, S., Fujii, Y.: Probabilistic enhancement of approximate indexing in metric spaces. Inf. Syst. **38**(7), 1007–1018 (2013). https://doi.org/10.1016/j.is.2012.05.012
19. Navarro, G.: Searching in metric spaces by spatial approximation. VLDB J. **11**(1), 28–46 (2002). https://doi.org/10.1007/s007780200060
20. Skopal, T., Pokornỳ, J., Snasel, V.: PM-tree: pivoting metric tree for similarity search in multimedia databases. In: ADBIS (Local Proceedings) (2004)
21. Socorro, R., Micó, L., Oncina, J.: A fast pivot-based indexing algorithm for metric spaces. Pattern Recogn. Lett. **32**(11), 1511–1516 (2011). https://doi.org/10.1016/j.patrec.2011.04.016
22. Uhlmann, J.K.: Metric trees. Appl. Math. Lett. **4**(5), 61–62 (1991). https://doi.org/10.1016/0893-9659(91)90146-M
23. Vidal Ruiz, E.: An algorithm for finding nearest neighbours in (approximately) constant average time. Pattern Recogn. Lett. **4**(3), 145–157 (1986). https://doi.org/10.1016/0167-8655(86)90013-9
24. Yianilos, P.N.: Data structures and algorithms for nearest neighbor search in general metric spaces. In: Proceedings of the 4th Annual ACM-SIAM Symposium on Discrete algorithms, pp. 311–321. Society for Industrial and Applied Mathematics, Philadelphia, PA, USA (1993)
25. Yianilos, P.N.: Excluded middle vantage point forests for nearest neighbor search. In: In DIMACS Implementation Challenge, ALENEX'99 (1999)

Some Branches May Bear Rotten Fruits: Diversity Browsing VP-Trees

Daniel Jasbick[1], Lucio Santos[2], Daniel de Oliveira[1], and Marcos Bedo[3(✉)]

[1] Institute of Computing, UFF, Niterói, RJ, Brazil
`danieljasbick@id.uff.br`, `danielcmo@ic.uff.br`
[2] Federal Institute of North of Minas Gerais, Montes Claros, MG, Brazil
`lucio.santos@ifnmg.edu.br`
[3] Fluminense Northwest Institute, UFF, St. A. Pádua, RJ, Brazil
`marcosbedo@id.uff.br`

Abstract. Diversified similarity searching embeds result diversification straight into the query procedure, which boosts the computational performance by orders of magnitude. While metric indexes have a hidden potential for perfecting such procedures, the construction of a suitable, fast, and incremental solution for diversified similarity searching is still an open issue. This study presents a novel index-and-search algorithm, coined _diversity browsing_, that combines an optimized implementation of the vantage-point tree (VP-Tree) index with the distance browsing search strategy and coverage-based query criteria. Our proposal maps data elements into VP-Tree nodes, which are incrementally evaluated for solving diversified neighborhood searches. Such an evaluation is based not only on the distance between the query and candidate objects but also on distances from the candidate to data elements (called _influencers_) in the partial search result. Accordingly, we take advantage of those distance-based relationships for pruning VP-Tree branches that are themselves _influenced_ by elements in the result set. As a result, _diversity browsing_ benefits from data indexing for _(i)_ eliminating nodes without valid candidate elements, and _(ii)_ examining the minimum number of partitions regarding the query element. Experiments with real-world datasets show our approach outperformed competitors GMC and GNE by at least 4.91 orders of magnitude, as well as baseline algorithm BRID_k in at least 87.51% regarding elapsed query time.

Keywords: Similarity searching · Result diversification · Metric spaces

1 Introduction

Similarity searching is a widely employed paradigm supporting modern computational applications that rely on data that are "alike" but not "equal",

M. Bedo—This study was financed in part by the Coordenação de Aperfeiçoamento de Pessoal de Nível Superior – Brasil (CAPES) - Finance Code 001 and Research Support Foundation of Rio de Janeiro State - G. E-26/010.101237/2018.

S. Satoh et al. (Eds.): SISAP 2020, LNCS 12440, pp. 140–154, 2020.
https://doi.org/10.1007/978-3-030-60936-8_11

e.g., instance-based learning, and content-based image retrieval [1,15]. An efficient approach for querying such data is the *metric space* model [9], where objects are mapped into a known domain and become comparable by a distance function. In practice, the most requested search is the neighborhood (or k-nearest neighbors – k-NN) query [3,12] that retrieves the k closest elements to a given object. Exact neighborhood searches are efficiently carried out by index-and-query algorithms [4], but they may present a semantic drawback in the querying of dense datasets. For instance, suppose a film student runs a search to recover the ten most similar movies to the "Leone's western *The Good, the Bad and the Ugly*" in a social repository of videos and retrieves remakes, versions, and parodies of the same motion picture. Although the answer is correct from a neighborhood viewpoint, the result is semantically redundant. A *diversified* neighborhood query, on the other hand, could consider not only the nearest films but also those of distinct styles, *i.e.*, *different collections* within the queried dataset.

Diversification strategies fit into one or more of the following categories: *(i) distance-based, (ii) novelty-based,* and *(iii) coverage-based* [18,20]. Approaches from the first and second groups require a two-phase execution in which an enlarged subset of candidates is selected and filtered to secure result diversification, while coverage-based methods separate candidates according to a similarity threshold at runtime and may be seamlessly embedded into low-level search routines. Thus far, however, no study has been conducted on designing an incremental algorithm to be coupled with index solutions for enhancing the execution of diversified similarity searches [6,7,20].

In this paper, we fill that gap by designing a browsing algorithm for *influence* and coverage-based query criteria [15,16]. We model *influences* as *closed balls* in *metric spaces* [9,16], in which every ball is centered at a single result set element with a radius defined as the distance between the query object and the ball center. In our diversified k-NN search, data are retrieved by order of proximity to the query element, and objects are outside the *influence* of the previous object. Such a model allows the pruning of *influenced* regions by matching them with disjoint and indexed partitions of the search space.

Accordingly, we define an incremental routine for inspecting query-matched regions with the concept of *distance browsing* [10] extended to *influence*, whose implementation is loosely coupled on top of the vantage-point tree (VP-Tree) indexing structure [19]. Our approach maintains two priority queues for *(i)* sorting VP-Tree internal nodes according to their minimum and maximal distances to the query element, and *(ii)* ordering leaf nodes' elements regarding the queried object. After the root insertion in the first queue, every incremental step examines only the queue whose top (partition or element) is the closest to the queried object. Non-influenced VP-Tree branches are inserted into the partitions' queue, whereas objects within examined leaf nodes are added to the second queue. A top element is selected as the next diversified neighbor only if it is not influenced by the partial result, and influence regions are constructed one entry at a time.

Experiments on real-world datasets indicate our approach outperformed novelty-based approaches *Greedy with Marginal Contribution* (GMC) and *Greedy Randomized with Neighborhood Expansion* (GNE) [18] by at least 4.91 orders of magnitude, as well as coverage-based approach *Better Result with Influence Diversification* (BRID_k) [16] by at least 87.51% regarding elapsed query time. Results also indicate balanced VP-Trees with pivots of maximal distance variance are most suitable than other VP-Tree settings for diversified k-NN searching.

The remainder of the paper is organized as follows. Section 2 describes the background and related work. Section 3 introduces diversity browsing. Section 4 presents the experimental findings, while Sect. 5 provides the conclusions.

2   Preliminaries and Related Work

A *metric space* is a pair $\langle \mathbb{O}, \delta \rangle$ for a given domain $\mathbb{O}$, and distance function δ that complies with properties of *(i)* <u>symmetry</u>, $\delta(o_i, o_j) = \delta(o_j, o_i)$; *(ii)* <u>non-negativity</u>, $\delta(o_i, o_j) \geq 0$; *(iii)* <u>identity</u>, $\overline{\delta(o_i, o_j)} = 0 \Leftrightarrow o_i = o_j$; and *(iv)* <u>triangle inequality</u>, $\delta(o_i, o_h) + \delta(o_h, o_j) \geq \overline{\delta(o_i, o_j)}$, for any objects $o_h, o_i, o_j \in \mathbb{O}$. Examples of distance functions include the L_p family, which operates on d-dimensional spaces, *i.e.*, $\mathbb{O} = \mathbb{R}^d$. Given a dataset $\mathcal{O} \subseteq \mathbb{O}$, a *rangequery($R_q$)* returns every element in $\mathcal{O}$ that is at most a given radius $\overline{\xi \in \mathbb{R}_+}$ from a query object $o_q \in \mathbb{O}$, *i.e.*, $R_q(o_q, \xi, \mathcal{O}, \delta) = \{o_i \mid o_i \in \mathcal{O}, \delta(o_i, o_q) \leq \xi\}$. Likewise, a *neighborhoodquery(k-NN)* retrieves $k \in \mathbb{N}$, elements in $\mathcal{O}$ whose distances to the query object $o_q \in \mathbb{O}$ are the smallest, *i.e.*, a range query with an initially unknown radius ξ so that $|R_q| = k$ [9]. The pair $\langle o_q, \xi \rangle$ defines a *closed ball* in $\mathcal{O}$ that covers more or fewer elements depending on ξ. Therefore, a k-NN query is an alternative to a range search whenever the choice of a suitable ξ is unclear. The side-effect, however, is the *sorting* of distances to filter the k closest elements, which can be boosted by the partitioning of the search space [5].

Pivot-based and ball-partitioning indexes adopt distinct criteria for splitting data elements into *partitions* so that, given a similarity query, only partitions related to the query ball are inspected [3,5,11]. *Vantage-point trees* (VP-Trees) are particularly versatile in this scenario since they organize the search space in a hierarchical and disjoint fashion [19]. The ball partitioning principle of VP-Trees is to use a *pivot* element p from a dataset $\mathcal{O} \subseteq \mathbb{O}$, the median μ_p of distances from elements $o_i \in \mathcal{O}$ to p, and the maximum distance M_p between p and any $o_i \in \mathcal{O}$ to divide the dataset into two disjoint partitions, namely *left* and *right* nodes. Elements whose distance to p fall inside $[0, \mu_p)$ interval are assigned to the left node, while the remaining objects are placed in the right node, within distance interval $[\mu_p, M_p]$. Left and right nodes are recursively divided until a user-provided number of elements per partition is reached. Therefore, VP-Tree parameters that fundamentally affect similarity searching are *(i)* the pivot selection criteria, which defines how pivots are chosen, *e.g.*, *random*, *convex-hull* or *maximal variance* [17,19], and *(ii)* overflow support in leaf nodes, which enforces tree balance for cases of non-unique medians [19].

2.1   The Distance Browsing Strategy

The order that partitions are traversed is crucial for the execution of k-NN queries since their query balls are dynamically constructed. The distance browsing strategy [10] employs two *priority queues* for retrieving one nearest neighbor at the time and limiting the total of examined partitions. The first queue sorts unvisited partitions and the second the elements from inspected partitions. Second queue entries are ordered by their *distances* to the query object, whereas partitions in the first queue are sorted by the *minimum* (δ_{min}) and *maximum* (tie-break – δ_{max}), distances between their boundaries and the query element.

At each iteration, the decision to select the potential next nearest neighbor is made upon the evaluation of both queues. If the top of the second queue is closer to the query object than the minimum distance to the partition in the head of the first queue, the top element in the second queue is returned as the next nearest neighbor. Otherwise, the top partition in the first queue is loaded from the disk, and its elements are inserted into the second priority queue.

2.2   Result Diversification

Unlike classical neighborhood searches, *diversity queries* ensure elements in the result set are of different collections, which ultimately generates new query criteria. Given a function div and a quantity $k \leq |\mathcal{O}|, \mathcal{O} \subset \mathbb{O}$, a diversified result set $\mathcal{R}$ is a subset of $\mathcal{O}$ that complies with Eq. 1 [20]. Diversification algorithms for such an optimization problem are categorized into *(i)* distance-based, *(ii)* coverage-based, and *(iii)* novelty-based.

$$\mathcal{R} = \mathrm{argmax}_{\mathcal{R}' \subseteq \mathcal{O}, |\mathcal{R}'|=k} div(\mathcal{R}') \tag{1}$$

Distance-based approaches formulate argument $div(\mathcal{R}')$ of Eq. 1 as an aggregation of distances, *e.g.*, *minimum sum* as in $div(\mathcal{R}') = \min_{o_i, o_j \in \mathcal{R}'} \delta(o_i, o_j)$. While such distance aggregation approaches can be solved as a variant of the p-dispersion problem [7,14], they target the exploration of the entire dataset $\mathcal{O}$ and disregard the perspective of a user-posed query object [15,16].

Coverage-based strategies exploit the perception of *coverage*, which can be modeled in metric spaces in terms of *coverage radii* from closed balls [8,9]. For instance, the r-DisC algorithm [8] that takes a user input $r, r \in \mathbb{R}_+$ for retrieving a set $\mathcal{R}$ of diversified elements in which $\forall\, o_i \in \mathcal{O}, \exists\, o_j \in \mathcal{R} \mid \delta(o_i, o_j) \leq r, o_i \neq o_j$, and $\forall\, o_i, o_j \in \mathcal{R}, \delta(o_i, o_j) > r$. Figures 1(a–b) compare k-NN and r-DisC queries for the same input. Notice r-DisC returns objects spaced among themselves but disregards both query element o_q and cardinality k.

Novelty-based methods employ a *score* function (f_{div}) for evaluating the *diversified* set $\mathcal{R}'$ regarding the distances of elements $o_i \in \mathcal{R}' \subseteq \mathcal{O} \subseteq \mathbb{O}$ and the query object $o_q \in \mathbb{O}$. The score is modeled as a linear combination of distance and diversity, whose weights are ruled by a $\boldsymbol{\lambda}$ **parameter**, as in Eq. 2.

$$f_{div}(o_q, \mathcal{R}') = (k-1) \cdot (1-\lambda) \cdot \sum_{o_i \in \mathcal{R}'} \delta(o_q, o_i) + 2 \cdot \lambda \cdot div(\mathcal{R}') \tag{2}$$

Since finding set $\mathcal{R} = \mathcal{R}'$ with the highest score is an NP-hard problem [2], *heuristics* are usually employed for discovering well-diversified result sets [7,18]. Those strategies reduce $\mathcal{R}'$ candidates by using *(i)* a sample $\mathcal{O}' \subseteq \mathcal{O}$ of the

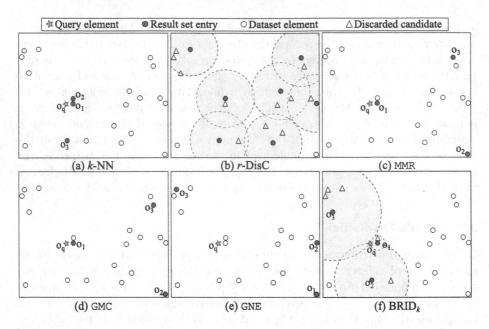

Fig. 1. Results produced by different search routines with setup $\delta = d = L_2$, $\lambda = .25$, and $r = .75$. BRID_k is the only approach requiring no extra user-provided parameters.

dataset, and *(ii)* a disconnection between *similarity* δ and *diversity* d metrics, which may be set as the same distance function, *i.e.*, $\delta = d$ [20].

A baseline solution to that rationale is the *Maximal Marginal Relevance* (MMR) algorithm. It includes the nearest neighbor to the query object in the partial result set $\mathcal{R}'$ and incrementally selects the remaining candidates $o_i \in \mathcal{O}'$ with the highest contribution $\text{MMR}(o_i, o_q) = (1 - \lambda) \cdot \delta(o_i, o_q) + 2 \cdot \lambda \cdot \sum_{o_j \in \mathcal{R}'} d(o_i, o_j)$ in $k - 1$ steps. Figure 1(a–c) shows the output of a diversified MMR query in contrast to that of a standard k-NN query. GMC [18] extends the MMR algorithm by weighting the impact of elements left *outside* the partial result set $\mathcal{R}'$. Such weighting is carried out by the GMC contribution function $\text{GMC}(o_i, o_q) = (1 - \lambda) \cdot \delta(o_i, o_q) + (\lambda/(k-1)) \cdot \left(\sum_{o_j \in \mathcal{R}'} d(o_i, o_j) + \sum_{j=1, o_j \in \mathcal{O}' \setminus \mathcal{R}'}^{j \leq k - |\mathcal{R}'|} d(o_i, o_j) \right)$ that considers the $k - |\mathcal{R}'|$ highest values for $d(o_i, o_j) : o_j \in \mathcal{O}' \setminus \mathcal{R}'$. Figure 1(d) shows the GMC output for the same query element in Fig. 1(c).

Another MMR extension is the GNE [18] algorithm, which replaces the incremental evaluation of $\mathcal{R}'$ by a GRASP-oriented heuristic. GNE uses the GMC contribution function, but instead of selecting the o_i entry with the highest score, a random element among the *top*-ranked objects is chosen. Such rationale leads to a two-phase execution in which a *candidate list* is kept in memory and is greedily refined by random picks. Figure 1(e) shows a GNE result set avoiding a couple of local maxima (border elements) that drew both MMR and GMC.

Better Results with Influence Diversification (BRID_k) [16] algorithm relies on the *influences* generated by the diverse neighbors in the search space for defining dynamic separation thresholds between *similarity* and *diversity* as in coverage-based approaches. The intuition is data elements mutually *influence* each other (in the sense of being similar) at a ratio proportional to their distance, and, formally, the mutual <u>influence</u> between two elements $o_i, o_j \in \mathbb{O}$ is given by $I(o_i, o_j) = 1/\delta(o_i, o_j)$. BRID_k assumes distances between result set and query elements are not independent, and the influences from a candidate element to both query object and result set entries constitute a ternary-relationship that determines whether or not the element is a valid *diversified* neighbor.

Given a dataset $\mathcal{O} \subseteq \mathbb{O}$, BRID sorts candidates $o_1, \ldots, o_n \in \mathcal{O}, n = |\mathcal{O}|$, according to their distances to a query object $o_q \in \mathbb{O}$, and traverses that ordered list targeting the elements that are more influenced by o_q than any other object in a (partial) result set $\mathcal{R}$. Since $\mathcal{R}$ is initially empty, the nearest object o_1 is added into the result as an *influencer* and defines the first *strong influence set* over $\mathcal{O}$. Strong influence sets enable discarding candidates $o_j \in \mathcal{O} \setminus \mathcal{R}$ that are more influenced by a result set entry $o_i \in \mathcal{R}$ than by the query object o_q itself, *i.e.*, the strong influence set I_{o_i, o_q} covers every candidate element that satisfies definition $\ddot{I}_{o_i, o_q} = \{o_j \in \mathcal{O} \setminus \mathcal{R} \mid I(o_i, o_j) \geq I(o_i, o_q) \wedge I(o_i, o_j) > I(o_j, o_q)\}$. Accordingly, BRID extends k-NN into diversified k-NN queries as follows.

Diversified k-NN Query. A diversified k-NN query retrieves the k non-influenced and nearest objects in $\mathcal{O}$ to o_q such that $\mathcal{R} = \{r_i \in \mathcal{O} \mid (\forall o_j \in \mathcal{R} \Rightarrow r_i \notin \ddot{I}_{o_j, o_q}) \wedge (\forall o_i \in \mathcal{O} \setminus \mathcal{R} \Rightarrow (\delta(r_i, o_q) \leq \delta(o_i, o_q) \vee \exists o_j \in \mathcal{R} \Rightarrow o_i \in \ddot{I}_{o_j, o_q})) \wedge (|\mathcal{R}| \leq k)\}$. Objects $r_i \in \mathcal{R}$ are *influencers*, while those in $\ddot{I}_{o_q, r_i}$ are *influenced elements*. Figure 1(f) shows an example of the result set returned by BRID_k in contrast to those of MMR, GMC, and r-Disc.

3   Diversity Browsing

This section describes *diversity browsing*, our approach for extending the distance browsing strategy into an algorithm that efficiently solves diversified k-NN queries. *Diversity browsing* shares diversity definitions with BRID_k, including those of *influencers* and *influenced elements*. We extend distance browsing because it guarantees that candidates $o_i \in \mathcal{O} \subseteq \mathbb{O}$ are incrementally retrieved according to their distances to the query object $o_q \in \mathbb{O}$. That way, we can naturally reduce *strong influence sets* into generic partitions defined by *closed balls* in the search space as in Lemma 1.

Lemma 1. Let the neighbor $o_i \in \mathcal{O}$ be the last *influencer* in result set $\mathcal{R}$, the problem of checking if the *next* distance browsing neighbor $o_{i+1} \in \mathcal{O} \setminus \cup_{j \in \mathcal{R}} \ddot{I}_{o_j, o_q}$ is within influence set $\ddot{I}_{o_i, o_q}$ is reduced to examine the closed ball $\langle o_i, \delta(o_i, o_q) \rangle$ in $\mathcal{O} \setminus \cup_{j \in \mathcal{R}} \ddot{I}_{o_j, o_q}$, *i.e.*, $\ddot{I}_{o_i, o_q} \equiv \langle o_i, \delta(o_i, o_q) \rangle$ regarding set $\mathcal{O} \setminus \cup_{j \in \mathcal{R}} \ddot{I}_{o_j, o_q}$.

Proof. Distance browsing ensures $\delta(o_i, o_q) \leq \delta(o_{i+1}, o_q)$ for any pair of consecutive neighbors $o_i, o_{i+1} \in \mathcal{O}$, so that $I(o_i, o_q) \geq I(o_{i+1}, o_q)$ is valid and inequalities $I(o_i, o_{i+1}) \geq I(o_i, o_q) \geq I(o_{i+1}, o_q)$ hold whenever $\delta(o_i, o_{i+1}) \leq \delta(o_i, o_q)$. ∎

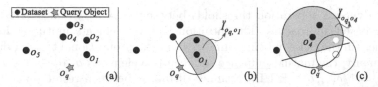

Fig. 2. Influence *vs.* closed ball partitions. (a–b) Querying set $\mathcal{O}$. (c) Querying $\mathcal{O}\backslash\ddot{I}_{o_1,o_q}$.

Figure 2 shows that the exclusion of $\cup_{j\in\mathcal{R}}\ddot{I}_{o_j,o_q}$ generates empty regions in $\mathcal{O}$. Consequently, rules derived from <u>triangle inequality</u> and <u>strong influence sets</u> can be applied for pruning both partitions and data elements, as follows.

Lemma 2. Let elements $o_1,\ldots,o_n \in \mathcal{O}$ be sorted by distances to o_q and influencer $o_i \in \mathcal{O}$ be the farthest diverse neighbor to o_q in $\mathcal{R}$, then $o_{i+1} \in \mathcal{O}$ can be safely assumed as the *next* nearest influencer whenever $\delta(o_{i+1},o_q) > 2\cdot\delta(o_i,o_q)$.

Proof. Since $\delta(o_{i+1},o_q) > 2\cdot\delta(o_i,o_q) > \delta(o_i,o_q)$ then o_{i+1} is an influencer to o_i, otherwise $\delta(o_i,o_q) \geq \delta(o_{i+1},o_i)$, i.e., $o_{i+1} \in \ddot{I}_{o_i,o_q}$. From the triangle inequality follows $2\cdot\delta(o_i,o_q) \geq \delta(o_{i+1},o_i)+\delta(o_i,o_q) \geq \delta(o_{i+1},o_q)$, which is a contradiction. Finally, $o_{i+1} \notin \ddot{I}_{o_j,o_q}, o_j \in \mathcal{R}\setminus\{o_i\}$ since $\delta(o_i,o_q) \geq \delta(o_j,o_q)$. ∎

Lemma 3. An index partition modeled as a closed ball $\langle p,\xi_p\rangle$ is within a strong influence set $\ddot{I}_{o_i,o_q}$ defined by result entry $o_i \in \mathcal{R}$ if $\delta(o_i,o_q) \geq \delta(p,o_i)+\xi_p$.

Proof. If closed ball $\langle o_i,\delta(o_i,o_q)\rangle$ in $\mathcal{O}\backslash\cup_{j\in\mathcal{R}}\ddot{I}_{o_j,o_q}$ covers $\langle p,\xi_p\rangle$ then $\delta(o_i,o_q) \geq \delta(o_i,p)$; $\delta(o_i,o_q) \geq \delta(o_i,o_j)$; and $\xi_p \geq \delta(p,o_j), \forall\ o_j \in \langle p,M_p\rangle$. From the triangle inequality follows $\delta(o_i,o_q) \geq \delta(o_i,p)+\xi_p \geq \delta(o_i,p)+\delta(p,o_j) \geq \delta(o_i,o_j)$. ∎

Lemma 4. If a partial and sorted result set $\mathcal{R}$ contains $k-1$ entries and the distance from diverse candidate $o_{k'} \in \mathcal{O}\backslash\cup_{o_j\in\mathcal{R}}\ddot{I}_{o_j,o_q}$ to o_q is known, then partition $\langle p,\xi_p\rangle$ can be pruned from the diversified search if $\delta_{min}(\langle p,\xi_p\rangle,o_q) > \delta(o_{k'},o_q)$.

Proof. Let $o_{k'} \in \mathcal{O}\backslash\cup_{o_j\in\mathcal{R}}\ddot{I}_{o_j,o_q}$ then $\delta(o_{k'},o_q) \geq \delta(o_i,o_q), \forall\ o_i \in \mathcal{R}$. Moreover, if $\delta_{min}(\langle p,\xi_p\rangle,o_q) > \delta(o_{k'},o_q)$ then $\delta(o_i,o_q) > \delta(o_{k'},o_q), \forall\ o_i \in \langle p,\xi_p\rangle$. Therefore, $\delta(o_j,o_q) > \delta(o_{k'},o_q) \geq \delta(o_i,o_q), \forall\ o_i \in \mathcal{R}$, and the sorted list of influencers is $o_1,\ldots,o_{k-1},o_{k'},o_h$, where every $o_h \in \langle p,\xi_p\rangle$ is after the kth candidate. ∎

We take full advantage of Lemmas 1–4 for designing our *diversity browsing* routine. We also loosely coupled our implementation on top of VP-Trees for experimental purposes, but other indexes can benefit from the search algorithm as well. The only requirement is defining how minimum (δ_{min}) and maximum (δ_{max}) distances from index partitions to query elements are calculated. In the case of VP-Trees, such values depend on determining where o_q is located with regards to a partition centered at pivot p: *(i)* o_q is outside the maximum distance

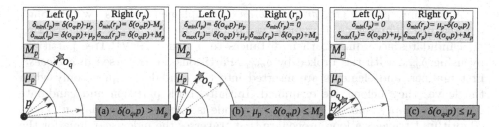

Fig. 3. Values δ_{min} and δ_{max} calculated for VP-Tree left and right nodes.

covered by the partition, *i.e.*, $\delta(o_q, p) > M_p$; *(ii)* o_q is within the right node, *i.e.*, $\mu_p < \delta(o_q, p) \leq M_p$; and *(iii)* o_q is within the left node, *i.e.*, $\delta(o_q, p) < \mu_p$. Figure 3 illustrates those scenarios and how minimum and maximum distances are obtained for the left $(\delta_{min}(l_p), \delta_{max}(l_p))$ and right $(\delta_{min}(r_p), \delta_{max}(r_p))$ nodes.

---

diversity_browsing(Root node p_{root}, query object o_q, diverse neighbors k);

---

1   $p_{root}.\delta_{min} \leftarrow 0.0$; $p_{root}.\delta_{max} \leftarrow \infty$; $\mathcal{R}' \leftarrow \emptyset$;
2   $nodeQ \leftarrow \{p_{root}\}$ /* Sorted queue for VP-Tree nodes   */
3   $candidateQ \leftarrow \emptyset$ /* Sorted queue for candidate elements   */
4   **while** $|nodeQ| > 0 \vee |candidateQ| > 0$ **do**
5   **if** $|candidateQ| = 0$ **then**
6          $node \leftarrow topAndPop(nodeQ)$;
7   **if** $isLeaf(node)$ **then** $candidateQ \leftarrow \{node.elements\}$;
8   **else**
             /* Set δ_{min} and δ_{max} for left and right partitions   */
9              $set\delta_{min}\delta_{max}(node.left)$; $set\delta_{min}\delta_{max}(node.right)$;
10             $nodeQ \leftarrow nodeQ \cup \{node.left\} \cup \{node.right\}$;
11   **else if** $|nodeQ| > 0 \wedge \delta_{min}(top(nodeQ)) < \delta(o_q, top(candidateQ))$ **then**
12          $node \leftarrow topAndPop(nodeQ)$;
         /* Pruning by Lemmas 3-4   */
13   **if** $node \not\subseteq \cup_{o_j \in \mathcal{R}'} \ddot{I}_{o_j, o_q} \wedge |\mathcal{R}'| < k$ **then**
14   **if** $isLeaf(node)$ **then**
15                  $candidateQ \leftarrow candidateQ \cup \{node.elements\}$;
16   **else**
17                  $set\delta_{min}\delta_{max}(node.left)$; $set\delta_{min}\delta_{max}(node.right)$;
18                  $nodeQ \leftarrow nodeQ \cup \{node.left\} \cup \{node.right\}$;
19   **else**
20          $o_i \leftarrow topAndPop(candidateQ)$;
         /* Pruning by Lemma 2   */
21   **if** $o_i \not\subseteq \cup_{o_j \in \mathcal{R}'} \ddot{I}_{o_j, o_q} \wedge |\mathcal{R}'| < k$ **then** $\mathcal{R}' \leftarrow \mathcal{R}' \cup \{o_i\}$;
22 **return** $\mathcal{R}'$;

Algorithm 1: The *diversity browsing* routine.

Algorithm 1 presents the *diversity browsing* implementation as a walkthrough inspection of dataset $\mathcal{O}$, in which two priority queues are used for ordering *(i)* candidates according to their distances to o_q, and *(ii)* VP-Tree partitions by using δ_{min}, with ties broken by δ_{max}. Partitions are traversed in a *closest-first* manner, and elements are inserted into the candidates' queue only when the leaves they belong are examined. Initially, both partition and candidate queues are empty, and the VP-Tree root node is inserted for evaluation. Then, Algorithm 1 triggers a loop procedure that traverses the index searching for the closest node to o_q while keeping the visited path of partitions and elements sorted by distance. Figure 4 illustrates how Algorithm 1 examines VP-Tree nodes in a diversified k-NN query example for a given query object o_q and $k = 3$.

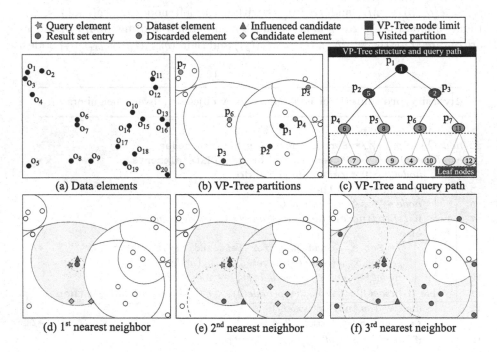

Fig. 4. Running of *diversity browsing* for a test-table query.

Figure 4(a–c) shows the two-dimensional queried set $\mathcal{O}$, and how data elements are organized in a VP-Tree index with distance function L_2 and support for two objects per leaf. In Fig. 4(d), Algorithm 1 splits root node p_1 into right and left nodes r_{p_1} and l_{p_1}, node r_{p_1} into r_{p_3} and l_{p_3}, and node l_{p_3} into r_{p_6} and l_{p_6}. Elements within l_{p_6} are, then, inserted into the candidates' queue. Since o_7 is closest to o_q than any possible element within l_{p_1}, r_{p_6} or r_{p_3} it is retrieved as the first diversified neighbor and included in the partial result set $\mathcal{R}'$. Candidate o_6 is the next nearest neighbor since it is closer to o_q than any node in the partitions' queue, but *diversity browsing* dismisses o_6 as *influenced* because

$\delta(o_7, o_q) \geq \delta(o_7, o_6)$. The algorithm does not consider element o_8 immediately because $\delta(o_8, o_q) > \delta_{min}(l_{p_1})$, which requires traversing the partitions' queue and loading objects o_{17}, o_{19}, o_{18}, o_{16} and o_{20} into the candidates' queue.

Candidate o_8 is, then, straightforwardly retrieved as the next influencer because $\delta(o_8, o_q) > 2 \cdot \delta(o_7, o_q)$, which complies with Lemma 2. The nearest candidate o_9 becomes closer to o_q than any uninspected partition, but it is also influenced by o_8 and discarded – Fig. 4(e). In the next step, Algorithm 1 evaluates the partitions' queue because $\delta(o_{17}, o_q) > \delta_{min}(r_{p_6})$ and loads objects o_4 and o_5 into the candidates' queue. The top of first queue becomes o_4, while the partition on top of second queue is r_{p_3} with $\delta(o_4, o_q) > \delta_{min}(r_{p_3})$.

Table 1. Description of queried datasets.

Name	$|\mathcal{O}|$	$|\mathcal{O}'|$	$|\mathcal{O}''|$	$\mathbb{R}^d$	δ	$\mathcal{D}$	Description
SINT10	$1M$	$70K$	$7K$	10	L_1	10	Random iid dimensions in the $[0, 1]$ interval
MNIST	$70K$	$70K$	$7K$	784	L_2	12	Handwritten digits in the $[0 - 9]$ interval
YAHOO	$2M$	$70K$	$7K$	400	L_2	08	Features from Yahoo® images
COPHIR	$10M$	$70K$	$7K$	282	L_1	15	Color features from photos

Accordingly, *diversity browsing* splits r_{p_3} into r_{p_7} and l_{p_7}, and r_{p_7} element o_{11} is inserted into the candidates' queue. At this point, Algorithm 1 retrieves o_4 as the next influencer and, then, discards entry l_{p_7} on top of the partitions' queue because $\delta(o_4, o_q) \geq \delta(o_4, p_7) + \mu_{p_7}$, which complies with Lemma 3. Finally, Algorithm 1 discards farther partitions l_{p_4} and l_{p_5} by Lemma 4 because $\delta_{min}(l_{p_5}) > \delta_{min}(l_{p_4}) > \delta(o_4, o_q)$. As a result, elements $\mathcal{R} = \{o_7, o_8, o_4\}$ are returned as the nearest *influencers* of the diversified k-NN query example.

4   Experiments

This section provides an empirical *diversity browsing* evaluation regarding one synthetic (SINT10) and three real-world (MNIST, YAHOO, and COPHIR) datasets. Table 1 describes the datasets, their original cardinality $|\mathcal{O}|$, sample sizes $|\mathcal{O}'|$ and $|\mathcal{O}''|$ dimensions $\mathbb{R}^d$, associated distance function δ, and data intrinsic dimension $\mathcal{D}$. The number of elements per leaf was set to 100 in every VP-Tree construction.

Data samples are required because current novelty-based competing algorithms examine a factorial-based number of combinations on data cardinality and, consequently, empirical runs would not finish within a reasonable time (months). Moreover, datasets were reduced to their intrinsic dimension through PCA for avoiding distance concentration [1, 13]. The intrinsic dimension was calculated by using the dataset distance distribution mean and standard deviation, as in [3]. All compared approaches were implemented under the same framework by using Java with JDK 13. The experiments were single-thread executed in our local cluster, a QLustar Debian-based server with two nodes, each one with 48

AMD Opteron 6320 hyper-thread processing cores, 96 GB of shared memory, and a local 01 TB SATA disk.

Three distinct aspects of *diversity browsing* are discussed in the following experiments. First, we examine possible tunings of VP-Tree parameters and their impact on diversified k-NN searches. Then, we compare *diversity browsing* on top of VP-Tree with novelty-based algorithms GMC and GNE, as well as with influence-driven approach BRID_k. Finally, we analyze VP-Tree construction costs and their effects on diversified k-NN search optimization.

VP-Tree Tuning. We evaluate distinct VP-Tree settings to verify if index construction parameters may enhance *diversity browsing* performance. Accordingly, we examine *(i)* the pivot selection criteria, which was set as one of the *random* (RND), *convex-hull* (CVX) or *maximal variance* (MAX) choices, and *(ii)* the presence (BAL) or absence (UBAL) of overflow support in leaf nodes.

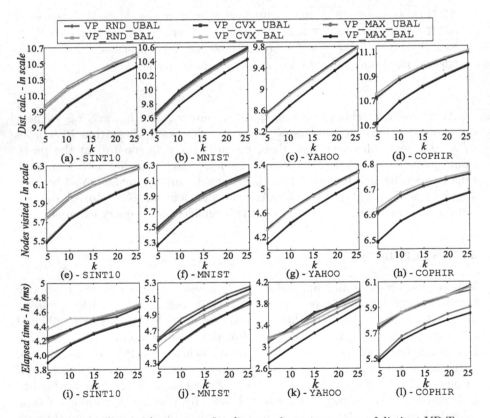

Fig. 5. Average (*ln* scale) measures for *diversity browsing* on top of distinct VP-Trees.

Figure 5 details the behavior of *diversity browsing* over distinct VP-Trees for diversified k-NN searches with values $k = \{5, 10, 15, 20, 25\}$. Such comparison is made with sampled datasets $\mathcal{O}'$ following a *holdout* strategy, in which 90% of elements in $\mathcal{O}'$ were indexed, and the remaining 10% were employed as query objects. Results indicate that the pivot selection criteria have a more significant impact on performance compared with overflow support since both VP-Trees with maximal variance pivots outperformed other VP-Trees.

In particular, VP_MAX_BAL outperformed VP_RND_BAL regarding distance calculation by, at least, 18.36%, 12.83%, 14.76%, and 9.26% in datasets SINT10, MNIST, YAHOO, and COPHIR, respectively. VP_MAX_BAL also outperformed VP_CVX_BAL by, at least, 15.76%, 12.21%, 12.92%, and 10.12% in the same scenario – Figs. 5(a–d). Similar findings were observed for the number of nodes examined during the search – Figs. 5(e–h). In this case, VP_MAX_BAL outperformed the closest non-maximal variance index VP_RND_UBAL by, at least, 17.57%, 13.56%, 19.28%, and 11.82% in datasets SINT10, MNIST, YAHOO, and COPHIR. Finally, no evidence of convex-based VP-Trees being systematically superior to those with random pivots was observed regarding either *distance calculations* or *nodes visited*.

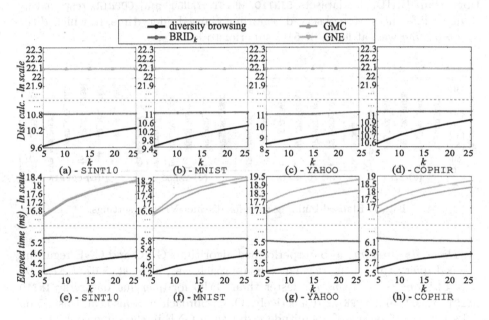

Fig. 6. Average (*ln* scale) measures for a diversified *k*-NN query executed by routines *diversity browsing*, BRID$_k$, and GMC and GNE with parameter $\lambda = .25$.

The average elapsed query time varied mostly in comparison to distances calculated, due to the cost of maintaining priority queues and the dynamical construction of strong influence sets by *diversity browsing*. Nevertheless, maximal variance indexes also outperformed competitors in every evaluated dataset. However, no significant differences were observed regarding overflow support in maximal variance structures, with VP_MAX_UBAL being slightly superior in for MNIST and VP_MAX_BAL providing better results for SINT10, YAHOO, and COPHIR. Therefore, we pick VP_MAX_BAL as a suitable VP-Tree tuning for *diversity browsing* to be evaluated against existing result diversification algorithms.

Diversity Browsing vs. Other Result Diversification Methods. We compare *diversity browsing*, $BRID_k$, GMC, and GNE approaches in the querying of datasets in Table 1. *Diversity browsing* and $BRID_k$ were evaluated over sampled datasets $\mathcal{O}'$, following the *holdout* strategy in which a random 90% of elements were indexed, and the remaining entries were used as query objects for diversified k-NN searches with $k = \{5, 10, 15, 20, 25\}$. Novelty-based GMC and GNE methods were tested over a smaller sample of candidates $\mathcal{O}''$, $|\mathcal{O}''| = 0.1 \times |\mathcal{O}'|$ with the same *holdout* validation strategy.

Figure 6 shows a comparison between the aforementioned approaches regarding average distance calculations and elapsed query time, in which *diversity browsing* outperformed the competitors for every value of k. In particular, our approach has required up to $2.01, 3.79, 14.71$, and $0.59\times$ less distance calculations than $BRID_k$ in datasets SINT10, MNIST, YAHOO, and COPHIR, respectively. Similar differences were observed regarding elapsed query time, in which *diversity browsing* was, at least, 87.51% superior to $BRID_k$.

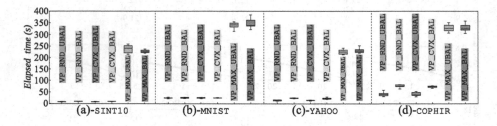

Fig. 7. Elapsed building time for distinct VP-tree settings.

Diversity Browsing also outperformed algorithms GMC and GNE regarding elapsed query time. In particular, our approach was, at least, $5.67, 5.41, 6.07$, and 4.91 *orders of magnitude* faster than GMC for querying datasets SINT10, MNIST, YAHOO, and COPHIR, respectively. Our approach was also, at least, 5.69, $5.42, 6.1$, and 5 orders of magnitude faster than GNE in the same context.

Index Construction Costs. In the last experiment, we evaluate the construction costs of VP-Trees and their impact on *diversity* browsing. Figure 7 reports building time box-plots (100 runs) for every dataset and VP-Tree examined in previous experiments. Results indicate that VP_MAX_BAL is the most expensive index to build, and VP_RND_UBAL is the cheapest structure to construct. Such findings reveal a *trade-off* for maximal variance pivots, *i.e.*, they boost diversified k-NN queries but are more expensive to build. In the case of *diversity browsing*, the construction of VP-Trees with maximal variance pivots pays off when the number of issued queries surpasses the costs of performing a $BRID_k$ search. For instance, *diversity browsing* accumulated query time with VP_MAX_BAL *plus* the index building cost becomes smaller, on average, than the total BRID_k elapsed time after $2379, 1104, 1061$, and 2351 queries (just 2.74% of the total of indexed elements) over datasets SINT10, MNIST, YAHOO, and COPHIR, respectively.

In real similarity-based applications, the ratio query/index is expected to be much higher and, hence, the use of maximal variance pivots with *diversity browsing* is recommended. Nonetheless, coupling our approach with low-cost VP-Trees may also be an alternative for the running of fewer queries. For instance, *diversity browsing* with the most inexpensive structure VP_RND_UBAL was still at least 14% faster than BRID_k in the experiments of Figs. 5 and 6.

5 Conclusions

This study has proposed *diversity browsing*, an algorithm that combines the distance browsing search strategy with metric indexes and influence-based query criteria to provide result diversification. *Diversity browsing* uses distance-based relationships for pruning indexes' branches that are themselves influenced by elements in the result set, according to their proximity to the query object. Empirical evaluations showed *diversity browsing* expressively outperformed approaches GMC and GNE, as well as the baseline diversity algorithm BRID_k. Such experimental gains are also related to the tuning of the VP-Trees, the underlying index we loosely coupled to *diversity browsing* in our implementation. Empirical findings pinpoint VP_MAX_BAL, *i.e.*, the use of maximal variance pivots, is the most suitable tuning for VP-Trees running *diversity browsing*.

Future works include the parallelization of *diversity browsing* and its extension to similarity-joins in a Map-Reduce framework, *e.g.*, Apache Spark.

References

1. Aggarwal, C.: Data Mining: The Textbook. Springer, Cham (2015). https://doi.org/10.1007/978-3-319-14142-8
2. Agrawal, R., Gollapudi, S., Halverson, A., Ieong, S.: Diversifying search results. In: ACM WSDM, pp. 5–14 (2009)
3. Chávez, E., Navarro, G., Baeza-Yates, R., Marroquín, J.: Searching in metric spaces. CSUR **33**(3), 273–321 (2001)
4. Chen, L., Gao, Y., Song, X., Li, Z., Miao, X., Jensen, C.: Indexing metric spaces for exact similarity search. arXiv preprint arXiv:2005.03468 (2020)
5. Chen, L., Gao, Y., Zheng, B., Jensen, C., Yang, H., Yang, K.: Pivot-based metric indexing. PVLDB **10**(10), 1058–1069 (2017)
6. Costa, V., Santos, R., Maconald, C., Ounis, I.: Sparse spatial selection for novelty-based search result diversification. In: Grossi, R., Sebastiani, F., Silvestri, F. (eds.) SPIRE 2011. LNCS, vol. 7024, pp. 344–355. Springer, Cham (2011). https://doi.org/10.1007/978-3-642-24583-1_34
7. Drosou, M., Jagadish, H., Pitoura, E., Stoyanovich, J.: Diversity in big data: a review. Big Data **5**(2), 73–84 (2017)
8. Drosou, M., Pitoura, E.: Multiple radii disc diversity: result diversification based on dissimilarity and coverage. ACM TODS **40**(1), 1–43 (2015)

9. Hetland, M.: The basic principles of metric indexing. In: Coello, C.A.C., Dehuri, S., Ghosh, S. (eds.) Swarm Intelligence for Multi-objective Problems in Data Mining. LNCS, vol. 242, pp. 199–232. Springer, Heidelberg (2009). https://doi.org/10.1007/978-3-642-03625-5_9

10. Hjaltason, G., Samet, H.: Index-driven similarity search in metric spaces. TODS **28**(4), 517–580 (2003)

11. Novak, D., Zezula, P.: PPP-codes for large-scale similarity searching. In: Hameurlain, A., Küng, J., Wagner, R., Decker, H., Lhotska, L., Link, S. (eds.) Transaction on Large-Scale Data-and Knowledge-Centered System. LNCS, vol. 9510, pp. 61–87. Springer, Cham (2016). https://doi.org/10.1007/978-3-662-49214-7_2

12. Padmanabhan, D., Deshpande, P.: Operators for Similarity Search - Semantics, Techniques and Usage Scenarios. Springer, Cham (2015). https://doi.org/10.1007/978-3-319-21257-9

13. Pestov, V.: Lower bounds on performance of metric tree indexing schemes for exact similarity search in high dimensions. Algorithmica **66**(2), 310–328 (2013)

14. Pisinger, D.: Upper bounds and exact algorithms for p-dispersion problems. Comput. Oper. Res. **33**(5), 1380–1398 (2006)

15. Santos, L., Blanco, G., Oliveira, D., Traina, A., Traina Jr., C., Bedo, M.: Exploring diversified similarity with kundaha. In: ACM CIKM, pp. 1903–1906 (2018)

16. Santos, L., Oliveira, W., Ferreira, M., Traina, A., Traina Jr., C.: Parameter-free and domain-independent similarity search with diversity. In: SSDBM, pp. 1–12 (2013)

17. Traina Jr., C., Santos, R., Traina, A., Vieira, M., Faloutsos, C.: The Omni-family of all-purpose access methods: a simple and effective way to make similarity search more efficient. VLDB J. **16**(4), 483–505 (2007)

18. Vieira, M., et al.: On query result diversification. In: IEEE ICDE, pp. 1163–1174. IEEE (2011)

19. Yianilos, P.: Data structures and algorithms for nearest neighbor search in general metric spaces. In: ACM-SIAM SDA, pp. 311–321. SIAM (1993)

20. Zheng, K., Wang, H., Qi, Z., Li, J., Gao, H.: A survey of query result diversification. Knowl. Inf. Syst. **51**(1), 1–36 (2016). https://doi.org/10.1007/s10115-016-0990-4

Continuous Similarity Search
for Evolving Database

Hisashi Koga$^{(\boxtimes)}$ and Daiki Noguchi

University of Electro-Communications, Chofu, Tokyo 182-8585, Japan
{koga,noguchi.d}@sd.is.uec.ac.jp

Abstract. Similarity search for data streams has attracted much attention recently in the area of information recommendation. This paper studies a continuous set similarity search which regards the latest W items in a data stream as an evolving set. So far, a top-k similarity search problem called CEQ (Continuous similarity search for Evolving Query) has been researched in the literature, where the query evolves dynamically and the database consists of multiple static sets. By contrast, this paper examines a new top-k similarity search problem, where the query is a static set and the database consists of multiple dynamic sets extracted from multiple data streams. This new problem is named as CED (Continuous similarity search for Evolving Database). Our main contribution is to develop a pruning-based exact algorithm for CED. Though our algorithm is created by extending the previous pruning-based exact algorithm for CEQ, it runs substantially faster than the one which simply adapts the exact algorithm for CEQ to CED. Our algorithm achieves this speed by devising two novel techniques to refine the similarity upper bounds for pruning.

Keywords: Data stream · Set similarity search · Sliding window · Pruning

1 Introduction

Similarity search for data streams has become significant these days, because it has many applications in the area of information recommendation. Typical similarity search problems [3,5,8] regard a single data stream as a pool of data, that is, a database which allows addition and/or removal of data. Their goals are to keep on seeking similar data to a query out of this dynamic database, i.e., the data stream.

Recently, another direction of similarity search which measures the similarity between a pair of data streams has attracted much attention. It views a single data stream as a set of data [2] and searches similar data streams by means of set similarity search. For instance, Efstathiades *et al.* [1] searched similar user pairs by measuring the similarity between their tweet sets. Wang *et al.* [4] studied how to generate memory-efficient sketches for streaming sets.

© Springer Nature Switzerland AG 2020
S. Satoh et al. (Eds.): SISAP 2020, LNCS 12440, pp. 155–167, 2020.
https://doi.org/10.1007/978-3-030-60936-8_12

Yang *et al.* [9] models a POI (Point of Interest) as a stream of visitors and labels a new POI semantically by searching similar known POIs. Xu *et al.* [6] formulated the Continuous similarity search problem for Evolving Queries (abbreviated as CEQ hereafter). In CEQ, one item is added to the data stream per time instant and its latest W items form the query set. Thus, every time a new item is added to the stream, the query evolves. The task of CEQ is to continuously find the top-k most similar sets to the query from the static database composed by n sets. For the evolving query, the CEQ demands to update the top-k most similar sets. The CEQ models a scenario in which recommendation systems must adapt to the changing preference of a user. The evolving query corresponds to the changing preference of the user, whereas the top-k most similar sets abstract the objects recommended to the user. For CEQ, Xu *et al.* [6] proposed a pruning-based exact algorithm.

In CEQ, the dynamic query originates from a single data stream, while the database consists of multiple static sets. By contrast, this paper examines a new top-k similarity search problem such that the query Q is a single static set and the database consists of multiple data streams with a fixed sliding window size W. This new problem purposes to continuously search the top-k data streams whose last W items are the most similar to the static query Q. Since the database evolves, we name this problem as the "Continuous similarity search problem for Evolving Database" abbreviated as CED hereafter. CED has realistic applications as follows. Consider the situation when, given a web ad, we would like to choose a group of users to whom the web ad should be delivered. Assume that the ad is assigned multiple keywords presenting the categories and that the preference of a user u is estimated from the history of web pages browsed by u. Then, the preference of u is viewed as a data stream. Here, if each web page is summarized with a few keywords, the sliding window of the data stream expresses the recent preference of u and forms a dynamic set of keywords. Thus, we can find the top-k users who should see the ad by solving the CED instance where the ad is given as the query.

In this paper, we develop a pruning-based exact algorithm for CED. Though we create it by extending the previous pruning-based exact algorithm for CEQ, we incorporate two novel techniques into it which refine the similarity upper bounds for pruning in order to reduce similarity computations. Experimentally, our algorithm runs substantially faster than the one which simply adapts the previous exact algorithm for CEQ to CED.

This paper is organized as follows. Section 2 reviews the known CEQ problem and the previous exact algorithms for CEQ. Section 3 formally defines the CED problem. Then, Sect. 4 presents our exact method for CED. Section 5 reports the experimental evaluation. Section 6 concludes this paper.

2   Continuous Similarity Search for Evolving Queries

This section reviews the CEQ problem formulated in [6]. Throughout this paper, let $\Phi = \{x_1, x_2, \cdots, x_{|\Phi|}\}$ be a set of alphabets. To the data stream, exactly one new item from Φ is added at every time instant. We denote an item added at time t by e_t. In CEQ, the latest W items in the data stream become the query Q_T at time T. That is, $Q_T = \{e_{T-W+1}, e_{T-W+2}, \cdots, e_T\}$. On the other hand, the database D stores n static sets of alphabets $\{S_1, S_2, \cdots, S_n\}$ which do not change over time. The task of CEQ is to search for the top-k most similar sets to Q_T in D at every T. Here, the similarity value between the query Q_T and a set S in D is measured with the Jaccard similarity

$$\text{sim}(S, Q_T) = \frac{|S \cap Q_T|}{|S \cup Q_T|}. \tag{1}$$

In case either S or Q_T is a multiset, the extended Jaccard similarity is used, where $|S \cap Q_T| = \sum_{i=1}^{|\Phi|} \min(s_i, q_i)$ and $|S \cup Q_T| = \sum_{i=1}^{|\Phi|} \max(s_i, q_i)$. s_i and q_i symbolize the numbers of the i-th alphabet x_i in S and Q_T respectively. When the time advances from T to $T + 1$, the top-k most similar sets to the query in D must be updated, because the query evolves from Q_T to Q_{T+1}.

CEQ is trivially solved if we compute the similarity values between the query and all the sets in D at every time instant. However, this brute-force solution is forced to compute the similarity values too many times. In the literature, two faster exact algorithms have been proposed. The first one is a pruning-based algorithm in [6] which will be explained in Sect. 2.1 in details.

The second algorithm EA-FIL in [7] makes use of the inverted lists. [7] proves that, if a set S_i contains neither e_T nor e_{T-W}, $\text{sim}(S_i, Q_T) = \text{sim}(S_i, Q_{T-1})$ and EA-FIL does not have to calculate $\text{sim}(S_i, Q_T)$ at T. EA-FIL uses the inverted lists to access quickly such sets having either e_T or e_{T-W} for which EA-FIL need to compute the similarity value at T. Though EA-FIL runs faster than GP, the inverted lists in EA-FIL exploit the property that the database is static.

2.1 Pruning-Based Exact Algorithm

Xu *et al.* [6] developed an exact algorithm named GP (General Pruning-based) which reduces the number of similarity computations between the query and the sets in D. For any set S_i in D, GP keeps on managing an upper bound of the similarity value $\text{sim}(S_i, Q_T)$ at every time instant T.

Denote the similarity value of the k-th most similar set at T by $\text{sim\_topk}(T)$. Then, GP omits the computation of $\text{sim}(S_i, Q_T)$, in case the corresponding upper bound falls below the lower bound of $\text{sim\_topk}(T)$.

2.1.1 Upper Bound of Similarity Value

GP derives the upper bound of $\text{sim}(S_i, Q_T)$ in the following way. Suppose that the similarity value for S_i was computed at time t_i for the last time before the current time T. Thus, $t_i < T$. Just after t_i, GP knows both $\text{sim}(S_i, Q_{t_i}) = \frac{|S_i \cap Q_{t_i}|}{|S_i \cup Q_{t_i}|}$ and $\text{sim\_topk}(t_i)$. From these two values, GP computes the minimum time steps $\min\_step_i$ after which the similarity value of S_i has a chance to exceed $\text{sim\_topk}(t_i)$. Since at most $\min\_step_i$ items are replaced after the $\min\_step_i$ steps, the similarity value of S_i never goes beyond $\frac{|S_i \cap Q_{t_i}| + \min\_step_i}{|S_i \cup Q_{t_i}| - \min\_step_i}$ for the time interval $(t_i, t_i + \text{min\_step}_i]$. Thus, min_step_i is defined as the smallest integer satisfying $\frac{|S_i \cap Q_{t_i}| + \text{min\_step}_i}{|S_i \cup Q_{t_i}| - \text{min\_step}_i} \geq \text{sim\_topk}(t_i)$. GP specifies $\frac{|S_i \cap Q_{t_i}| + \text{min\_step}_i}{|S_i \cup Q_{t_i}| - \text{min\_step}_i}$ as the upper bound of $\text{sim}(S_i, Q_T)$ for any T in $(t_i, t_i + \text{min\_step}_i]$,

2.1.2 Extraction of Top-k Most Similar Sets at Time Instant T

At T, GP first chooses a group of sets $R \subset D$ such that $R = \{S_i | t_i + \text{min\_step}_i = T\}$. R contains at least k sets, because the top-k most similar sets at $T-1$ belong to R. Next, GP determines the top-k most similar sets $C = \{R_1, R_2, \cdots R_k\}$ in R. C is going to memorize the candidates of the top-k most similar sets at T. Let τ be the minimum similarity value of the k sets in C. Because C holds k sets, τ works as the lower bound of $\text{sim\_topk}(T)$.

For a set S_i in the remaining $D \backslash R$, if the upper bound of $\text{sim}(S_i, Q_T)$ exceeds τ, GP computes $\text{sim}(S_i, Q_T)$. Otherwise, GP omits computing $\text{sim}(S_i, Q_T)$. When $\text{sim}(S_i, Q_T)$ is computed and $\text{sim}(S_i, Q_T) > \tau$, GP replaces the least similar set in the current C with S_i and increases τ accordingly. After all the sets in $D \backslash R$ are processed, the k sets in C are returned as the final top-k most similar sets to Q_T at T. In this way, GP shortens the execution time by omitting the similarity computations for some sets in $D \backslash R$.

3 Problem Statement

In our CED problem, a single static set of alphabets serves as a query Q. The database D consists of n data streams $\{S_1, S_2, \cdots, S_n\}$. Each data stream S_i manages a sliding window which stores the last W items added to S_i. We denote the contents stored in the sliding window of S_i at time T by S_i^T. At every time T, CED demands to find the top-k most similar sets to the query set Q from the n sets $D_T = \{S_1^T, S_2^T, \cdots, S_n^T\}$, where the similarity between S_i^T and Q is evaluated with the Jaccard similarity $\text{sim}(S_i^T, Q) = \frac{|S_i^T \cap Q|}{|S_i^T \cup Q|}$. Because D_T evolves with T, we must search the top-k most similar sets continuously. To simplify the problem, we assume that the n data streams are synchronized so that exactly one new item from Φ is added to every data stream at every time instant.

This paper is interested only in pruning-based exact algorithms. Algorithms based on the inverted lists are out of scope, because the inverted lists look improper for the dynamic database in CED: Because one item enters and another item leaves every set in the database at every time instant, the inverted lists have to handle many insertions and deletions of sets all the time. Thus, the inverted lists will suffer from enormous management costs.

In this research, we implement the sliding window of a data stream S_i as a fifo list, because the fifo list is the easiest to implement. For example, the fifo list is the optimal in terms of memory usage in storing the sliding window S_i^T and can update it in $O(1)$ time at every time instant. However, the fifo list is alphabetically unordered. It is known that the time complexity to compute the Jaccard similarity between two sets of size W amounts to a big $O(W^2)$ time, when they are represented as unordered lists. To cope with this issue, we decide to manage

S_i^T as an unordered list usually and to sort it temporarily in the alphabetical order just before $\text{sim}(S_i^T, Q)$ need be computed. This approach reduces the time complexity to compute the Jaccard similarity up to an $O(W \log W + 2W)$ time, where the $W \log W$ term covers the sorting of S_i^T and the $2W$ term presents the time spent to compute the Jaccard similarity between two ordered lists.

4   Our Exact Algorithm for CED

This section explains our exact algorithm for CED. First, Sect. 4.1 claims that the known exact algorithm GP [6] for CEQ can be adapted to CED. We name this adapted version as GPD (General Pruning-based method for dynamic Database). Then, Sects. 4.2 and 4.3 introduce two novel techniques that decrease the frequency of similarity computations by tightening the similarity upper bounds. We call the first technique as the incremental update, while the second one is termed as the common element method.

4.1   Extension of Exact Algorithm for CEQ to CED

Whereas the previous CEQ problem considers a dynamic query and the static database, our CED problem treats a static query and the dynamic database. Therefore, CED is the same as CEQ in that one of the query and the database is static and the other is dynamic. Thus, we can make an exact algorithm for CED by modifying the known exact algorithm GP for CEQ. What we have to do is to replace (1) the dynamic query Q_T in GP with the static query Q and (2) the static database $D = \{S_1, S_2, \cdots, S_n\}$ in GP with the dynamic database $D_T = \{S_1^T, S_2^T, \cdots, S_n^T\}$. We name this exact algorithm for CED as GPD.

For example, suppose that GPD computes $\text{sim}(S_i^{t_i}, Q) = \frac{|S_i^{t_i} \cap Q|}{|S_i^{t_i} \cup Q|}$ at time t_i. GPD first derives the minimum time steps min_step_i which satisfies

$$\frac{|S_i^{t_i} \cap Q| + \text{min\_step}_i}{|S_i^{t_i} \cup Q| - \text{min\_step}_i} \geq \text{sim\_topk}(t_i). \tag{2}$$

Then, GPD specifies $\frac{|S_i^{t_i} \cap Q| + \text{min\_step}_i}{|S_i^{t_i} \cup Q| - \text{min\_step}_i}$ as the upper bound of similarity value between Q and the stream S_i in the time period $(t_i, t_i + \text{min\_step}_i]$. Like GP, GPD uses this similarity upper bound for S_i for the whole period $(t_i, t_i + \text{min\_step}_i]$.

The left-hand side of Eq. (2) is derived by replacing S_i and Q_{t_i} in the upper bound formula for GP with $S_i^{t_i}$ and Q respectively. In the same way, the procedure of GP to select the top-k most similar sets at time T in Sect. 2.1.2 can be migrated to CED by replacing S_i with S_i^T and Q_T with Q.

4.2   Incremental Update

Without updating the similarity upper bound for the period $(t_i, t_i + \text{min\_step}_i]$, GPD minimizes the overhead of managing the similarity upper bounds.

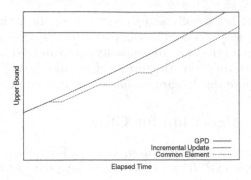

Fig. 1. Similarity upper bounds

Instead, the similarity upper bound in GPD is rather loose as follows: Let j be an integer satisfying $1 \leq j \leq \text{min\_step}_i$. At $t_i + j$, an upper bound of

$$\frac{|S_i^{t_i} \cap Q| + j}{|S_i^{t_i} \cup Q| - j} \leq \frac{|S_i^{t_i} \cap Q| + \text{min\_step}_i}{|S_i^{t_i} \cup Q| - \text{min\_step}_i}. \tag{3}$$

is available, because j items have changed inside the sliding window of S_i before $t_i + j$.

By updating the similarity upper bound incrementally at every time instant, our incremental update technique uses $\frac{|S_i^{t_i} \cap Q| + j}{|S_i^{t_i} \cup Q| - j}$ at time $t_i + j$. Figure 1 illustrates how the similarity upper bounds differs between GPD and the incremental update. While GPD always keeps the similarity upper bound around the top-k similarity value at t_i, the incremental update increases it gradually from $\text{sim}(S_i^{t_i}, Q)$.

We expect the incremental update to accelerate the similarity search by lessening the similarity computations in exchange for the extra overhead to update similarity upper bounds. Fortunately, we can update the upper bound in $O(1)$ time per time instant by remembering $|S_i^{t_i} \cap Q|$ and $|S_i^{t_i} \cup Q|$ at t_i. Therefore, the overhead to update the upper bound once becomes much less than that incurred in computing the Jaccard similarity.

Besides, the incremental update allows us to exclude the variable min_step_i, because $\frac{|S_i^{t_i} \cap Q| + j}{|S_i^{t_i} \cup Q| - j}$ does not contain this variable. To remove "min_step_i" completely, we need to define R as the top-k most similar streams at time $T - 1$.

4.3   Common Element Method

Here, we refine the incremental update in Sect. 4.2 so that the similarity upper bound may lower more. This technique is named as the *common element method*. We would point out that both the incremental update and GPD assume pessimistically that items which will depart from the sliding window in near future are unknown. However, in practice, any similarity search algorithm can recognize

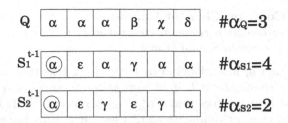

Fig. 2. Common elements

them, because they exist in the current sliding window. In particular, we exploit the property that, if the item which departs at time t is a common element between S_i^{t-1} and Q, $|S_i^t \cap Q| \le |S_i^{t-1} \cap Q|$ so that $\mathrm{sim}(S_i^t, Q) \le \mathrm{sim}(S_i^{t-1}, Q)$.

4.3.1 Common Elements

When S_i^{t-1} are multisets, which items in S_i^{t-1} are common with Q is unclear. Thus, we first define the common elements rigorously: Consider the scenario such that the similarity is computed between S_i and Q at time t_i. At t_i, the sliding window $S_i^{t_i} = \{e_{t_i-W+1}, e_{t_i-W+2}, \cdots, e_{t_i-1}, e_{t_i}\}$. To ease the exposition, we rewrite $\{e_{t_i-W+1}, e_{t_i-W+2}, \cdots, e_{t_i-1}, e_{t_i}\}$ as $\{a_1, a_2, \cdots, a_{W-1}, a_W\}$ by substituting a_l for e_{t_i-W+l} for $1 \le l \le W$. a_l is going to depart the sliding window at $t_i + l$. Definition 1 below defines the common elements.

Definition 1 (Common Element). *If the intersection size between S_i and Q is decreased by 1 when a_l departs at $t_i + l$, a_l is a common element with Q.*

If a_j is a common element, $|S_i^{t_i+l-1} \cap Q| \le |S_i^{t_i+l} \cap Q|$ without regard to the arrival of an element at $t_i + l$, because the departure at $t_i + l$ decreases the intersection size by 1.

Suppose a_l has an alphabet label, say α. Whether a_l is a common element depends on the numbers of α characters in Q and $S_i^{t_i+l-1}$. We denote the number of α characters in Q by $\#\alpha_Q$ and that in $S_i^{t_i+l-1}$ by $\#\alpha_{S_i}^l$. a_l is a common element if and only if $\#\alpha_{S_i}^l \le \#\alpha_Q$. See Fig. 2 as an example. While Q holds 3 instances of α, S_1^{t-1} and S_2^{t-1} have 4 and 2 instances of α respectively. Thus, $|S_1^{t-1} \cap Q| = 3$ and $|S_2^{t-1} \cap Q| = 2$. After the circled elements at the head left S_1^{t-1} and S_2^{t-1}, only S_2 reduces the intersection size from 2 to 1.

According to the above discussion, we may judge if a_l is a common element only at $t_i + l - 1$ by comparing $\#\alpha_{S_i}^l$ with $\#\alpha_Q$. However, this scheme is hard to implement efficiently, when dynamic sets are represented with unordered lists as mentioned in Sect. 3, because we must scan the list $S_i^{t_i+l-1}$ to obtain $\#\alpha_{S_i}^l$. In addition, this scheme requires to scan one list every time instant.

So that we may estimate the similarity upper bound without troubled by this difficulty, we introduce a notion of common element candidates as follows.

Definition 2 (Common Element Candidate). *At time* $t < t_i + l$, *if* a_l *still has a potential to become a common element at* $t_i + l$, a_l *is called a common element candidate between* S_i *and* Q *at* t.

Again, suppose that a_l has a label of α. Definition 2 tells that a_l is no longer a common element candidate at t, if at least $\#\alpha_Q$ instances of α appear *behind* a_l in S_i^t, because it guarantees that $\#\alpha_{S_i}^l > \#\alpha_Q$ at $t_i + l - 1$. By contrast, a_l is still a common element candidate at t, if less than $\#\alpha_Q$ instances of α exist behind a_l in S_i^t.

The point of the common element candidate is that, instantly $\text{sim}(S_i^{t_i}, Q)$ is computed at t_i, we can decide if each element a_i $(1 \le i \le W)$ in the sliding window is a common element candidate by scanning only one unordered list $S_i^{t_i}$ from back to front.

4.3.2 Estimation of Upper Bounds

This subsection describes our method to estimate the similarity upper bound. By scanning the unordered list $S_i^{t_i}$ in an $O(W)$ time, our method first identifies all the common element candidates in $S_i^{t_i}$ at t_i, just when $\text{sim}(S_i^{t_i}, Q)$ is computed for the stream S_i. After that, we can create an array AR_i such that $AR_i[j]$ records how many common element candidates appear in the prefix of $S_i^{t_i}$ whose length equals j for $1 \le j \le W$.

Theorem 1 below relates the number of common element candidates in the prefix of $S_i^{t_i}$ to the intersection size.

Theorem 1. *Let* $j \le W$ *and* $\{a_1, a_2, \cdots, a_j\}$ *be a prefix of* $S_i^{t_i}$. *If the number of common element candidates in this prefix equals* v, $|S_i^{t_i+j} \cap Q| \le |S_i^{t_i} \cap Q| + (j - v)$.

Theorem 1 immediately derives that $\text{sim}(S_i^{t_i+j}, Q) \le \dfrac{|S_i^{t_i} \cap Q| + (j - v)}{|S_i^{t_i} \cup Q| - (j - v)}$. Thus, we got a similarity upper bound tighter than that for the incremental update in Eq. (3). See Fig. 1 for illustration. Unlike the incremental update, the common element method occasionally does not increase the similarity upper bound.

Proof. The period $[t_i + 1, t_i + j]$ accompanies exactly j item departures and j item arrivals. Here, one item departure either decreases the intersection size by 1 or keeps it unchanged. On the other hand, one item arrival either increases the intersection size by 1 or keeps it unchanged. Therefore, $|S_i^{t_i+j} \cap Q|$ takes the maximum value of $|S_i^{t_i} \cap Q| + j$, if all the j departures keep the intersection size unchanged and all the j arrivals increase it by 1.

Conversely, let f_1 be the number of departures that decrease the intersection size and f_2 be the number of arrivals which keep the intersection size unchanged. Then, it holds that $|S_i^{t_i+j} \cap Q| = |S_i^{t_i} \cap Q| + j - (f_1 + f_2)$. Thus, Theorem 1 is correct if we can prove $f_1 + f_2 \ge v$.

Suppose that v' out of the v common element candidates finally departed as common elements, and the remaining $v - v'$ candidates departed as non-common elements. Obviously $v' = f_1$ from the definition of common elements.

Input: TOPK^{T-1}: top-k most similar data streams at $T-1$

1: **foreach** $S_i \in \text{TOPK}^{T-1}$ **do**
2:   Compute $\text{sim}(S_i^T, Q)$, while searching the common element candidates in S_i^T.
3:   Record # of common element candidates in the prefixes of S_i^T into array AR_i.
4: Set τ to the similarity value of the least similar stream in TOPK^{T-1}.
5: **foreach** $S_i \notin \text{TOPK}^{T-1}$ **do**
6:   Update the upper bound of $\text{sim}(S_i^T, Q)$ incrementally by referring to AR_i.
7:   **if** Upper bound $> \tau$ **then**
8:     Compute $\text{sim}(S_i^T, Q)$, while searching the common element candidates in S_i^T.
9:     Record # of common element candidates in the prefixes of S_i^T into AR_i.
10:     Increase τ to the k-th largest similarity value computed at T.
11: Select the k streams whose similarity values are the highest as TOPK^T.

Fig. 3. Behavior of GP_CE at T

Next, consider the latter $v - v'$ candidates. Let a_j be any one of them and have a label of α. Since a_j was a common element candidate at t_i and left the sliding window as non-common elements at $t_i + j$, there must be a time instant $t' \in [t_i + 1, t_i + j - 1]$ when a_j becomes no longer a candidate. At t', the number of α behind a_j increased from $\#\alpha_Q - 1$ to $\#\alpha_Q$. Therefore, the item arriving at t' must have a label of α. Note that this fact also tells that there exist $\#\alpha_Q - 1$ instances of α behind a_j at $t' - 1$ in the sliding window. Because these $\#\alpha_Q$ instances including a_j never departs before $t_i + j$, they all remain in the sliding window at t' regardless of the item departure at t'. Thus, the item arrival at t' cannot increase the intersection size. In this way, with each of the $v - v'$ candidates, we can associate one distinct item arrival which does not increase the intersection size. Hence, $f_2 \geq v - v'$.

Thus, $f_1 + f_2 \geq v' + (v - v') = v$ and Theorem 1 is proved.   □

4.4   Whole Image of Our Algorithm

Figure 3 illustrates the way how our exact algorithm returns the top-k most similar data streams to Q at time instant T in the pseudocode style. Our algorithm is named as GP_CE (General Pruning supported by Common Elements).

At the beginning, GP_CE calculates the similarity values for the k data streams which were the most similar to Q at $T - 1$. In the 4th line, the least similarity value for these k streams is set to τ as the initial lower bound of $\text{sim\_topk}(T)$, i.e., the similarity value of the k-th most similar stream at T.

Then, after updating the similarity upper bounds for all of the remaining streams incrementally in the 6th line, GP_CE actually computes similarity values only for such streams whose similarity upper bounds exceed the current lower bound of $\text{sim\_topk}(T)$. In the 2nd and 8th lines, when the Jaccard similarity is computed, GP_CE checks the number of common candidate elements that help to tighten the similarity upper bounds.

5   Experiments

We evaluate our GP_CE experimentally with synthetic datasets and two real datasets. The experimental platform is a PC with Intel Core i7-4770 CPU@ 3.40 GHz, 16 GB memory. As the baselines for comparison, we also implement (1) the brute-force method BFM which computes the Jaccard similarity n times at each time instant and (2) GPD in Sect. 4.1 which simply adapts the pruning-based exact algorithm in [6] for CEQ to CED. To examine the effect of common element method, we also developed an algorithm which only makes use of the incremental update in Sect. 4.2. This algorithm is called "IU-only". Since BFM was much slower than GPD, BFM will be displayed in only one graph reporting the execution time in the subsequence.

5.1   Results with Synthetic Datasets

We construct the synthetic database D with n data streams as follows. After fixing the sliding window size W, we first make a pool P of 1000 random sets which choose their elements from the alphabet Φ with the IBM Quest data generator, where the average set cardinality in P is set to W. Then, we prepare n long sequences of alphabets by randomly concatenating the sets in P. One data stream is simulated by passing one element from a long sequence at each time instant. The query Q is specified by pick up one set in the pool P randomly.

We evaluate the efficiency of an algorithm by measuring the total processing time to perform the continuous top-k similarity search 1000 times for the time period from $T = 1$ to $T = 1000$. Since Q is randomly chosen from P, we report the average processing time over 10 different queries.

Here, we fix n to 10000 and W to 100, whereas we vary the alphabet size $|\phi|$ and k, since $|\phi|$ and k highlight the property of GP_CE well. Intuitively, $|\Phi|$ controls the number of common elements between Q and the dynamic sets. As $|\phi|$ becomes larger, the number of common elements tends to decrease. Experiment (1) When $|\Phi|$ is changed.

With fixing k to 10, we change $|\Phi|$ in the range $\{200, 400, 800, 1600, 3200, 6400\}$. Figure 4(a) displays the total processing time.

For any $|\Phi|$, our GP_CE runs more than 4 times faster than GPD. Interestingly, as $|\Phi|$ becomes larger, GPD and IU-only behave more differently. We infer that this result is caused by the difference of similarity upper bound between GPD and IU-only: Roughly speaking, GPD sets the similarity upper bound for a data stream S_i to sim_topk(t_i) from the beginning, whereas IU-only gradually increases it from sim($S_i^{t_i}, Q$) towards sim_topk(t_i). For large $|\Phi|$ values, the expectation of sim($S_i^{t_i}, Q$) becomes small, because $E[|S_i^{t_i} \cap Q|]$ reduces. Thus, GPD is to specify a much higher similarity upper bound than IU-only. As the result, GPD has to compute the similarity values more often than IU-only.

By contrast, for small Φ values, $E[|S_i^{t_i} \cap Q|]$ becomes larger so that the expectation of sim($S_i^{t_i}, Q$) gets closer to sim_topk(t_i). Thus, when Φ is small, the effect of incremental update is restricted. Remarkably even when Φ is small,

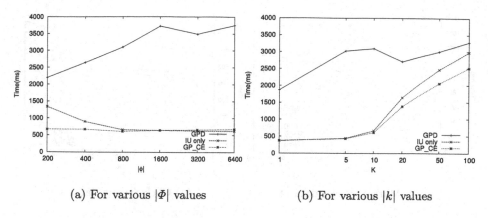

(a) For various $|\Phi|$ values   (b) For various $|k|$ values

Fig. 4. Processing time for synthetic datasets

GP_CE retains its advantage over GPD successfully, because the common element method makes a big contribution when $E[|S_i^{t_i} \cap Q|]$ is large and covers the deterioration of the incremental update.

Experiment (2) When k is changed.

We change k in the range $\{1, 5, 10, 20, 50, 100\}$ for $|\Phi| = 800$. Figure 4(b) shows the total processing time for various k values. Also in this experiment, GP_CE outperforms GPD for any k value. The relative gain of IU-only to GPD shrinks as k increases. This result can be also attributed to the difference of similarity upper bound between GPD and IU-only. As k becomes larger, $\text{sim\_topk}(t_i)$ decreases and gets nearer to $\text{sim}(S_i^{t_i}, Q)$, so that IU-only performs more similarly to GPD. Interestingly, GP_CE definitely defeats GPD even for large k values thanks to the common element method that does not depend much on k. Thus, the common element method complements the incremental update again.

5.2   Results with Real Datasets

Next, we evaluate GP_CE with two real datasets: a Market Basket dataset[1] and a Click Stream dataset[2]. Both of them were also used in the previous work [6].

The Market Basket dataset is a collection of itemsets each of which presents a set of items purchased by one customer. This dataset consists of 88162 itemsets, where $|\Phi| = 16470$ denotes the number of distinct item kinds. To apply the Market Basket dataset to GPD, we emulate each of $n = 10000$ data streams in D by concatenating itemsets in the dataset randomly. The sliding window size W is set to 10, since the average set cardinality equals 10.3. The query is generated by choosing one itemset randomly from the dataset.

The Click Stream dataset is a collection of sequences of web pages (URLs) visited by some users. By associating each web page with some predefined category such as "news" and "tech", a sequence of web pages visited by a user is

---

[1] http://fimi.ua.ac.be/data/.

[2] http://www.philippe-fournier-viger.com/spmf/index.php?link=datasets.php.

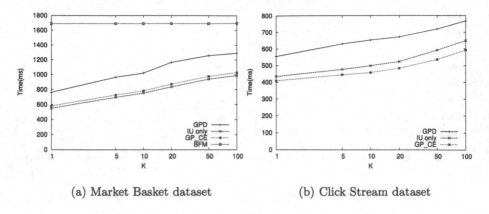

(a) Market Basket dataset   (b) Click Stream dataset

Fig. 5. Processing time for real datasets

represented as a multiset of alphabets, where $|\Phi| = 17$ corresponds to the number of categories. This dataset consists of 31790 multisets. With this dataset, 10000 data streams in D are synthesized completely in the same way as the Market Basket dataset. W is set to 13 which almost equals the average set cardinality. One multiset chosen from this dataset serves as a query.

The dynamic sets in D have few common elements with Q for the Market Basket dataset whose $|\phi|$ is quite huge. By contrast, they share many common elements with Q for the Click Stream dataset as $|\Phi|$ is only 17. We evaluate an algorithm with the total time to process the continuous top-k similarity search over the time period of length 1000. We vary k in the range $\{1, 5, 10, 20, 50, 100\}$. Again, we report the average processing time over 10 different queries.

Figure 5(a) shows the processing time for the Market Basket dataset. While GP_CE outperforms GPD, it runs slightly slower than IU-only. This is because the common element method in GP_CE is not rewarded, when the dynamic sets in D rarely share common elements with Q. However, the extra overhead caused by the common element method is moderate since the gap between GP_CE and IU-only is slight. Figure 5(b) shows the processing time for the Click Stream dataset. Again, GP_CE is superior to GPD for any k value. For this dataset, GP_CE runs faster than IU-only, because the common element method can handle the dynamic sets having many common elements with Q efficiently.

6   Conclusion

This paper investigates the Continuous similarity search for Evolving Database (CED) for the first time, where the query is a static set and the database consists of dynamic sets which correspond to the sliding window of data streams. In the past, the previous researches considered only the Continuous similarity search for Evolving Queries (CEQ) problem which deals with the dynamic query and the database of static sets.

Especially, we design a pruning-based exact algorithm GP_CE for the CED. Here, we not only adapt the known pruning-based exact algorithm for CEQ to CED, but also incorporate two techniques into GP_CE: (1) the incremental update and (2) the common element method both of which improve the efficiency of pruning by refining the similarity upper bound. The incremental update lowers the upper bound by increasing it stepwise according to the elapsed time since the last similarity calculation. The common element method reduces the upper bound further by monitoring the common elements in the current sliding window.

We show experimentally that GP_CE always runs faster than the baseline algorithm which simply adapts the previous exact algorithm for CEQ to CED. Interestingly, the common element method often complements the incremental update, especially when the incremental update cannot decrease the upper bound well for the reason that the top-k similarity value is not far from the average similarity value of dynamic sets.

One interesting future direction of this research is to decide whether or not to apply the common element method adaptively on a set-by-set basis.

Acknowledgments. This work was supported by JSPS KAKENHI Grant Number JP18K11311, 2020.

References

1. Efstathiades, C., Belesiotis, A., Skoutas, D., Pfoser, D.: Similarity search on spatio-textual point sets. In: Proceedings of the 19th International Conference on Extending Database Technology, EDBT, pp. 329–340 (2016)
2. Mann, W., Augsten, N., Jensen, C.S.: SWOOP: top-k similarity joins over set streams. CoRR abs/1711.02476 (2017). http://arxiv.org/abs/1711.02476
3. Leong Hou, U., Zhang, J., Moruatidis, K., Li, Y.: Continuous top-k monitoring on document streams. IEEE Trans. Knowl. Data Eng. **29**(5), 991–1003 (2017)
4. Wang, P., et al.: A memory-efficient sketch method for estimating high similarities in streaming sets. In: Proceedings of the 25th ACM SIGKDD International Conference on Knowledge Discovery and Data Mining, pp. 25–33 (2019)
5. Wang, X., Zhang, Y., Zhang, W., Lin, X., Huang, Z.: Skype: top-k spatial-keyword publish/subscribe over sliding window. Proc. VLDB Endow. **9**(7), 588–599 (2016)
6. Xu, X., Gao, C., Pei, J., Wang, K., Al-Barakati, A.: Continuous similarity search for evolving queries. Knowl. Inf. Syst. **48**(3), 649–678 (2015). https://doi.org/10.1007/s10115-015-0892-x
7. Yamazaki, T., Koga, H., Toda, T.: Fast exact algorithm to solve continuous similarity search for evolving queries. In: Sung, W.K., et al. (eds.) AIRS 2017. LNCS, vol. 10648, pp. 84–96. Springer, Cham (2017). https://doi.org/10.1007/978-3-319-70145-5_7
8. Yang, D., Shastri, A., Rundensteiner, E.A., Ward, M.O.: An optimal strategy for monitoring top-k queries in streaming windows. In: Proceedings of the 14th International Conference on Extending Database Technology, pp. 57–68 (2011)
9. Yang, D., Li, B., Cudré-Mauroux, P.: POIsketch: semantic place labeling over user activity streams. In: Proceedings of the IJCAI 2016, pp. 2697–2703 (2016)

Taking Advantage of Highly-Correlated Attributes in Similarity Queries with Missing Values

Lucas Santiago Rodrigues[✉], Mirela Teixeira Cazzolato,
Agma Juci Machado Traina, and Caetano Traina Jr.

Institute of Mathematics and Computer Sciences, University of São Paulo (USP),
São Carlos, Brazil
{lucas_rodrigues,mirelac}@usp.br, {agma,caetano}@icmc.usp.br

Abstract. Incompleteness harms the quality of content-based retrieval and analysis in similarity queries. Missing data are usually evaluated using exclusion and imputation methods to infer possible values to complete gaps. However, such approaches can introduce bias into data and lose useful information. Similarity queries cannot perform over incomplete complex tuples, since distance functions are undefined over missing values. We propose the SOLID approach to allow similarity queries in complex databases without the need neither of data imputation nor deletion. First, SOLID finds highly-correlated metric spaces. Then, SOLID uses a weighted distance function to search by similarity over tuples of complex objects using compatibility factors among metric spaces. Experimental results show that SOLID outperforms imputation methods with different missing rates. SOLID was up to 7.3% better than the competitors in quality when querying over incomplete tuples, reducing 16.42% the error of similarity searches over incomplete data, and being up to 30.8 times faster than the closest competitor.

Keywords: Missing data · Similarity search · Complex · Metric spaces

1 Introduction

Advances in data collection and sharing have substantially increased the amount of available data in the last decades. Datasets of complex data such as images, time series, and audios have been the target of several studies [3,11]. Similarity queries explore complex objects using low-level data representations and compare them using distance functions to retrieve the relevant objects by their content [3]. However, problems in data acquisition, recording, and management may lead to missing values in the data. As data quality poses significant challenges for content-based retrieval, analysis, and management, missing data is a relevant issue [10]. A common approach to query over incomplete datasets is to remove the tuples or attributes with missing values. Applications can also maintain the

© Springer Nature Switzerland AG 2020
S. Satoh et al. (Eds.): SISAP 2020, LNCS 12440, pp. 168–176, 2020.
https://doi.org/10.1007/978-3-030-60936-8_13

attribute and ignore only the missing values [2], but removing tuples from the dataset may also discard relevant information.

Figure 1 shows an example of a database of complex attributes extracted from drawn characters, with missing values. Distance functions can not measure the similarity of incomplete tuples. Imputation approaches replace missing values following predetermined criteria, filling the data, for traditional analyses. Mean imputation [4] is one of the most common approaches, replacing missing values with the corresponding dimension's average or mode. Other techniques include the K-Nearest Neighbor (kNN) imputation [2], which estimates missing values based on the most similar objects considering the existing dimensions, fuzzy sets, interpolation functions, or regression methods [5]. However, ignoring or imputing values can introduce bias in the data and harm similarity searches [10].

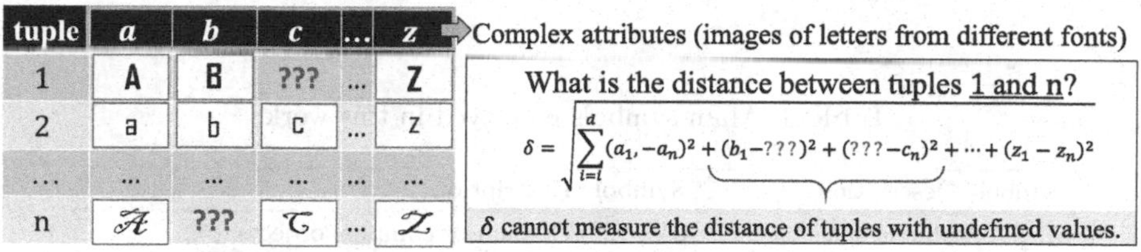

tuple	a	b	c	...	z
1	A	B	???	...	Z
2	a	b	c	...	z
...	...	...	...	...	...
n	$\mathscr{A}$	???	$\mathcal{C}$	...	$\mathcal{Z}$

Complex attributes (images of letters from different fonts)

What is the distance between tuples 1 and n?

$$\delta = \sqrt{\sum_{i=1}^{d}(a_1, -a_n)^2 + (b_1 - ???)^2 + (??? - c_n)^2 + \cdots + (z_1 - z_n)^2}$$

δ cannot measure the distance of tuples with undefined values.

Fig. 1. Example of the problem of missing values in similarity searches.

In this work, we propose the SOLID method to retrieve the most similar tuples comparing complex attributes over incomplete databases. SOLID implements our CorDiS method to compute the correlation among metric spaces defined over complex attributes, and generates compatibility factors to yield more importance to the most correlated attributes. Experiments show that SOLID is three orders of magnitude better than the baseline approach when querying over incomplete tuples, being up to 7.3% better than the Decision Tree Imputation method, on average. In datasets with large amounts of missing values, SOLID reduced the error by 26.3% and 16.4% compared with the baseline approach and the closest competitor, respectively. Also, SOLID was faster than all competitors.

Paper Outline. The remaining sections of this work are organized as follows: Sect. 2 describes the background and related work. Section 3 proposes the SOLID method. Section 4 shows the experiments. Section 5 concludes this work.

2   Background and Related Work

Database Management Systems (DBMS) compare scalar data using identity and order operators, but those are not adequate to compare complex data. Similarity-based comparisons of complex data rely on the data representation obtained by **Feature Extraction Methods (FEM)**, and object comparisons carried out by **Distance Functions** (δ). Table 1 shows the symbols employed in this work.

FEMs extract discriminative features from complex data, often represented by numerical feature vectors. There are several FEMs dedicated to images, such as the ones defined by the MPEG-7 [9]. Distance functions (δ) measure the similarity between pairs of feature vectors, and output a real value R^+, where the smaller the distance, the greater the similarity. Euclidean and Manhattan are examples of widely used distance functions [11]. A distance function is called a metric when it meets the properties of metric spaces, following Definition 1.

Definition 1. Metric Space (M). *A Metric Space is a pair* $M = \langle S, \delta \rangle$, *where* S *is the domain of complex data and* $\delta : \mathbb{R}^m \times \mathbb{R}^m$ *is a distance function. A* δ *is a metric if, for any complex objects* $s_i, s_j, s_k \in S$, *it meets the following properties: (i)* **Non-Negativity:** $\delta(s_i, s_j) \geqslant 0$; *(ii)* **Identity of the indiscernible:** $\delta(s_i, s_i) = 0 \Rightarrow s_i = s_j$; *(iii)* **Symmetry:** $\delta(s_i, s_j) = \delta(s_j, s_i)$; *(iv)* **Triangular Inequality:** $\delta(s_i, s_j) \leqslant \delta(s_i, s_k) + \delta(s_k, s_j)$.

Table 1. Main symbols employed in this work.

Symbol	Description	Symbol	Description
M	Metric space	S	Domain of complex objects
$\mathfrak{M}$	Set of metric spaces	d	Dimensionality of $\mathcal{D}$
δ	Distance function	ω	Compatibility factor of a metric space
s_i	Complex attribute	Ω	Set of compatibility factors
ϕ	Correlation function	ψ	SOLID's distance function
S	Feature vector set	γ	Threshold for the compatibility factors
$\mathcal{D}$	Dataset	Υ	Set of consolidated compatibility factors
n	Cardinality of $\mathcal{D}$	ϖ	Consolidated compatibility factors

Let $\mathcal{D}$ be a dataset of complex objects $S \in S$, and δ be a metric distance function. The **Range** and **k-Nearest Neighbor** are the most employed similarity queries. Let s_q and s_i be two complex attributes in domain S, where s_q is the query center. A Range Query (R_q) retrieves every element $s_i \in S$ whose similarity to s_q is less or equal than a similarity threshold ξ, that is $\delta(s_q, s_j) \leqslant \xi$. Given an amount k of objects and a query center s_q, a k-NN Query (Knn$_q$) retrieves the k objects $s_i \in S$ most similar to s_q measured by δ.

Traditional distance functions cannot measure the similarity between incomplete complex objects, drastically reducing both the available data and the efficiency of similarity queries. Many existing methods treat incompleteness based on data deletions or imputation [5]. The Mean Imputation is the simplest approach, which infers an average value of the available attributes in a tuple. Regression Imputation involves incorporating knowledge, such as data correlation, for inferring missing values [5]. The kNN Imputation searches for the k-nearest neighbors to the missing value, computing and inserting the average of the candidates into the missing spot [2]. Imputation methods based on Decision Trees

Imputation partition the data based on its correlation, looking for the best candidates for each partition to infer values [7]. However, imputation methods may lead to biased results when it increases the rate of missing values [10].

3   The Proposed Method

SOLID (*Search Over Correlated and Incomplete Data*) allows similarity queries over incomplete databases by taking advantage of *highly correlated search spaces*. The method does not discard nor replace missing values. Instead, it weights compatible metric spaces to give an approximation of the missing complex object location in each search space. We describe each step of SOLID next.

Defining Metric Spaces. SOLID extracts features from the d complex attributes in a dataset $\mathcal{D}$ with the chosen FEM. SOLID defines a metric space $\mathbb{M}_i$ for every complex attribute, *i.e* every image, using the corresponding distance function. Let $\mathfrak{M} = \mathbb{M}_1, \mathbb{M}_2, \cdots, \mathbb{M}_d$ be the set d of metric spaces defined over the attributes of $\mathcal{D}$. We represent the ith metric space defined over $\mathcal{D}$ as $\mathbb{M}_i = \langle \mathbb{S}, \delta \rangle \in \mathfrak{M}$, $1 \leq i \leq d$, and δ is the corresponding distance function.

Measuring the Correlation. SOLID implements CorDiS (Algorithm 1) to map the correlation between pairs of metric spaces, working with a sample $\mathcal{D}'$ from $\mathcal{D}$ and the set of extracted features (Line 2). Function *GetSetOfAverageDistances* computes the average distances (Lines 3–4), comparing each object to every other element in the feature set (Lines 8–10), and returning the set of mean distances of each element in the feature set (Lines 7–12). After obtaining the

---

Algorithm 1: CorDiS (*Correlation of Distance Spaces*)

Input : A pair of metric spaces ($< \mathbb{M}_i, \mathbb{M}_j >$), the distance function (δ), the correlation function (ϕ)

Output: The correlation map $\mathcal{F}_{\text{corr}}$

1 **begin**
    // Generate sets of average distances $\mathbb{A}_i$ and $\mathbb{A}_j$
2   Let S_i and S_j be the feature vector sets from $\mathbb{M}_i$ and $\mathbb{M}_j$;
3    $\mathbb{A}_i \leftarrow GetSetOfAverageDistances(S_i, \delta)$;
4    $\mathbb{A}_j \leftarrow GetSetOfAverageDistances(S_j, \delta)$;
5    $\mathcal{F}_{\text{corr}} \leftarrow \phi(\mathbb{A}_i, \mathbb{A}_j)$; // Compute the correlation
6   **return** $\mathcal{F}_{\text{corr}}$;

7 **Function** GetSetOfAverageDistances(S, δ)
8   **foreach** *object* s_i *in* S **do**
9   **foreach** *object* s_j *in* S **do**
10            $D_i \leftarrow D_i \cup [\ \delta(s_i, s_j)\ ]$; // Get distances to s_i
11        $\mathbb{A} \leftarrow \mathbb{A} \cup \textbf{\textit{MEAN}}(D_i)$; // Compute the mean distance
12   **return** $\mathbb{A}$;

distance of every object to all elements in $\mathcal{D}'$, CorDiS computes and returns the correlation $\mathcal{F}_{corr}$ between both metric spaces (Line 5–6).

Generating Compatibility Factors. SOLID's mapping step obtains a correlation matrix relating all metric spaces. SOLID normalizes each row of the matrix's correlation values, which corresponds to the compatibility factors of that attribute. The compatibility factor is given by Definition 2.

Definition 2. *Compatibility Factor* (ω). *Let a and b be two attributes of $\mathcal{D}$, such that $0 \leq a, b \leq d$. The compatibility factor $\omega_{a,b}$ is the complement to one of the normalized correlation considering the metric spaces $< \mathrm{M}_a, \mathrm{M}_b >$.*

Accordingly, Ω_a is the set of compatibility factors of attribute a over each attribute from $\mathcal{D}$, and $\omega_{a,b} \in \Omega_a$ is the compatibility factor of a over attribute b. A minimum threshold parameter γ allows removing low correlation values.

Querying over Incomplete Tuples. To compare a pair of tuples $< t_q, t_i >$, $1 \leq i \leq n$, SOLID considers that zero or more attributes in each tuple may have missing values. Algorithm 2 looks for the attributes with missing values in each tuple. It accumulates the compatibility factors ω corresponding to the missing dimensions in t_i (Lines 3 to 5), adding it to the consolidated set of factors Υ_i. The algorithm returns the set of consolidated compatibility values Υ_i of t_i (Line 7). Let v be a feature vector. SOLID compares each pair of tuples using the SOLID-dist distance ψ, according to Definition 3.

Definition 3. *SOLID-dist* (ψ). *Let $< v_{a,q}, v_{a,i} >$ be a pair of feature vectors from attribute a, $1 \leq a \leq d$. Let Υ_q, Υ_i be the sets of consolidated compatibility factors of t_q and t_i, respectively. The distance between $< t_q, t_i >$ is given by:*

$$\psi(t_q, t_i) = \sum_{a=1}^{d} \varpi_{q,a} \times \varpi_{i,a} \times \delta(v_{a,q}, v_{a,i}), \tag{1}$$

where $\varpi_{q,a} \in \Upsilon_q$ and $\varpi_{i,a} \in \Upsilon_i$ are the consolidated compatibility factors for a.

Algorithm 2: SOLID (Consolidated Compatibility Factors of a Tuple)

Input : t_i: A tuple
Output: Υ: The set of consolidated compatibility factors of t_i.

1 **begin**
2   Let P_i be the set of missing attributes in t_i;
3   **foreach** *attribute a in the t_i* **do**
4   **foreach** *missing attribute p in p_i* **do**
5              $W_p[a] += \omega_{a,p}$;
6          $\Upsilon_i.\mathrm{add}(W_p[a])$;
7   **return** Υ_i;

---

4   Experimental Analysis

Material and Methods. We employed the four image datasets described in Table 2, composed of tuples of complex attributes, ensuring alignment of features when comparing tuples. We explored the Euclidean distance and the FEMs Local Binary Pattern, Edge Histogram, Haralick, Zernike, Color Layout, Scalable Color, Color Structure, Texture Spectrum, and Color Histogram. CorDiS ran over ten random combinations of FEMs, and choose those with the highest correlation sum as the features of every complex attribute. We employed the Pearson coefficient, since it presented higher correlation values than the Spearman coefficient. The source code, datasets and the detailed description of features are provided in the Git repository github.com/lsrusp/SOLID-Method.

Our competitors are: **Deletion** (baseline), which removes missing attributes; **Mean** Imputation; k**NN** Imputation [2]; kNN-**Regression** Imputation [5]; and the **Decision-Tree** Imputation [7]. We split the datasets into train and test: 70%/30% split proportion for DS-LibraGestures, and 50%/50% for the remaining datasets. The correlation threshold was $\gamma = 0.5$ for all experiments.

Quantitative Analysis. First, we analyzed *how well SOLID retrieves information from complete datasets using a query tuple with missing values.* We randomly placed missing values into the query center, with 10%, 20%, 30%, 40%, and 50% missing rates. We measured the query quality using the Jaccard coefficient between the Knn_q results posed over the complete tuples and the results of each approach. Figure 2 shows the results for different k values, where the higher the values, the better. SOLID presented the best results in most cases, being up to 3 orders of magnitude better than the Deletion approach, and up to 7.3% better than the Decision Tree Imputation, on average. SOLID is better and faster at reducing the missingness impact when performing queries.

Table 2. Datasets used in the experiments

Dataset	Imgs	n	d	Description
DS-MSTSpine [8]	540	54	10	Lumbar muscles and vertebral bodies MRIs
DS-HandPD [6]	594	66	9	Handwritten image to detect Parkinsons diseases
DS-LibraGestures [1]	4800	120	40	Images of hand gestures
DS-Letters [a]	15340	295	52	Font types of alphabetic letters

[a] https://www.kaggle.com/killen/bw-font-typefaces?select=BRLNSR

Next, we evaluated *whether SOLID reduces the error of similarity queries posed over databases containing high amounts of missing data.* We randomly inserted missing values into the database, with 10%, 20%, 30%, 40%, and 50% of missing rates. The error was computed as one minus the Jaccard coefficient between query results over the complete and the incomplete databases for each approach. Figure 3 shows the results, where the lower the values, the better.

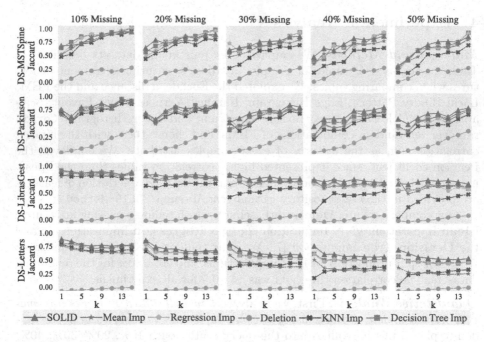

Fig. 2. Incomplete query tuple (**the higher, the better**): SOLID ties or outperforms the competitors in most experiments, for every percentage of missing values.

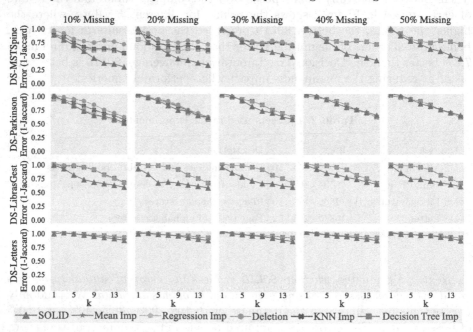

Fig. 3. Incomplete databases (**the lower the error, the better**): SOLID presented the lowest error rates in most experiments.

SOLID reduced the error in most of the analyzed scenarios because it takes advantage of correlated attributes given by the metric spaces' compatibility factors. Our method outperformed its competitors for high values of k, reducing the error by 26.3% on average when compared to the Deletion approach, and being 16.4% more precise than the Decision Tree Imputation.

Performance Analysis. We computed the average time of 100 runs of each approach over a dataset with 15% of missing values. Figure 4 shows the execution time results of Knn_q with $k = 15$, with a 50/50 train and test division. The fastest approach was the Deletion, which drops the objects with missing values. SOLID was the fastest approach among the remaining ones. It was up to 30.8 times faster than its direct competitor, the Decision Tree Imputation.

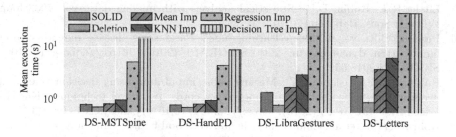

Fig. 4. Execution time of all approaches, for each dataset.

5   Conclusions

In this work, we proposed the SOLID approach to answer similarity queries over complex attributes with missing values, without discarding elements. SOLID computes the correlation of metric spaces by implementing our CorDiS correlation method. The approach gives higher importance to attributes that more likely present correlated spatial distribution of data concerning the missing attribute. Experimental results over four representative datasets show that SOLID improved the similarity search quality by 7.3% regarding its best competitor. Even in datasets with missing rates as high as 50%, SOLID reduced the error by up to 16.4%, and was up to 30.8 times faster than other approaches. Thus, SOLID has proved to be well-fitted for similarity search over incomplete data.

Acknowledgments. This research was financed in part by the Coordenação de Aperfeiçoamento de Pessoal de Nível Superior - Brasil (CAPES) - Finance Code 001, by the São Paulo Research Foundation (FAPESP, grants No. 2016/17078-0, 2018/24414-2, 2020/10902-5, 2020/07200-9), and the National Council for Scientific and Technological Development (CNPq).

References

1. Bastos, I.L.O., Angelo, M.F., Loula, A.C.: Recognition of static gestures applied to Brazilian sign language (libras). In: 28th SIBGRAPI (2015). https://doi.org/ 10.1109/SIBGRAPI.2015.26
2. Batista, G.E.A.P.A., Monard, M.C.: A study of K-nearest neighbour as an imputation method. His **87**(251–260), 48 (2002)
3. Figueroa, K., Reyes, N.: Permutation's signatures for proximity searching in metric spaces. In: Amato, G., Gennaro, C., Oria, V., Radovanović, M. (eds.) SISAP 2019. LNCS, vol. 11807, pp. 151–159. Springer, Cham (2019). https://doi.org/10.1007/ 978-3-030-32047-8_14
4. Hunt, L.A.: Missing data imputation and its effect on the accuracy of classification. In: Palumbo, F., Montanari, A., Vichi, M. (eds.) Data Science. SCDAKO, pp. 3–14. Springer, Cham (2017). https://doi.org/10.1007/978-3-319-55723-6_1
5. Little, R.J., Rubin, D.B.: Statistical analysis with missing data, vol. 793. John Wiley & Sons, Hoboken (2019)
6. Pereira, C.R., et al.: Deep learning-aided Parkinson's disease diagnosis from handwritten dynamics. In: 29th SIBGRAPI (2016). https://doi.org/10.1109/ SIBGRAPI.2016.054
7. Rahman, M.G., Islam, M.Z.: Missing value imputation using decision trees and decision forests by splitting and merging records: two novel techniques. Knowl.-Based Syst. **53**, 51–65 (2013). https://doi.org/10.1016/j.knosys.2013.08.023
8. Rohrmeier, A., et al.: Lumbar muscle and vertebral bodies segmentation of chemical shift encoding-based water-fat MRI: the reference database MyoSegmenTUM spine. BMC Musculoskelet. Disord. **20**, 152 (2019). https://doi.org/10.1186/ s12891-019-2528-x
9. Salembier, P., Sikora, T., Manjunath, B.: Introduction to MPEG-7: Multimedia Content Description Interface. John Wiley & Sons, Hoboken (2002)
10. Traina, A.J., et al.: Querying on large and complex databases by content: challenges on variety and veracity regarding real applications. Inf. Syst. **86**, 10–27 (2019). https://doi.org/10.1016/j.is.2019.03.012
11. Zabot, G.F., Cazzolato, M.T., Scabora, L.C., Traina, A.J.M., Traina-Jr., C.: Efficient indexing of multiple metric spaces with spectra. In: 2019 IEEE ISM, pp. 169–1697 (2019). https://doi.org/10.1109/ISM46123.2019.00038

Similarity Between Points in Metric Measure Spaces

Evgeny Dantsin[✉] and Alexander Wolpert[✉]

Roosevelt University, Chicago, IL, USA
{edantsin,awolpert}@roosevelt.edu

Abstract. This paper is about similarity between objects that can be represented as points in metric measure spaces. A metric measure space is a metric space that is also equipped with a measure. For example, a network with distances between its nodes and weights assigned to its nodes is a metric measure space. Given points x and y in different metric measure spaces or in the same space, how similar are they? A well known approach is to consider x and y similar if their neighborhoods are similar. For metric measure spaces, similarity between neighborhoods is well captured by the Gromov-Hausdorff-Prokhorov distance, but it is NP-hard to compute this distance even in quite simple cases. We propose a tractable alternative: the *radial distribution distance* between the neighborhoods of x and y. The similarity measure based on the radial distribution distance is coarser than the similarity based on the Gromov-Hausdorff-Prokhorov distance but much easier to compute.

Keywords: Metric measure space · Gromov-Hausdorff-Prokhorov distance · Radial distribution

1 Introduction

A *metric measure space* is a metric space that is also equipped with a measure. Such spaces play an important role in geometry, especially after Gromov's works [8], and they have proven to be useful in other areas of mathematics, for example, in optimization theory [2] and in probability theory [4]. Metric measure spaces are also used to model real-world systems and processes, for example, in image recognition [11], in genetics [15], in machine learning [3], etc.

A natural example of a metric measure space is given by a connected graph G in which all vertices and edges are labeled with numbers: the number assigned to a vertex is its "weight" and the number assigned to an edge is its "length". The corresponding metric space is formed by the set of all vertices with the shortest path metric in G: the distance between two vertices is the length of a shortest path between them. A measure on this space is defined on all subsets of the vertices: the measure of a subset A is the total weight of all vertices of A. The weights and lengths can be interpreted in various ways. For example, if G is a communication network, then the weight of a vertex can describe the traffic at

© Springer Nature Switzerland AG 2020
S. Satoh et al. (Eds.): SISAP 2020, LNCS 12440, pp. 177–184, 2020.
https://doi.org/10.1007/978-3-030-60936-8_14

this vertex. Another example: if G is a propagation network in epidemic models, then the weight can be the number of infected individuals.

Consider objects that can be modeled by points in metric measure spaces: suppose one object is represented by a point x in a space $\mathcal{X}$ and another object is represented by a point y in a space $\mathcal{Y}$. How similar are these objects? How can we measure similarity between them if the only information we have is the pairs $(\mathcal{X}, x)$ and $(\mathcal{Y}, y)$? Such pairs are called *rooted metric measure spaces* or *rooted mm spaces* for short, see Sect. 2 for precise definitions. In this paper we address the question of similarity between objects modeled by rooted mm spaces.

The most obvious type of similarity between $(\mathcal{X}, x)$ and $(\mathcal{Y}, y)$ is an *isomorphism* between them, which means that there is a bijection from $\mathcal{X}$ to $\mathcal{Y}$ that maps x to y and preserves the metric and measure. This is an "all or nothing" measure of similarity: any two rooted mm spaces are either similar or not. Clearly, this measure is not a good solution for applications because real-world objects, like social, biological, or technological networks, are very rarely, if ever, isomorphic to one another.

Can we improve the isomorphism-based approach to make it more flexible? How could we measure to what extent $(\mathcal{X}, x)$ and $(\mathcal{Y}, y)$ look isomorphic? The concept of "approximate isomorphism" between rooted mm spaces can be implemented using the idea proposed by Edwards [6] and Gromov [7]. To compare $(\mathcal{X}, x)$ and $(\mathcal{Y}, y)$, we embed $\mathcal{X}$ and $\mathcal{Y}$ into another metric measure space $\mathcal{Z}$ and compare their images in $\mathcal{Z}$. More exactly, we take embeddings f and g that preserve the metric and measure and compare the images $f(\mathcal{X})$ and $g(\mathcal{Y})$ in $\mathcal{Z}$. We consider $(\mathcal{X}, x)$ and $(\mathcal{Y}, y)$ *similar* if their images are close to each other in the following sense:

- the point $f(x)$ is close to the point $g(y)$ in the space $\mathcal{Z}$;
- the set of points of $f(\mathcal{X})$ is close to the set of points of $g(\mathcal{Y})$ in the space $\mathcal{Z}$;
- the measures induced by f and g in $\mathcal{Z}$ are close to one another.

The second condition is formalized using the *Hausdorff distance* and the third condition is formalized using the *Lévy-Prokhorov distance*, see Sect. 2. Taking the infimum over all possible spaces $\mathcal{Z}$ and embeddings f and g, we obtain a distance function on rooted mm spaces called the *Gromov-Hausdorff-Prokhorov distance* (the *GHP distance* for short).

Both the isomorphism-based similarity measure and the GHP-based similarity measure have the following disadvantage for applications. Most real-world systems have the distance decay effect, also called the gravity model, which is often expressed as "all things are related, but near things are more related than far things". For example, when comparing points x and y in metric spaces, the role of their local neighborhoods is more important than the role of points that are far away from x and y. However, neither the isomorphism approach nor the GHP distance capture this effect: all points are considered equally important, independently of their distance from x and y.

This disadvantage is eliminated using the distance defined in [1]. Loosely speaking, this distance between two rooted mm spaces combines the GHP distance with an exponential decay: a point is taken into account with a weight that exponentially decreases with increasing its distance from the root. We call it the *neighborhood-based* distance and describe it in Sect. 2.

Intuitively, the neighborhood-based distance is the best approach to capture similarity between points in metric measure spaces. To put this distance to work in practical applications, we need to compute it efficiently. However, it is NP-hard to compute the neighborhood-based distance, which follows from [12]. Moreover, under standard complexity-theoretic assumptions, it is not possible to approximate it with a reasonable factor in polynomial time [16].

In Sect. 3, we propose a tractable alternative to the neighborhood-based distance: namely, we define the *radial distribution distance* between rooted mm spaces. This distance can be viewed as a coarser variant of the neighborhood-based distance or, more exactly, as a lower bound on the neighborhood-based distance. The advantage of the radial distribution distance is that it can be computed efficiently: a straightforward algorithm that computes this distance between finite rooted mm spaces takes time quasilinear in the total number of points.

What is the idea of radial distribution distance? First, we view a point x in a metric measure space $\mathcal{X}$ as the center of a ball of radius r around x. This ball has its own measure (its "mass") and we consider how such masses change with increasing r. Second, we consider this change of masses with an exponential distance decay, which means that the "contribution" of points exponentially decreases with increasing their distance from x. The radial distribution distance between $(\mathcal{X}, x)$ and $(\mathcal{Y}, y)$ is basically the distance between two functions that describe the change of masses around x in $\mathcal{X}$ and around y in $\mathcal{Y}$.

2 Neighborhood-Based Distance

A metric measure space is usually defined as a complete separable metric space with a Borel measure on this metric space. However, in this paper, we deal with only compact metric spaces and finite measures. Therefore, to simplify terminology, we use the term "metric measure space" to refer to this restricted case.

Definition 1 (mm space). *Let X be a set and d be a metric on X such that (X, d) is a compact metric space. Let μ be a Borel measure on this metric space. The triplet (X, d, μ) is called a* metric measure space *(an mm space for short).*

We use the following notation for balls in metric spaces: for every number $r \in [0, \infty)$ and every point $x \in X$,

- $B_r(x) = \{y \in X \mid d(x, y) < r\}$ is an *open ball* of radius r around x;
- $\overline{B}_r(x) = \{y \in X \mid d(x, y) \leq r\}$ is a *closed ball* of radius r around x.

For every number $r > 0$ and every nonempty subset $S \subseteq X$, the r-*neighborhood* of S, denoted by $N_r(S)$, is the union of open balls of radius r around the points of S. Let S_1 and S_2 be nonempty subsets of X. The *Hausdorff distance* between them is defined by

$$d_H(S_1, S_2) = \inf\{r \in [0, \infty) \mid S_1 \subseteq N_r(S_2) \text{ and } S_2 \subseteq N_r(S_1)\}.$$

Given two subsets of points in a metric space, the Hausdorff distance shows how close they are. Given two measures on a metric space, how close are they? This question is answered by the *Lévy-Prokhorov distance*, a measure-theoretic analogue of the Hausdorff distance, defined as follows. Let μ_1 and μ_2 be finite Borel measures on a metric space (X, d). Let $\mathcal{B}$ be the Borel σ-algebra on (X, d). The *Lévy-Prokhorov distance* (sometimes called the *Prokhorov distance*) between these measures, denoted by π, is given by

$$\pi(\mu_1, \mu_2) = \inf\{r \in [0, \infty) \mid \mu_1(S) \leq \mu_2(N_r(S)) + r \text{ and}$$
$$\mu_2(S) \leq \mu_1(N_r(S)) + r \quad \text{for all } S \in \mathcal{B}\}.$$

Measuring similarity between points in different mm spaces or in the same space, it is convenient to deal with *rooted mm spaces* (sometimes called *pointed mm spaces*).

Definition 2 (rooted mm space). *A* rooted mm space *is a pair* $(\mathcal{X}, o)$ *where* $\mathcal{X}$ *is an mm space and o is a designated point in $\mathcal{X}$ called the* origin *of $\mathcal{X}$.*

A general approach to comparing rooted mm spaces was outlined in Sect. 1. This approach combines the idea of Gromov-Hausdorff distance with the idea of Lévy-Prokhorov distance. The combination was introduced in [13] and has slight variations. The following definition is the version from [9].

Definition 3 (GHP distance). *Let* $(\mathcal{X}_1, o_1)$ *and* $(\mathcal{X}_2, o_2)$ *be rooted mm spaces with* $\mathcal{X}_1 = (X_1, d_1, \mu_1)$ *and* $\mathcal{X}_2 = (X_2, d_2, \mu_2)$. *The* Gromov-Hausdorff-Prokhorov *distance (the* GHP distance *for short), denoted* d_{GHP}, *is defined by*

$$d_{GHP}((\mathcal{X}_1, o_1), (\mathcal{X}_2, o_2)) =$$
$$\inf_{\mathcal{Y}, f_1, f_2}\{d(o_1, o_2) + d_H(f_1(X_1), f_2(X_2)) + \pi\left(\mu_1 \circ f_1^{-1}, \mu_2 \circ f_2^{-1}\right)\}$$

where the infimum is taken over all mm spaces $\mathcal{Y}$ and all measurable isometries $f_1 : \mathcal{X}_1 \to \mathcal{Y}$ *and* $f_2 : \mathcal{X}_2 \to \mathcal{Y}$. *The distances d, d_H, and π denote respectively the distance, the Hausdorff distance, and the Lévy-Prokhorov distance in $\mathcal{Y}$. The measures $\mu_1 \circ f_1^{-1}$ and $\mu_2 \circ f_2^{-1}$ are the push-forward measures.*

As noted in Sect. 1, the GHP distance has the disadvantage that d_{GHP} does not capture the distance decay effect occurring in most real-world systems. This disadvantage is eliminated in the distance defined in [1]. We define a simplified version of this distance below and call it *neighborhood-based distance*. Its idea can be informally described as follows. When comparing rooted mm spaces $(\mathcal{X}_1, o_1)$ and $(\mathcal{X}_2, o_2)$, we consider the restrictions of $\mathcal{X}_1$ and $\mathcal{X}_2$ to closed balls of radius r around o_1 and o_2. For every radius r, we consider the GHP distance between the corresponding restrictions and sum up these distances over all values of r with an exponential decrease when r increases.

Let $\overline{B}$ be a closed ball in an mm space $\mathcal{X} = (X, d, \mu)$. This ball, along with the metric and measure obtained by restricting d and μ to $\overline{B}$, form the mm space called the *restriction* of $\mathcal{X}$ to $\overline{B}$.

Definition 4 (neighborhood-based distance). *Let $(\mathcal{X}_1, o_1)$ and $(\mathcal{X}_2, o_2)$ be rooted mm spaces. For every number $r \in [0, \infty)$, let δ_r denote the GHP distance between the restriction of $\mathcal{X}_1$ to $\overline{B}_r(o_1)$ and the restriction of $\mathcal{X}_2$ to $\overline{B}_r(o_2)$. The* neighborhood-based distance, *denoted d_{nb}, is defined by*

$$d_{nb}((\mathcal{X}_1, o_1), (\mathcal{X}_2, o_2)) = \int_0^\infty e^{-r} \delta_r \, dr$$

It follows from the definition that

$$d_{nb}((\mathcal{X}_1, o_1), (\mathcal{X}_2, o_2)) = 0$$

for isomorphic $(\mathcal{X}_1, o_1)$ and $(\mathcal{X}_2, o_2)$ even if they are different. Therefore, d_{nb} is not a metric but, as shown in [1], d_{nb} is a pseudometric.

3   Radial Distribution Distance

Can we compute the neighborhood-based distance d_{nb} efficiently? Note that an efficient algorithm for computing d_{nb} could also be used to compute the Gromov-Hausdorff distance efficiently. However, as shown in [10,12], it is NP-hard to compute the Gromov-Hausdorff distance for finite metric spaces (see Proposition 3.16 in [12]). Moreover, under standard complexity-theoretic assumptions, there is no polynomial-time approximation algorithm with a reasonable factor for computing this distance [16].

In this section, we define another distance between rooted mm spaces called the *radial distribution distance* and denoted by d_{rd}. On the one hand, d_{rd} is coarser than d_{nb}, more exactly, d_{rd} is a lower bound on d_{nb}. On the other hand, d_{rd} can be computed efficiently: computing the radial distribution distance between finite rooted mm spaces takes time quasilinear in the total number of points.

Definition and Properties. Consider rooted mm spaces $(\mathcal{X}_1, o_1)$ and $(\mathcal{X}_2, o_2)$ where $\mathcal{X}_1 = (X_1, d_1, \mu_1)$ and $\mathcal{X}_2 = (X_2, d_2, \mu_2)$. To define the *radial distribution distance* between them, we first define the following functions m_1 and m_2 from $[0, \infty)$ to itself: for every number $r \in [0, \infty)$,

$$m_1(r) = \mu_1\left(\overline{B}_r(o_1)\right) = \mu_1\left(\{x \in X_1 \mid d_1(x, o_1) \leq r\}\right)$$
$$m_2(r) = \mu_2\left(\overline{B}_r(o_2)\right) = \mu_2\left(\{x \in X_2 \mid d_2(x, o_2) \leq r\}\right)$$

That is, $m_1(r)$ is the measure (we could call it "mass" or "weight") of the ball of radius r around the origin o_1 and, similarly, for $m_2(r)$. The functions are non-decreasing and bounded. Also, m_1 and m_2 are càdlàg functions, i.e., they are right continuous with left limits, see Lemma 2.8 in [1]. Therefore, the distance function in the definition below is well defined.

Definition 5 (radial distribution distance). *Let* $(\mathcal{X}_1, o_1)$ *and* $(\mathcal{X}_2, o_2)$ *be rooted mm spaces. The* radial distribution distance, *denoted* d_{rd}, *is defined by*

$$d_{rd}((\mathcal{X}_1, o_1), (\mathcal{X}_2, o_2)) = \int_0^\infty e^{-r} |m_1(r) - m_2(r)| \, dr$$

We use the term "radial distribution distance" keeping in mind radial distribution functions from physics where they are used to measure the probability of finding a particle at distance r from a certain "origin" particle. In the setting of metric measure spaces, a radial distribution function describes how the total "mass" ("weight") of points at distance r from a center x changes when r increases. The radial distribution distance is basically a combination of the L^1 distance between the cumulative versions of the radial distribution functions and an exponential distance decay. Theorems 1 and 2 show basic properties of the radial distribution distance.

Theorem 1. *The function* d_{rd} *is a pseudometric on the set of rooted mm spaces.*

Proof. It is obvious that d_{rd} is a distance function. The triangle inequality for d_{rd} can be seen from the following two facts:

- The functions $e^{-r} m_1(r)$ and $e^{-r} m_2(r)$ are measurable functions on $[0, \infty)$ such that the integrals $\int_0^\infty e^{-r} m_1(r) \, dx$ and $\int_0^\infty e^{-r} m_2(r) \, dx$ are finite.
- The distance function

$$d(f_1, f_2) = \int_0^\infty |f_1(x) - f_2(x)| \, dx$$

  is the L^1 metric on the set of measurable functions f such that the integral $\int_0^\infty |f(x)| \, dx$ is finite.

$\square$

Theorem 2. *For all rooted mm spaces* $(\mathcal{X}_1, o_1)$ *and* $(\mathcal{X}_2, o_2)$,

$$d_{rd}((\mathcal{X}_1, o_1), (\mathcal{X}_2, o_2)) \le d_{nb}((\mathcal{X}_1, o_1), (\mathcal{X}_2, o_2)). \tag{1}$$

Proof. Let $\overline{B}_r(o_1)$, $\overline{B}_r(o_2)$, and δ_r be the same as in Definition 4. Let μ_1 and μ_2 be the measures in $\mathcal{X}_1$ and $\mathcal{X}_2$ respectively. We prove the following inequality that implies claim (1):

$$|\mu_1(\overline{B}_r(o_1)) - \mu_2(\overline{B}_r(o_2))| \le \delta_r. \tag{2}$$

By the definition of the GHP distance, we have

$$\delta_r \ge \pi(\mu_1 \circ f_1^{-1}, \mu_2 \circ f_2^{-1}) \tag{3}$$

where f_1 and f_2 are arbitrary measurable isometries of the corresponding restrictions to an arbitrary mm space $\mathcal{Y}$ and π is the Lévy-Prokhorov distance in $\mathcal{Y}$. By the definition of π,

$$\pi(\mu_1 \circ f_1^{-1}, \mu_2 \circ f_2^{-1}) = \gamma \tag{4}$$

where γ is the infimum of all $\epsilon \geq 0$ such that

$$\begin{cases} \mu_1\left(\overline{B}_r(o_1)\right) \leq \mu_2\left(\overline{B}_{r+\epsilon}(o_2)\right) + \epsilon \\ \mu_2\left(\overline{B}_r(o_2)\right) \leq \mu_1\left(\overline{B}_{r+\epsilon}(o_1)\right) + \epsilon \end{cases}$$

Both inequalities above hold if we take

$$\epsilon = |\mu_1(\overline{B}_r(o_1)) - \mu_2(\overline{B}_r(o_2))|$$

and, therefore, we have

$$\gamma \geq |\mu_1(\overline{B}_r(o_1)) - \mu_2(\overline{B}_r(o_2))|. \tag{5}$$

Now, combining (3)–(5), we obtain inequality (2). □

The theorem above shows that d_{rd} is a lower bound on d_{nb}. This bound is strict, which can be seen from the following simple example. Consider a rooted mm space on a set of three points: $\{a, b, c\}$ where a is the origin. The distance between any two points is 1. The measure assigned to each point is 1. Consider another rooted mm space that differs from the first one in only the measure of b and c: in the second space, the measure of b is 0.5 and the measure of c is 1.5. It is easy to see that the radial distribution distance between these spaces is 0, while the neighborhood distance is not zero.

Computing the Radial Distribution Distance. There is a straightforward algorithm that computes the radial distribution distance between finite rooted mm spaces efficiently, namely taking $O(n \log n)$ steps where n is the number of points in the input spaces. However, due to space limitation for this paper, we have to omit its description here.

Extension for Feature Vectors. The radial distribution distance d_{rd} can be used to measure similarity between points in a metric space if points are described with a single feature: a value of this feature for a given point is viewed as the point's "weight". However, it is more typical that a point is described by a feature vector rather than a single feature. Each component of the vector corresponds to a measure on the metric space. For example, consider a recommender system for movies that uses item-item collaborative filtering. A metric on movies is based on the similarity between them calculated using people's ratings. In addition to the metric, each movie is described by a feature vector that can include, for example, the number of reviews, budget, box office, etc. [14].

How can we compare points in a metric space if points are described using feature vectors? Suppose a feature vector consists of k components that correspond to measures $\mu_1, \ldots, \mu_k$. Each measure μ_i determines the radial distribution distance $d_{rd}^{(i)}$ and we can consider their sum

$$d_{rd}^* = d_{rd}^{(1)} + \ldots + d_{rd}^{(k)}. \tag{6}$$

The value $d_{rd}^*((\mathcal{X}, x), (\mathcal{Y}, y))$ essentially shows the distance between the neighborhoods of x and y if we compare points by their feature vectors. Note that the sum of pseudometrics is also a pseudometric. Also note that instead of the sum in (6), we could take any other norm, for example, the Euclidean norm or the maximum.

References

1. Abraham, R., Delmas, J.-F., Hoscheit, P.: A note on the Gromov-Hausdorff-Prokhorov distance between (locally) compact metric measure spaces. Electron. J. Probab. **18**(14), 1–21 (2013)
2. Ambrosio, L., Gigli, N., Savaré, G.: Gradient Flows. In: Metric Spaces and in the Space of Probability Measures. Lectures in Mathematics. Birkhäuser (2005)
3. Arjovsky, M., Chintala, S., Bottou, L.: Wasserstein generative adversarial networks. In: Proceedings of the 34th International Conference on Machine Learning, ICML 2017, pp. 214–223 (2017)
4. Athreya, S., Löhr, W., Winter, A.: Invariance principle for variable speed random walks on trees. Ann. Probab. **45**(2), 625–667 (2017)
5. Burago, D., Burago, Y., Ivanov, S.: A Course in Metric Geometry, volume 33 of Graduate Studies in Mathematics. American Mathematical Society (2001)
6. Edwards, D.A.: The structure of superspace. In: Studies in Topology, pp. 121–133. Academic Press (1975)
7. Gromov, M.: Groups of polynomial growth and expanding maps (with an appendix by Jacques tits). Publications Mathématiques de l'IHÉS **53**, 53–78 (1981)
8. Gromov, M.: Metric structures for Riemannian and non-Riemannian spaces, volume 152 of Progress in Mathematics. Birkhäuser (1999). Based on the 1981 French original
9. Lei, T.: Scaling limit of random forests with prescribed degree sequences. Bernoulli **25**(4A), 2409–2438 (2019)
10. Mémoli, F.: On the use of Gromov-Hausdorff distances for shape comparison. In: Proceedings of the Symposium on Point Based Graphics, Prague 2007, pp. 81–90 (2007)
11. Mémoli, F.: Gromov-Wasserstein distances and the metric approach to object matching. Found. Comput. Math. **11**(4), 417–487 (2011)
12. Mémoli, F., Smith, Z., Wan, Z.: Gromov-Hausdorff distances on p-metric spaces and ultrametric spaces. ArXiv e-prints (2019)
13. Miermont, G.: Tessellations of random maps of arbitrary genus. Annales Scientifiques de L'École Normale Supérieure **42**(5), 725–781 (2009)
14. Ning, X., Desrosiers, C., Karypis, G.: A comprehensive survey of neighborhood-based recommendation methods. In: Recommender Systems Handbook, pp. 37–76. Springer, Boston (2015)
15. Rabadán, R., Blumberg, A.: Topological Data Analysis for Genomics and Evolution: Topology in Biology. Cambridge University Press (2019)
16. Schmiedl, F.: Computational aspects of the Gromov-Hausdorff distance and its application in non-rigid shape matching. Discrete Comput. Geom. **57**(4), 854–880 (2017)

High-Dimensional Data and Intrinsic Dimensionality

GTT: Guiding the Tensor Train Decomposition

Mao-Lin Li[1]([✉]), K. Selçuk Candan[1], and Maria Luisa Sapino[2]

[1] Arizona State University, Tempe, AZ, USA
{maolinli,candan}@asu.edu
[2] University of Turin, Turin, Italy
mlsapino@di.unito.it

Abstract. The demand for searching, querying multimedia data such as image, video and audio is omnipresent, how to effectively access data for various applications is a critical task. Nevertheless, these data usually are encoded as *multi-dimensional* arrays, or *Tensor*, and traditional data mining techniques might be limited due to the *curse of dimensionality*. Tensor decomposition is proposed to alleviate this issue, commonly used tensor decomposition algorithms include CP-decomposition (which seeks a diagonal core) and Tucker-decomposition (which seeks a dense core). Naturally, Tucker maintains more information, but due to the denseness of the core, it also is subject to exponential memory growth with the number of tensor modes. Tensor train (*TT*) decomposition addresses this problem by seeking a sequence of three-mode cores: but unfortunately, currently, there are no guidelines to select the decomposition sequence. In this paper, we propose a *GTT* method for guiding the tensor train in selecting the decomposition sequence. GTT leverages the data characteristics (including number of modes, length of the individual modes, density, distribution of mutual information, and distribution of entropy) as well as the target decomposition rank to pick a decomposition order that will preserve information. Experiments with various data sets demonstrate that GTT effectively guides the TT-decomposition process towards decomposition sequences that better preserve accuracy.

Keywords: Low-rank embedding · Tensor train decomposition

1 Introduction

Tensors are commonly used to represent multi-dimensional sets and tensor decomposition operations, such as CP [5,10] and Tucker [24] form the basis of

This work has been supported by: NSF grants #1633381, #1909555, #1629888, #2026860, #1827757, DOD grant W81XWH-19-1-0514, a DOE CYDRES grant, and a European Commission grant #690817. Experiments for the paper were conducted using NSF testbed: "Chameleon: A Large-Scale Re-configurable Experimental Environment for Cloud Research".

S. Satoh et al. (Eds.): SISAP 2020, LNCS 12440, pp. 187–202, 2020.
https://doi.org/10.1007/978-3-030-60936-8_15

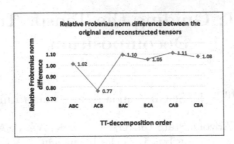

Fig. 1. Effect of the decomposition order on the accuracy for a 3-mode tensor from the Wisconsin Diagnostic Breast Cancer data [9]: $ID(mode_A)$, $Diagnosis(mode_B)$ and $Radius(mode_C)$. See Sect. 6 for more details

many dimensionality reduction techniques for multi-modal data sets to support similarity search and retrieval [4,11]. In the Tucker-decomposition, for example, given a tensor with d modes, each entry in the resulting $r_1 \times r_2 \times \ldots \times r_d$ dense core encodes the strength of the d-way relationship among the groups consisting of elements of the individual modes.

Tucker decomposition has been shown to be highly effective in many applications [4,11,18,25], but due to the denseness of the core, it also is subject to exponential memory growth with the number of tensor modes. The tensor train (TT) decomposition addresses this problem, by seeking a sequence of 3-mode cores [23]: while, collectively, this sequence (or "train") of cores capture the high-modal information, they require fewer resources. Consequently, the TT-decomposition has been used in various applications of similarity search and retrieval, including deep learning [6,21], crowdsourcing [16], and recommendation systems [22].

1.1   Impact of the Decomposition Order

One critical challenge with the TT-decomposition, however, is the fact that finding an optimal TT representation is non-trivial [27]. Figure 1 illustrates this issue: given a 3-mode ($mode_A$: ID, $mode_B$: $Diagnosis$ and $mode_C$: $Radius$) tensor from the Wisconsin Diagnostic Breast Cancer data set in UCI Machine Learning Repository [9]; the figure compares the relative Frobrenius norm difference (ratio of the norm of the difference tensor to the norm of the original tensor) between the input tensor and the reconstructed tensor for different TT-decomposition orders. As the figure shows, the ordering of the TT-decomposition has a significant impact on the ability of the final representation in preserving the original information: in this case, the order ACB is $(0.77 - 1.02)/1.02 = 24.5\%$ better than the closest alternative.

1.2   Our Contributions

In this paper, we propose a novel approach for *guiding the tensor train* (GTT) in selecting the mode sequence for tensor train decompositions:

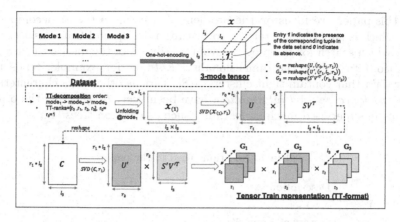

Fig. 2. TT-decomposition for converting a 3-mode tensor $\mathcal{X}^{l_1 \times l_2 \times l_3}$

- we identify significant relationships among various data characteristics and the accuracies of different tensor train decomposition orders;
- we propose four order selection strategies, (a) *aggregate mutual information (AMI)*, (b) *path mutual information (PMI)*, (c) *inverse entropy (IE)*, and (d) *number of parameters (NP)*, for tensor train decomposition; and
- we show that good tensor train orders can be selected through a *hybrid (HYB)* strategy that takes into account multiple characteristics of the data.

Experiments reported in Sect. 6 show that the proposed HYB strategy provides an effective order selection strategy, without any additional decomposition time overhead.

2   Related Work

Tensor decomposition has been shown to be effective in multi-aspect data analysis and similarity search by capturing high-order structure in high-dimensional data [4,11,18]. However, a major challenge is its high computational complexity and large memory overhead [12,14,15]. Tensor-train decomposition [23] provides a memory-saving representation called *TT-format*, with linear space complexity (see Fig. 2). TNrSVD [2] adapts the randomized SVD to implement TT-decomposition, and FastTT [19] computes the TT-decomposition of a sparse tensor by its sparsity. However, as discussed in the introduction, TT-decomposition involves strictly sequential multi-linear products over latent cores and this makes it difficult to search for best TT representation for a given tensor. [20] and [27] extended TT-decomposition by adding auxiliary variables to obtain an alternative data structure, Tensor Ring (*TR*), which provides *circular dimensional permutation invariance* – the sequence can be shifted circularly without changing the result [1], however, it does not eliminate the need to pick a (*circularly-arranged*) permutation of modes.

In this paper, we focus on the problem of selecting a tensor decomposition order, which is superficially related to feature ordering and feature selection [3, 17,26], which search for the most relevant attributes of the data set (for a given application) to reduce the dimensionality. *Entropy*, for example, tends to be low for data that contain tight clusters [7,8]. Various other data characteristics, such as *variance, mutual information*, have been used for selecting the order of decisions in supervised machine learning, such as decision trees [4].

Table 1. Notations used in the paper

	Description
$\mathcal{X}$	A tensor
$\rho_{\mathcal{X}}$	Density of tensor $\mathcal{X}$
$\mathcal{X}_{(i)}$	A mode-i unfolding matrix of a tensor $\mathcal{X}$
l_i	Length of mode i in a tensor
m_i	The mode i in a tensor
r_i	TT-rank of mode i
r_{max}	Given maximum TT-rank
X	A discrete random variable with possible values $\{x_1, \ldots, x_n\}$
U	Left factor matrix
S	Singular matrix
V	Right factor matrix
G_i	3-mode core for mode i of TT-decomposition
H_i	Shannon entropy of random variable for mode i
$H_{i\|j}$	Conditional entropy for mode i given mode j
$H_{(i,j)}$	Averaged conditional entropy for $H_{i\|j}$ and $H_{j\|i}$
$MI_{(i,j)}$	Mutual information between mode i and mode j

Algorithm 1. TT-SVD (adapted from [23])

Input:
A d-mode tensor $\mathcal{X} \in \mathbb{R}^{l_1 \times l_2 \times \cdots \times l_d}$; a target tt-rank, r; a permutation, Π
Output:
TT-format with TT-cores $G_1, G_2, \ldots G_d$.
- $numel(C)$: number of elements in C.
- $reshape(A, [d_1, \ldots, d_k])$: reshape a tensor A into shape $d_1 \times d_2 \times \cdots \times d_k$.
- $min(a, b)$: return a if $a < b$, else return b.

1: **procedure** TT-SVD($\mathcal{X}, r, \Pi$)
2:    Initialize $r_0 = r_d = 1$, $C = \mathcal{X}$.
3:    **for** $k \leftarrow 1$ to $d - 1$ **do**
4:       $C \leftarrow reshape(C, [r_{k-1} \times l_{\pi_k}, \frac{numel(C)}{r_{k-1} \times l_{\pi_k}}])$.
5:       $U, S, V = SVD(C, r_k = min(r_{max}, l_{\pi_k}))$.
6:       $G_k \leftarrow reshape(U, [r_{k-1}, l_{\pi_k}, r_k])$.
7:       $C \leftarrow SV^T$.
8:    **end for**
9:    $G_d \leftarrow C$.
10:    return TT-format with TT-cores $G_1, \ldots, G_d$.
11: **end procedure**

3 Preliminaries

Table 1 summarizes the key notations. Intuitively, the tensor model maps a schema with d attributes to a d-modal array (where each potential tuple is a tensor cell). TT-decomposition [23] is obtained by applying a sequence of singular value decompositions (SVD) to approximate the original tensor: given (i) an input tensor, $\mathcal{X} \in \mathbb{R}^{l_1 \times l_2 \times \cdots \times l_d}$, (ii) a permutation, $\Pi = \langle \pi_1, \pi_2, \ldots, \pi_d \rangle$, of modes, and (iii) a sequence of decomposition ranks, $\langle r_0, r_1, r_2, \ldots, r_d \rangle$, where $r_0 = r_d = 1$, the tensor train decomposition approximates the input tensor, $\mathcal{X}$, with a sequence of tensor cores $G_k \in \mathbb{R}^{r_{k-1} \times l_{\pi_k} \times r_k}$, $k = 1 \ldots d$, where $\mathcal{X} \approx \hat{\mathcal{X}}_\Pi = G_1 \cdot G_2 \cdots G_d$. In this paper, without loss of generality, we will assume that all ranks (except $r_0 = r_d = 1$) have the same value, r. Note that, while there are several non-parametric decomposition techniques, such as [13] which can learn also the appropriate rank, this is outside of the scope of this paper – most tensor decomposition (in fact most latent semantic search) literature takes the number of latent-semantics as input. Algorithm 1 presents the pseudocode and Fig. 2 visualizes the TT-SVD process for a 3-mode tensor $\mathcal{X} \in \mathbb{R}^{l_1 \times l_2 \times l_3}$.

Accuracy. To evaluate the accuracy, we use the Frobenius norm of the difference between mode-i unfolding $\mathcal{X}_{(i)}$, of the original tensor and mode-i unfolding $\hat{\mathcal{X}}_{\Pi_{(i)}}$ of the reconstructed tensor, $\hat{\mathcal{X}}_\Pi$: $Error(\hat{\mathcal{X}}_\Pi, \mathcal{X}) = \|\mathcal{X}_{(i)} - \hat{\mathcal{X}}_{\Pi_{(i)}}\|_{Frob}$. This term gives the same value independently of the mode i selected for matrix unfolding.

4 Problem Statement

In this paper, we aim to seek a decomposition sequence that minimizes the reconstruction error:

Problem 1 (Tensor Train Decomposition Sequence Selection). *Let us be given a d-dimensional tensor, $\mathcal{X} \in \mathbb{R}^{l_1 \times l_2 \times \cdots \times l_d}$, and a target decomposition rank, r. Our goal is to find a permutation, $\Pi = \langle \pi_1, \pi_2, \ldots, \pi_d \rangle$, which minimizes the approximation error; i.e., $argmin_{\Pi \in \mathfrak{P}} \left(Error(\hat{\mathcal{X}}_\Pi, \mathcal{X}) \right)$, where $\mathfrak{P}$ denotes the set of all possible d! permutations.*

5 GTT: Guiding the Tensor Train

In this paper, we propose a novel approach to *guide the tensor trains* (GTT) in selecting the decomposition sequence. GTT leverages the various characteristics/statistics of the input data tensor (sparse or dense) to identify and recommend a mode ordering for the TT-decomposition process.

5.1 Data Characteristics

In this subsection, we describe data characteristics, or *features*, relevant for tensor train mode sequence selection. Note that these data characteristics are very general and can be computed for any data set with categorical entries. We leave the extension to non-categorical data to future work.

Mode Length. Given a d-mode tensor, $\mathcal{X} \in \mathbb{R}^{l_1 \times l_2 \times \cdots \times l_d}$, we compute the average of the mode lengths, along with the absolute and relative standard deviations:

$$\mu_{length}(\mathcal{X}) = average(l_1, l_2, \ldots, l_d), \tag{1}$$

$$\sigma_{length}(\mathcal{X}) = stdev(l_1, l_2, \ldots, l_d), \tag{2}$$

$$\phi_{length}(\mathcal{X}) = \sigma_{length}/\mu_{length}. \tag{3}$$

Intuitively, the larger the lengths of the modes, the larger will be the number of parameters to be sought. The absolute and relative standard deviations indicate how discriminative the mode length feature is in the given tensor.

Mode Entropy. Given a data set with d modes, let X_i be a discrete random variable with possible values $\{x_1, \ldots, x_{n_i}\}$ for mode i. Given this, we can compute the Entropy for mode i as $H_i = H(X_i) = -\sum_{j=1}^{n_i} p_i(j) \log_2 p_i(j)$, where $p_i(j)$ represents the probability that x_j occurs in the given mode i. Given the entropy statistics for each mode of the tensor, we then compute the average and standard deviation statistics as follows:

$$\mu_{entropy}(\mathcal{X}) = average(H_1, H_2, \ldots, H_d), \tag{4}$$

$$\sigma_{entropy}(\mathcal{X}) = stdev(H_1, H_2, \ldots, H_d), \tag{5}$$

$$\phi_{entropy}(\mathcal{X}) = \sigma_{entropy}/\mu_{entropy}. \tag{6}$$

Intuitively, entropy indicates how easy it is to have a low-rank approximation of a tensor along a given mode and the absolute and relative standard deviations indicate how discriminative the mode entropy feature is.

Tensor Density. Note that the above definition of entropy is meaningful especially for sparse tensors[1]. Therefore, we also compute a *density* statistic. Given a d-mode tensor $\mathcal{X} \in \mathbb{R}^{l_1 \times l_2 \times \cdots \times l_d}$, we compute the density ρ of $\mathcal{X}$ as

$$\rho(\mathcal{X}) = \frac{\# \ of \ nonzero \ values \ in \ \mathcal{X}}{l_1 \times l_2 \times \cdots \times l_d}. \tag{7}$$

Pairwise Average Conditional Entropy. The tensor train representation links consecutive modes in the sequence; therefore, pairwise statistics may also be needed. Given a data set with a d-mode tensor, let X_i be a discrete random variable with possible values $\{x_1, \ldots, x_{n_i}\}$ for mode i. The conditional entropy of X_i given X_j is defined as:

$$H_{i|j} = H(X_i|X_j) = \sum_{h=1}^{n_j} p_j(x_h) H(X_i|X_j = x_h). \tag{8}$$

We compute average pairwise conditional entropy as $ACE_{(i,j)} = \frac{H_{i|j} + H_{j|i}}{2}$. Given this, we can then compute the average and standard statistics for ACE as follows:

$$\mu_{ace}(\mathcal{X}) = average(ACE_{(i,j)} \mid i \neq j), \tag{9}$$

$$\sigma_{ace}(\mathcal{X}) = stdev(ACE_{(i,j)} \mid i \neq j), \tag{10}$$

$$\phi_{ace}(\mathcal{X}) = \sigma_{ace}/\mu_{ace}. \tag{11}$$

---

[1] Alternative definitions of entropy may be used for dense tensors.

Note that at each step of the TT-decomposition process, the algorithm creates a core that links two modes of the tensor. Intuitively, the average pairwise entropy (ACE) indicates the ease with which one can obtain the low-rank decomposition of a pair of modes. The average and standard deviation statistics then indicate how significant this feature is in the data and how discriminative the feature is to help select pairs of modes to consider in sequence.

Pairwise Mutual Information. A related measure to conditional entropy is the pairwise mutual information. Given a d-mode tensor, let X_i be a discrete random variable with possible values $\{x_1, \ldots, x_{n_i}\}$ for mode i. The mutual information of X_i and X_j is defined as

$$MI_{(i,j)} = \sum_{x \in X_i} \sum_{y \in X_j} p_{(X_i, X_j)}(x,y) \log\left(\frac{p_{(X_i, X_j)}(x,y)}{p_{X_i}(x) p_{X_j}(y)}\right) \tag{12}$$

$$= H_i - H_{i|j} = H_j - H_{j|i}. \tag{13}$$

where $p_{(X_i, X_j)}$ is the joint probability mass function of X_i and X_j. We then compute that average and standard statistics for mutual information as follows:

$$\mu_{mi}(\mathcal{X}) = average(MI_{(i,j)} \mid i \neq j), \tag{14}$$

$$\sigma_{mi}(\mathcal{X}) = stdev(MI_{(i,j)} \mid i \neq j), \tag{15}$$

$$\phi_{mi}(\mathcal{X}) = \sigma_{mi}/\mu_{mi}. \tag{16}$$

Intuitively, mutual information can be used to measure how closely related the rows and columns of a given matrix are; the more closely related two modes are, the better are the chances to obtain a more accurate decomposition.

5.2 GTT-NP: Number of Parameters

Consider the TT-decomposition process depicted in Fig. 2. Here a 3-mode input tensor $\mathcal{X} \in \mathbb{R}^{l_1 \times l_2 \times l_3}$ is being converted into TT-format with a given decomposition sequence ($mode_1 \rightarrow mode_2 \rightarrow mode_3$) following Algorithm 1. In this example, the total number of parameters that the two SVD algorithms involved in the process have to solve for is the sum of the number of variables for U, SV^T, U' and SV'^T, which is $(r_0 \times l_1 \times r_1) + (r_1 \times l_2 \times l_3) + (r_1 \times l_2 \times r_2) + (r_2 \times l_3)$. It is easy to generalize this to

$$NP_{\Pi}(\mathcal{X}) = \sum_{i=1}^{d-1} \left(\underbrace{r_{i-1} \times l_{\pi_i} \times r_i}_{U} + \underbrace{r_i \times \prod_{j=i+1}^{d} l_{\pi_j}}_{SV^T} \right).$$

GTT-NP computes the number, $NP_{\Pi}(\mathcal{X})$ of parameters for each possible permutation, Π, and selects an order with the least number of parameters.

5.3 GTT-AMI and GTT-PMI: Mutual Information

Aggregate Mutual Information (AMI). Mutual information (Eq. 12) can be seen as a measure of dependency between the two variables. *GTT-AMI* guides the

TT-decomposition process based on the *aggregate mutual information* each mode has with the rest of the modes in the tensor. More specifically, given a d-mode tensor, the AMI value for mode i is computed as $AMI_i = \sum_{j=1}^{d} MI_{(i,j)}$. A potential strategy to guide the ordering of the modes in the TT-decomposition would be to (a) first find the mode with the largest AMI value and (b) then select this as the first mode. The process is, then, continued by (c) recomputing the AMI values among the remaining modes, (d) finding the mode with the largest (updated) AMI value among the remaining modes, and (e) selecting this as the next mode in the sequence. The process is repeated until all the modes have been ordered (when only two modes remain, the order is picked randomly). Figure 3 illustrates an example for a 3-mode (Mode 1: m_1, Mode 2: m_2, Mode 3: m_3) categorical data set. First, we compute AMI for each mode, which are: $AMI_1 = 1.5 + 0.2 = 1.7$, $AMI_2 = 1.5 + 0.7 = 2.1$, and $AMI_3 = 0.2 + 0.7 = 0.9$. In this case, AMI strategy described above would select mode m_2 as the first mode followed by m_1 or m_3. Intuitively, this process ensures that, at each step of the process, we consider and factorize a matrix where the rows have the highest statistical dependency with the columns.

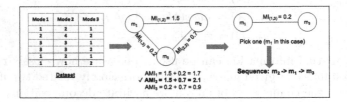

Fig. 3. GTT-AMI computation for a 3-mode tensor

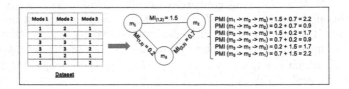

Fig. 4. GTT-PMI computation for a 3-mode tensor

Path Mutual Information (PMI). Note that the above process, which first picks the mode with the highest aggregate mutual information with the rest of the modes, is likely to lead to orderings where the total mutual information along the sequence is low: Figure 4 illustrates an example, where $MI_{(1,2)} = 1.5$, $MI_{(1,3)} = 0.2$, and $MI_{(2,3)} = 0.7$. With a total MI of $(1.5 + 0.7) = 2.2$, the orders $m_1 \to m_2 \to m_3$ and $m_3 \to m_2 \to m_1$ have the highest total mutual information. In fact, *surprisingly*, permutations with a low total MI tend to lead to higher accuracies than orders with a high total MI. This somewhat counter-intuitive result (which we experimentally validate in the "Experimental Results" section), indicates that the accuracies of initial decomposition steps are very important in obtaining high accuracy in TT-decompositions. We refer to this strategy as *path mutual information (GTT-PMI)*.

5.4   GTT-IE: (Inverse) Entropy

Remember that at the first step of the process, we matricize the given tensor $\mathcal{X}$ and then apply SVD to obtain U and SV^T matrices: here U represents clusters along the first selected mode and SV^T represents tensor $\mathcal{X}$ except the first mode. In the following steps of the algorithm, we apply several other clustering steps on the remaining matrix SV^T. It is therefore important that the matrix SV^T lends itself to a good clustering. One indicator of this is the entropy: if SV^T has high entropy, it is likely that it will lead to better clusters. Since the overall entropy in $\mathcal{X}$ is fixed, this implies that the matrix U should ideally have low entropy.

This leads to a third strategy, *GTT-IE*, which guides the TT-decomposition process based on the *(inverse) entropy of* each mode: at each step the algorithm selects the mode with the lowest entropy among the remaining modes. Again, Fig. 5 depicts an example of *GTT-IE*, given a 3-mode (m_1, m_2, m_3) categorical data set, IE strategy computes the entropy for each mode (H_1, H_2, H_3), and then decides a TT-decomposition sequence base on entropy in ascending order.

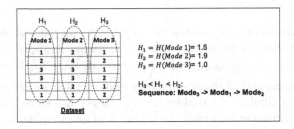

Fig. 5. GTT-IE computation for a 3-mode tensor

5.5   GTT-HYB: Hybrid Strategy

In Table 3, we list the data sets we use in our experiments along with along with the (non-hybrid) strategy with the best accuracy performance. As we see in the table, none of the strategies lead to a universally accurate order. While this is initially disappointing, the facts that different strategies work well for different data sets and that, often, where one strategy fails to lead to an accurate decomposition, another strategy excels, indicate that a hybrid strategy which carefully switches between the different approaches can lead to a better accuracy than any of the individual strategies.

To show the feasibility of such a hybrid technique, for each strategy[2], $\mathfrak{S}$, we have considered the data characteristics described earlier Sect. 5.1 as features and train a (linear) SVM classifier (with L1-regularization) that separates the data sets for which the strategy provides better accuracies than the rest (i.e., strategy $\mathfrak{S}$ vs. rest). In particular, for each scenario we consider the top-20% of the tensor instances for which the given strategy returns the best results against the lowest-20% of the tensor instances for which the given strategy returns the worst results. Intuitively, the separator can be interpreted as a feature selector that describes the data characteristics that best matches the given strategy. For each decomposition scenario, we then select the strategy that is recommended collectively by the trained separators; for any scenario for which the classifiers recommend more than one strategy, we pick the strategy that has the largest margin from the corresponding separator.

5.6   Complexity of GTT Decomposition

Let $\mathcal{X}$ be a d-mode input tensor and $t = |\mathcal{X}|$ indicates the number of non-zero entries in data set. Let al.so n_i denote the size of mode d_i and n denote the average mode size.

Guidance Step. The time complexities for the various strategies are as follows:

- *GTT-AMI* makes a pass over t data and for each it computes its contribution to the mutual information among $\frac{d(d-1)}{2}$ mode pairs; therefore its cost is $O\left(t \times \frac{d(d-1)}{2}\right)$.
- *GTT-PMI* also computes mutual information for all pairs of modes, but then it further computes a minimum path on the resulting graph with d nodes and $\frac{d(d-1)}{2}$ edges; therefore its cost is $O\left(\left(t \times \frac{d(d-1)}{2}\right) + \left(\frac{d(d-1)}{2} + d \log d\right)\right)$.
- *GTT-NP* enumerates $d!$ many sequences and, for each sequence computes the corresponding number of variables at $O(d)$ time – therefore it costs $O(d! \times d)$.
- *GTT-IE* requires one pass over the entire data for computing all of the mode entropies – i.e., its cost is $O(t)$.

Note that, as we experimentally show in the next section (Table 4), the time complexity for statistics collection is negligible relative to the time needed to decompose the tensor.

Decomposition Step. GTT provides a decomposition order which is then fed into TT-SVD to obtain the actual decomposition. The decomposition time complexity is therefore equal to that of TT-SVD [23], which is $O(dnr^3)$ and the number of parameters will be $O(dnr + (d-2)r^3)$.

---

[2] Note that the two mutual information based strategies, *GTT-AMI* and *GTT-PMI*, are hard to separate; since, as we see in Tables 4 and 5 in Sect. 6, *GTT-PMI* is overall more accurate among the two, we omit emphGTT-AMI in hybrid selection.

Table 2. Data sets [9]

Data set	#Inst.	#Modes	Data set	#Inst.	#Modes
Dermatology	366	34	Flare	1395	11
Mushroom	8124	23	House-votes	435	17
Soybean	307	36	Tic-tac-toe	958	10
Breast	699	10	Nursery	12960	9
Balance-scale	625	5	Primary-tumor	339	18
Hayes-roth	160	6	Lymphography	148	19
Car	172	7	Spect	267	23
Chess	3196	37			

6   Experimental Results

Here, we present experimental evaluations of the proposed GTT strategies[3]. Note that (once the decomposition order is selected) the data tensors are decomposed using TT-SVD [23] on a 4-core CPU (2.7 GHz each) machine, with 16 GB RAM.

6.1   Competitors

We compare five order selection strategies (GTT-AMI, GTT-PMI, GTT-IE, GTT-NP, GTT-HYB) and a baseline strategy, ARB, which represents the "average" decomposition performance of uninformed (i.e. arbitrary) order selection.

Evaluation Criteria. For accuracy, we adapt the reconstruction error introduced in Sect. 3. We report and compare average reconstruction errors for each strategy and the percentage improvement over ARB:

- Given a d-mode tensor, we enumerate *ALL* ($d!$) permutations and compute error for each permutation.
- We use the mean of all these $d!$ reconstruction errors as the (average) error for arbitrary selection, ARB.

In addition to the absolute values of reconstruction errors, we also report percentages of decompositions with better than (B) and worse than (W) the average ranking by arbitrary selection, ARB. We further report the ratio $gain = B/W$ – the value of *gain* indicates how well a given strategy promotes good decomposition, while avoiding the bad ones.

   We also report the average decomposition times for the decomposition orders selected by the various strategies.

---

[3] Our implementation and data sets can be found: https://shorturl.at/DMOSY.

Table 3. The relative average ranking against ARB for each data set (we normalize average ranking of ARB strategy as 1, and the bold number means the best ranking within four proposed strategies - **the lower, the better**) and the percentage improvement in reconstruction error (RE % impr.) against ARB using the *GTT-HYB* strategy - **the higher, the better**. *Inst. weighted average = (# of instances for a data set * Relative average ranking or RE % impr. for a strategy)/(total # of instances).

Data set	Relative average ranking (Lower, the better)								RE % impr. using HYB (Higher, the better)	
	r = 3				r = 5				r = 3	r = 5
	IE	NP	PMI	AMI	IE	NP	PMI	AMI	% impr.	% impr.
Tic-tac-toe	**0.68**	1.04	0.97	0.93	**0.67**	1.04	0.98	0.94	45%	49%
Balance-scale	**0.56**	1.12	0.97	0.71	**0.55**	1.02	0.93	0.71	16%	16%
Breast	**0.75**	1.01	0.99	0.92	**0.74**	1.03	0.97	0.94	10%	12%
Hayes-roth	**0.66**	0.75	0.75	0.75	**0.65**	0.73	0.74	0.69	4%	10%
Primary-tumor	0.97	**0.68**	0.92	0.93	0.98	**0.66**	0.90	0.93	10%	5%
Nursery	0.96	**0.79**	0.92	0.82	1.00	0.90	0.93	**0.81**	4%	5%
Dermatology	**0.82**	0.96	1.01	1.12	**0.84**	0.96	0.99	1.13	4%	4%
Spect	0.98	1.00	0.96	**0.95**	0.97	1.00	0.95	**0.94**	8%	4%
Flare	**0.83**	0.86	0.96	1.10	**0.84**	0.85	0.96	1.11	4%	3%
House_votes	**0.92**	1.05	0.99	1.21	**0.89**	1.06	1.00	1.22	2%	3%
Soybean	0.92	**0.88**	0.97	1.02	0.96	**0.92**	0.97	0.99	2%	3%
Mushroom	1.00	0.90	**0.87**	1.12	1.05	**0.85**	0.90	1.13	3%	3%
Lymphography	**0.77**	0.88	0.97	1.03	**0.77**	0.92	0.97	1.03	5%	1%
Car	1.10	0.87	0.98	**0.66**	1.15	1.14	0.99	**0.71**	0%	−2%
Chess	**0.91**	0.97	0.98	1.02	**0.90**	0.96	0.98	1.02	−16%	−6%
Average	**0.86**	0.92	0.95	0.95	**0.86**	0.94	0.94	0.95	6.7%	7%
*Inst. weighted Average	**0.88**	0.91	0.96	1.00	**0.89**	0.93	0.96	1.01	7%	7%

Data Sets. Table 2 lists the 15 data sets we use in these experiments. The data sets are taken from the UCI Machine learning repository [9]. From each data set, we extracted randomly selected 3-, 4-, and 5-mode tensor instances (up to 100 each, as allowed by the dimensionality of the data set). The total number of tensors extracted from these data sets and used in the experiments is 3632.

Target Ranks. Here, we consider two target ranks, 3 and 5. As discussed in Sect. 3, we assume the target TT-rank is given and fixed for each mode. While there are several non-parametric decomposition techniques, such as [13] which can learn also the appropriate rank, this is outside of the scope of this paper. We leave this to the future works.

6.2   Evaluations and Analysis

Accuracy. In Table 3, we first list the relative average ranking for each proposed strategy against ARB (lower, the better), as we can see, the best single strategy can vary from data set to data set – this motivates the need for a hybrid strategy (*GTT-HYB*) to select an effective combined strategy. As shown in Table 3, *GTT-HYB* provides improvements for all data set except the *car* and *chess* data sets.

Table 4. Average reconstruction error, rate of improvement against arbitrary selection (ARB) and average decomposition time.

	Average reconstruction error (Lower, the better)		Rate of improvement (Higher, the better)		Avg. Dec. Time (ms)	
Method	r = 3	r = 5	r = 3	r = 5	r = 3	r = 5
ARB	5.18	5.37	–	–	82.9	85.3
IE	4.94	5.1	4.7%	5.0%	**78.6**	**81.8**
NP	5.07	5.29	2.1%	1.5%	84.4	82.5
PMI	5.1	5.29	1.6%	1.5%	81.5	84.3
AMI	5.14	5.34	0.8%	0.4%	82.1	82.3
HYB	**4.86**	**5.05**	**6.2%**	**6.0%**	80.7	82.3

Table 5. Percentages of decompositions with better than (B) and worse than (W) the rank of decomposition returned on average by an uniformed, arbitrary ARB selection strategy)

Method	r = 3			r = 5		
	B	W	Gain	B	W	Gain
IE	54.0	34.0	1.6	53.0	35.0	1.5
NP	38.0	25.0	1.5	37.0	27.0	1.4
PMI	46.0	34.0	1.4	46.0	34.0	1.4
AMI	45.0	43.0	1.0	45.0	42.0	1.1
HYB	54.5	29.2	**1.9**	52.4	30.8	**1.7**

To get a more general view of the benefit of proposed strategies, in Table 4, we aggregate all data sets and report average reconstruction errors and percentage of improvements against the baseline (ARB). As we see in the table, all proposed GTT strategies improve reconstruction performance against ARB, with *GTT-IE* providing the highest improvement among the single criterion strategies. The table also shows that the hybrid strategy (*GTT-HYB*, described in Sect. 5.5) provides the highest overall improvement in accuracy. Table 3 also depicts the percentage improvement of reconstruction error (RE) against ARB using the *GTT-HYB* strategy for each data set, and we further see that the proposed hybrid strategy is indeed beneficial for 13 out of 15 of the considered data sets.

Again, with aggregating all data sets, in Table 5, we report the percentage of tensors for which each strategy returns better than (B) and worse than (W) the arbitrary selection, ARB, and the overall gain (*gain = B/W*). As we see, the *GTT-IE* strategy provides the largest *gain* among the four strategies and as before *GTT-HYB* strategy provides the best overall gain for both target tt-ranks.

Note that, among the two mutual information, based strategies, *GTT-PMI* is more effective than *GTT-AMI* in terms of both reconstruction error (Table 4) and *gain* (Table 5). Therefore, as reported in Sect. 5.5, we do not consider GTT-AMI, when constructing a hybrid strategy.

Decomposition Time. Table 4 reports the average decomposition times for different strategies. As we discussed in Sect. 5.6, the proposed strategies do not add any overhead to the decomposition time over arbitrary selection, ARB. In fact, the hybrid strategy, *GTT-HYB*, appears to reduce the decomposition time over ARB, we plan to explore this further in future work.

Table 6. Three major contributors to the *GTT-IE*, *GTT-NP*, and *GTT-PMI* strategies (positive values indicate positive, negative values indicate negative contribution)

IE	$\sigma_{ace}[3.9]$; $\phi_{ace}[-2.7]$; $\rho[-1.9]$
NP	$\phi_{length}[3.1]$; $\rho[-2.0]$; $\sigma_{entropy}[-1.1]$
PMI	$\sigma_{ace}[3.2]$; $\phi_{ace}[-2.0]$; $\mu_{length}[1.8]$

Top-Contributors to Each Strategy. In Table 6, we present the top-3 positive and/or negative contributors (among the various statistics considered in Sect. 5.1) for the *GTT-IE*, *GTT-NP*, and *GTT-PMI* strategies:

– For *GTT-IE*, the two main contributors are σ_{ace} and ϕ_{ace}. This echos the argument in Sect. 5.4: *GTT-IE* prefers that the entropies of the modes are considered in ascending order and thus *GTT-IE* is more effective when the discriminatory power of ACE is high.
– As discussed in Sect. 5.2, the number of parameters that needs to be learned depends on the length of the modes and the more discriminative the mode length parameter is, the more effective *GTT-NP* – this explains the positive contribution of ϕ_{length} to the *GTT-NP* selection criterion.
– For the mutual information based strategy, *GTT-PMI*, the higher the spread of ACE, the higher the impact of *GTT-PMI*. This confirms our discussion in Sect. 5.3: mutual information can be considered as a measure of dependency and, since the entropy of a mode is fixed, its dependency with the adjacent mode (mutual information) is constrained by the conditional entropy between them. Hence, the more the parameter ACE is (i.e., the larger is the value of σ_{ace}), the higher the benefits of *GTT-PMI*.

7 Conclusion

While the TT-decomposition promises a good trade-off between accuracy and resource requirements, the final accuracy is highly dependent on the order of the tensor modes in the tensor train. In this paper, we proposed a novel approach for *guiding the tensor train* (GTT) in selecting the decomposition sequence.

We have shown that we can leverage the various characteristics of the given data set to identify an effective order strategy. In particular, we proposed three order selection strategies and have shown that a *hybrid (HYB)* strategy that combines these three strategies taking into account the specific characteristics of the given data set can lead to decomposition sequences with high accuracy.

References

1. Batselier, K.: The trouble with tensor ring decompositions. CoRR abs/1811.03813 (2018)
2. Batselier, K., Yu, W., Daniel, L., Wong, N.: Computing low-rank approximations of large-scale matrices with the tensor network randomized SVD. SIAM J. Matrix Anal. Appl. **39**(3), 1221–1244 (2018)
3. Bedo, M.V.N., Ciaccia, P., Martinenghi, D., de Oliveira, D.: A k-skyband approach for feature selection. In: SISAP 2019, Newark, NJ, USA, Proceedings (2019)
4. Candan, K.S., Sapino, M.L.: Data Management for Multimedia Retrieval. Cambridge University Press, USA (2010)
5. Carroll, J.D., Chang, J.J.: Analysis of individual differences in multidimensional scaling via an n-way generalization of "eckart-young" decomposition. Psychometrika **35**(3), 283–319 (1970). https://doi.org/10.1007/BF02310791
6. Chen, Y., Jin, X., Kang, B., Feng, J., Yan, S.: Sharing residual units through collective tensor factorization to improve deep neural networks. In: IJCAI-18
7. Dash, M., Choi, K., Scheuermann, P., Liu, H.: Feature selection for clustering - A filter solution. In: 2002 IEEE ICDM, 2002, Proceedings, pp. 115–122 (2002)
8. Dash, M., Liu, H., Yao, J.: Dimensionality reduction of unsupervised data. In: IEEE International Conference on Tools with Artificial Intelligence (Nov 1997)
9. Dua, D., Graff, C.: UCI machine learning repository (2017)
10. Harshman, R.: Foundations of the parafac procedure: Models and conditions for an "explanatory" multi-modal factor analysis. UCLA Working Papers in Phonetics, vol. 16 (1970)
11. Houle, M.E., Kashima, H., Nett, M.: Fast similarity computation in factorized tensors. In: Navarro, G., Pestov, V. (eds.) SISAP 2012. LNCS, vol. 7404, pp. 226–239. Springer, Heidelberg (2012). https://doi.org/10.1007/978-3-642-32153-5_16
12. Huang, S., Candan, K.S., Sapino, M.L.: Bicp: block-incremental CP decomposition with update sensitive refinement. In: CIKM 2016. ACM, New York (2016)
13. Imaizumi, M., Maehara, T., Hayashi, K.: On tensor train rank minimization: Statistical efficiency and scalable algorithm. In: NIPS, pp. 3930–3939 (2017)
14. Jeon, I., Papalexakis, E.E., Kang, U., Faloutsos, C.: Haten2: Billion-scale tensor decompositions. In: 2015 IEEE 31st ICDE, pp. 1047–1058 (2015)
15. Kim, M., Candan, K.S.: Decomposition-by-normalization (DBN): leveraging approximate functional dependencies for efficient CP and tucker decompositions. Data Min. Knowl. Disc. **30**(1), 1–46 (2016)
16. Ko, C.Y., Lin, R., Li, S., Wong, N.: Misc: Mixed strategies crowdsourcing. In: IJCAI-19, pp. 1394–1400 (2019). https://doi.org/10.24963/ijcai.2019/193
17. Kohavi, R., John, G.H.: Wrappers for feature subset selection. Artif. Intell. **97**(1), 273–324 (1997). Relevance
18. Kolda, T.G., Bader, B.W.: Tensor decompositions and applications. SIAM Rev. **51**(3), 455–500 (2009). https://doi.org/10.1137/07070111X

19. Li, L., Yu, W., Batselier, K.: Faster tensor train decomposition for sparse data. ArXiv (2019)
20. Mickelin, O., Karaman, S.: Tensor ring decomposition. CoRR abs/1807.02513 (2018)
21. Novikov, A., Podoprikhin, D., Osokin, A., Vetrov, D.: Tensorizing neural networks. In: NIPS 2015, pp. 442–450. MIT Press, Cambridge (2015)
22. Novikov, A., Trofimov, M., Oseledets, I.: Exponential Machines (2016). arXiv e-prints arXiv:1605.03795
23. Oseledets, I.: Tensor-train decomposition. SIAM J. Sci. Comput. **33**(5), 2295–2317 (2011). https://doi.org/10.1137/090752286
24. Tucker, L.: Some mathematical notes on three-mode factor analysis. Psychometrika **31**(3), 279–311 (1966)
25. Yamaguchi, Y., Hayashi, K.: Tensor decomposition with missing indices. In: IJCAI 2017, pp. 3217–3223. AAAI Press (2017)
26. Yu, L., Liu, H.: Feature selection for high-dimensional data: A fast correlation-based filter solution. In: ICML 2003, vol. 2, pp. 856–863 (2003)
27. Zhao, Q., Zhou, G., Xie, S., Zhang, L., Cichocki, A.: Tensor ring decomposition. CoRR abs/1606.05535 (2016)

Noise Adaptive Tensor Train Decomposition for Low-Rank Embedding of Noisy Data

Xinsheng Li[1(✉)], K. Selçuk Candan[1], and Maria Luisa Sapino[2]

[1] Arizona State University, Tempe, AZ 85281, USA
{lxinshen,candan}@asu.edu
[2] University of Torino, 10149 Turin, Italy
marialuisa.sapino@unito.it

Abstract. Tensor decomposition is a multi-modal dimensionality reduction technique to support similarity search and retrieval. Yet, the decomposition process itself is expensive and subject to dimensionality curse. Tensor train decomposition is designed to avoid the explosion of intermediary data, which plagues other tensor decomposition techniques. However, many tensor decomposition schemes, including tensor train decomposition is sensitive to noise in the input data streams. While recent research has shown that it is possible to improve the resilience of the tensor decomposition process to noise and other forms of imperfections in the data by relying on probabilistic techniques, these techniques have a major deficiency: they treat the entire tensor uniformly, ignoring potential non-uniformities in the noise distribution. In this paper, we note that noise is rarely uniformly distributed in the data and propose a *Noise-Profile Adaptive Tensor Train Decomposition* (NTTD) method, which aims to tackle this challenge. NTTD leverages a model-based noise adaptive tensor train decomposition strategy: any *rough* priori knowledge about the noise profiles of the tensor enable us to develop a sample assignment strategy that best suits the noise distribution of the given tensor.

1 Introduction

Tensors and tensor decomposition (such as CP [16] and Tucker [36]) are increasingly being used for data-intensive tasks, including anomaly detection, correlation analysis [34], pattern discovery [21,22], and similarity retrieval [19].

1.1 Tensor Train Decomposition

A common problem faced by tensor decomposition techniques, such as Tucker, which generates dense core tensors, is that, even when the input is sparse, the

This work has been supported by NSF grants #1633381, #1909555, #1629888, #2026860, #1827757, a DOE CYDRES grant, and a European Commission grant #690817. Experiments for the paper were conducted using NSF testbed: "Chameleon: A Large-Scale Re-configurable Experimental Environment for Cloud Research".

© Springer Nature Switzerland AG 2020
S. Satoh et al. (Eds.): SISAP 2020, LNCS 12440, pp. 203–217, 2020.
https://doi.org/10.1007/978-3-030-60936-8_16

intermediary and final steps in the decomposition may lead to very large datasets. Recent research has shown that several generalizations of higher order tensors' low-rank decompositions, such as hierarchical Tucker (HT) [20] and the Tensor Train (TT) [32] format, are effective solutions to this problem. Intuitively, the TT decomposition (which can be interpreted as a special case of HT, without a recursive formulation) avoids the creation of a high-modal dense core, by splitting the core into a sequence of low (3) modal cores (Fig. 1). Since, computation and storage are exponential in the number of modes, TT is widely used for decomposition in various applications [35, 38].

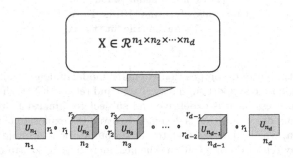

Fig. 1. Illustration of the Tensor Train (TT) decomposition

1.2   Challenge: Noisy Data

The problem the tensor train decomposition faces is that the overall decomposition process can be negatively affected by the noise and low quality in the data, which is especially a concern for sparse web and web-based user data [6,39]. Recent research has shown that it may be possible to avoid such over-fitting by relying on probabilistic techniques [37]. Unfortunately, existing probabilistic approaches have one major deficiency: they treat the entire tensor uniformly, ignoring possible non-uniformities in the distribution of noise in the given tensor. [30] has shown that if available, even *rough* a priori knowledge about the noise profiles of the tensor may enable CP-based decomposition strategies that are robust against noise, but these *uni-core* techniques are not applicable to the *multi-core* tensor train decomposition process, which results in a sequence of low-modal cores.

1.3   Noise-Profile Adaptive Tensor Train Decomposition (NTTD)

In this paper, we propose a *Noise-Profile Adaptive Tensor Train Decomposition (NTTD)* method, which leverages *rough* a prior information about noise in the data (which may be user provided or obtained through automated techniques) to improve decomposition accuracy. NTTD decomposes each mode matricization probabilistically through Bayesian factorization – the resulting factor matrix are then reconstructed to obtain the tensor approximations. Most importantly,

Algorithm 1. PTTD

Input: d dimensional tensor $\mathcal{X}$, Rank $R = \{r_1, ..., r_{d-1}\}$

Output: Decomposed factors $U^{(1)}, ..., U^{(d)}$ of

TT-approximation $\tilde{\mathcal{X}}$

1. generate the appropriate sampling number for each mode ,$S = \{s_1, ..., s_{d_1}\}$ with **intelligent sampling assignment strategy**
2. Temporary Tensor: $M = \mathcal{X}$
3. for k = 1 to d-1 do
   (a) $M := reshape(M, [r_{k-1}n_k, \frac{numel(M)}{r_{k-1}n_k}])$
   (b) apply the Probabilistic Matrix Factorization (PMF) on the matrix M with pre-given rank r_k and sampling number s_k to get the $U^{(k)} and V^{(k)}$
   (c) New core: $U^{(k)} = reshape(U, [r_{k-1}, n_k, r_k])$
   (d) $M := SV^{(k)T}$
4. $U_d = M$
5. Return tensor $\tilde{\mathcal{X}}$ in TT-format with scores $U^{(1)}, ..., U^{(d)}$

NTTD provides a resource allocation strategy, which accounts for the impacts of (a) the noise density of each mode and (b) inherent approximation error of the Tensor Train decomposition process, on the overall decomposition accuracy of the input tensor.

In other words, a priori knowledge about noise distribution on the tensor and the inherently approximate nature of the tensor train decomposition process are both considered to obtain a decomposition strategy, which involves (a) the order of the modes and (b) the number of Gibbs samples allocated to each step of the decomposition process, that best suits the noise distribution of the given tensor.

2   Background and Notations

The tensor model maps a schema with N attributes to an N-modal array and the decomposition process generalizes the matrix decomposition process to tensors. The two most popular tensor decomposition algorithms are the Tucker [36] and the CANDECOMP/PARAFAC (CP) [16] decompositions.

2.1   Tensor Train Decomposition

A major difficulty with the Tucker decomposition is that the dense core can be prohibitively large and expensive for high-modal tensors. While several tensor network approaches [7,8,18,28,29,31,32] (where network have been proposed to avoid large, dense core tensors, the tensor train (TT) format [32,35], which creates a linear tensor network (or a matrix product state, MPS [15]) avoids the deficiencies of many other complex decomposition structures:

Definition 1 (Tensor Train (TT) Format). *Let $\mathcal{X} \in \mathbb{R}^{n_1 \times n_2 \times \cdots \times n_d}$ be a tensor of order d. As we see in Fig. 1 (in Introduction), the tensor train decomposition decomposes $\mathcal{X}$ into d matrices $\mathbf{U}_{n_1}, \mathbf{U}_{n_2}, ..., \mathbf{U}_{n_d}$ such that,*

$$\mathcal{X} \approx \tilde{\mathcal{X}} = \mathbf{U}_{n_1} \circ \mathbf{U}_{n_2} \circ \cdots \circ \mathbf{U}_{n_d}, \tag{1}$$

where $\mathbf{U}_{n_1} \in \mathbb{R}^{n_1 \times r_1}$, $\mathbf{U}_{n_i} \in \mathbb{R}^{r_{i-1} \times n_i \times r_i} (i = 2, \ldots, d-1)$, *and* $\mathbf{U}_{n_d} \in \mathbb{R}^{r_{d-1} \times n_d}$.

$\Diamond$

Tensor train decomposition proceeds with matricizations along one mode at a time: at each step, a low-dimensional core corresponding to the current mode is obtained and the remainder of the data (which now has one less mode) is passed to the next step in the process [32].

3   Probabilistic TT Decomposition (PTTD)

Given the limitations of SVD on sparse and noisy tensors, the first step is to introduce a *probabilistic TT decomposition* scheme (PTTD), which extends TT tensor train decomposition framework with probabilistic matrix factorization [33].

PTTD replaces the SVD decomposition step in tensor train decomposition with probabilistic matrix factorization, in order to avoid over-fitting due to data sparsity and noise. More specifically, we first matricize the tensor $\mathcal{X}$ and we apply probabilistic matrix factorization on the resulting matrix, M. Under a Bayesian formulation, the prior distributions over U and V are assumed to be Gaussian:

$$p(U|\mu_U, \Lambda_U) = \prod_{i=1}^{n_1} \mathcal{N}(U_i|\mu_U, \Lambda_U^{-1}) \tag{2}$$

A similar formulation holds for V. The resulting factor matrix, U, is assigned as the first TT factor matrix. The matrix V is reshaped into the matrix M_{next} to be factorized in the next step. This probabilistic factorization and reshape processes are repeated until the decomposition is completed. The pseudo code of the algorithm is presented in Algorithm 1. Note that since the exact evaluation of the probabilistic factorization process is intractable, we instead seek to approximate the solution. While variational methods [17,23] are possible, they can produce inaccurate results because they tend to involve overly simple approximations to the posterior. MCMC-based methods [40], however, where the factor matrix $\{U_i^{(k)}, V_j^{(k)}\}$ are sampled by running a Markov chain, have been shown to asymptotically approach the exact results.

It is important to note that, in and of itself, PTTD does not leverage **a priori** knowledge about noise distribution and internal decomposition interaction, but it provides the framework in which noise-profile based adaptation can be implemented. More specifically, each row of factor matrices U and V follows a Gaussian distribution and this Gaussian is related to the uncertainty in the corresponding element and, thus, provides an opportunity to discover the distribution of data noise across the tensor, as we discuss in the next section.

4   Noise Adaptive Probabilistic Tensor Train Decomposition (NTTD)

One key advantage of the probabilistic decomposition framework presented above is that it can simultaneously uncover (Gaussian) noise while obtaining

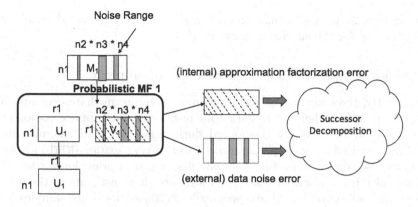

Fig. 2. Two types of errors propagate to downstream matricizations in tensor train decomposition: (*internal*) approximate factorization error and (*external*) data noise error

the decomposition [32]. Yet, it fails to account for the *potentially available (user-provided or automatically discovered) knowledge* about (a) the distribution of the <u>external</u> data noise across the tensor and (b) the noise generated *internally* due to the inherent imperfections in the decomposition process at the different steps of the tensor train network.

4.1   External and Internal Noise

External (Data) Noise. In this paper, we define, (external data) *noise density* as the ratio of the cells that are subject to noise. Without loss of generality, we assume noise exists only on cells that have values (i.e., the observed values can be faulty, but there are no spurious observations) and, thus, we formalize noise density as the ratio of the non-null cells that are subject to noise. Note that noise may impact the observed values in the tensor in different ways: in *value-independent noise* [30], the correct data may be overwritten by a completely random new value, whereas in *value-correlated noise*, existing values may be perturbed (often with a Gaussian noise, defined by a standard deviation, σ). We refer to the amount of perturbation as the *noise intensity*.

Internal (Decomposition) Noise. As we discussed in Sect. 2.1, in the tensor train format, the network structure acts as a "train" or "chain" of tensors: the core tensors only interact with their neighboring cores as illustrated in Fig. 1. The corresponding tensor train decomposition relies on sequential projections (formulated as sequential matrix factorizations) and *the decomposition accuracy of the intermediate matrix, M_k, depends on the accuracy of the previous matrix M_{k-1}'s (approximate) decomposition; similarly, the factorization error of M_k propagates to the following sequence of (intermediate) matrices, $M_{k+1}, ..., M_N$, of the chain.* This implies that a predecessor matrix which is poorly decomposed due to data noise or approximation error may negatively impact decomposition accuracies also for the successor matrices.

4.2   Noise Adaptation Through Sample Assignment

Consequently, the inaccuracies resulting from each intermediate decomposition along the chain (whether due to data noise or factorization approximation error, Fig. 2) need to be carefully considered during planning and resource allocation. The proposed *noise-profile adaptive tensor decomposition* (NTTD) algorithm adapts to (user provided or automatically discovered) a priori knowledge about noise by selecting a resource assignment strategy that best suits to the internal and external noise profiles. More specifically, NTTD assigns Gibbs samples to the decompositions of the various individual matricizations in a way that maximizes the overall decomposition accuracy of the whole tensor.

4.3   Gibbs Sampling and (Internal) Decomposition Error

As we discussed in Sect. 3, the probabilistic tensor train decomposition process consists of several sequential probabilistic matrix decompositions. Consequently, any inaccuracies generated in any of the upstream decompositions will *propagate* to the downstream matrix decompositions along the "train" structure. In this section, we ignore the external data noise and focus on the impact of this internal noise generated due to decomposition inaccuracies. More specifically, we aim to investigate how to allocate Gibbs samples in a way that is sensitive to (a) the internal noise generated by the individual matrix factorizations, (b) the downstream (internal) noise propagation, and (c) their impacts to the accuracy of the overall tensor train decomposition.

As discussed in Sect. 3, Gibbs sampling is used for tackling the challenge of evaluating the predictive distribution of the posterior by approximating the expectation by an average of samples drawn from the posterior distribution through a Markov Chain Monte Carlo (MCMC) technique. As we see in Fig. 2, for each intermediate matrix decomposition in PTTD, two factor matrices are generated: The U factor matrix is used to construct the core tensor corresponding to the current mode, whereas the V factor matrix is re-shaped as an input matrix for the successor decomposition step. Therefore,

- the accuracy of the U_k matrix has *direct* impact on the accuracy of one of the cores, whereas
- the accuracy of the V_k matrix *indirectly* influences accuracies of all downstream cores,

This observation, along with the observation that more samples can help provide better accuracy (*to certain degree*) in matrix factorization, can be used to improve the overall decomposition accuracy, to help allocate Gibbs samples to the different steps in tensor decomposition. More specifically, we argue that the number of samples for an intermediate matrix, M_k, should be allocated proportional to the size of the factor matrix, $size(U_k) + size(V_k)$, which reflects the number of unknowns to be discovered during the factorization of matrix M_k. In other words, the *internal decomposition error* sensitive sampling number, $L_{i\_err}(M_k)$, for matrix M_k can be computed as

$$L_{i\_err}(M_k) = L_{min}(M_k) + \lceil \gamma_{i\_err} \times (size(U_k) + size(V_k)) \rceil,$$

where γ_{i_err} is a scaling parameter such that the sum of all the sample counts is equal to the total number, $L_{i\_err(total)}$, of samples allocated for dealing with *internal decomposition errors* for the whole tensor decomposition:

$$L_{i\_err(total)} = \sum_{k=1}^{d-1} L_{i\_err}(M_k) \tag{3}$$

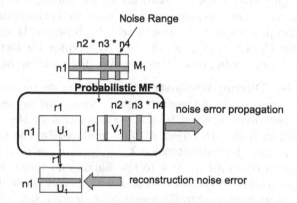

Fig. 3. Illustration of the decomposition noise error propagation and reconstruction noise error for the first decomposition step

4.4   Gibbs Sampling and (External) Noise

Equation 7, above, helps allocate samples across intermediate decomposition phases. However, it ignores one crucial piece of information that may be available: *distribution of the noise across the input tensor.*

The basic probabilistic tensor train decomposition (Sect. 3) assumes the noise is uniformly distributed across the tensor. In the real world, however, noise is rarely *uniformly* distributed along the entire tensor. More often, we would expect that noise would be clustered across slices of the tensor (corresponding, for example, to unreliable information sources or difficult to obtain data). In many cases, even if we do not have precise knowledge about the cells that are subject to such noise or the amount of noise they contain, we may have a rough idea about the distribution of noise across the different modes [30]. As we experimentally show in Sect. 5, there is a direct relationship between the noise distribution across the tensor and the number of Gibbs samples it requires for accurate decomposition. Consequently, given a tensor with non-uniform noise distribution across different modes, uniform assignment of the number samples, $L_{n\_err}(M_k) = \frac{L_{n\_err(total)}}{d-1}$ (where $L_{n\_err(total)}$ is the total number of Gibbs samples for tackling the impact of noise) becomes ineffective. Therefore, in this section, we aim to answer the question

> *can we leverage rough information that may be available about noise distribution in improving the accuracy of the overall tensor train decomposition?*

Noise taints accuracy through two distinct mechanisms: (a) impact of noise during decomposition and, for applications (such as recommendation and prediction) that involve the recovery of missing entries in the tensor, (b) impact of noise during reconstruction. As we see in Fig. 3, the noise in the input matrix partitions itself into the resulting factor matrices U and V. The factor matrix V_i is reshaped as input matrix for the following tensor train decomposition steps, therefore is involved in the propagation of the noise to downstream steps during the decomposition process (Fig. 4). The matrix, U_i, however, is separated into a factor matrix (for U_1) or more generally to a core tensor for factor $i > 1$, and thus impacts accuracy during reconstruction. We discuss these next.

Impact of Noise During Reconstruction. The noise reconstruction error taints the overall accuracy in the reconstruction process due to the matrix tensor multiplication operations involved in the recomposition of the (approximate) tensor. For example, if the ith object of U_k is polluted by the noise, after the reconstruction process, the complete slice $\tilde{\mathcal{X}}_{*,\ldots,*,k(i),*,\ldots,*}$ will be tainted by the noise pollution from column $U_{k(i)}$ due to the matrix and tensor multiplication. Consequently, to account for the noise reconstruction error, the number of Gibbs samples should be proportional to the mode noise density, nd_k.

Impact of Noise During Decomposition. A naive approach to allocate the number of samples for a noisy matrix, M_k, is to allocate it proportional to its noise density, nd_k. However, since the probabilistic tensor train decomposition process follows a "train" structure, errors propagate downstream as shown in Fig. 4. Consequently, allocating sampling number proportional to the noise density maybe not the best strategy.

As mentioned earlier, the Gibbs sampling algorithm cycles through the latent variables, sampling each one from its distribution conditional on the current values of all other variables. Due to the use of conjugate priors for the parameters and hyperparameters in the Bayesian PMF model, the conditional distributions derived from the posterior distribution are easy to sample from. In particular, the conditional distribution over the feature vector U_i, conditioned on the other features V_i, observed matrix cell value M_i, and the values of the hyperparameters are Gaussian:

$$p(U_i|M, V, \Theta_U, \alpha) = \mathcal{N}(U_i|\mu_i, \Lambda_i^{-1})$$

$$\approx \prod_{i=1}^{n_1} \left([\mathcal{N}(\widetilde{M}_{i,j}|U_i^T, V_i, \alpha^{(-1)})]^{n_{i,j}} \right. \tag{4}$$

$$\left. \times p(U_i|\mu_U, \Lambda_U^{-1}) \right).$$

Note that the conditional distribution over the latent feature matrix U factorizes into the product of conditional distributions over the individual feature vector:

$$p(U|M, V, \Theta_U) = \prod_{i=1}^{n_1} p(U_i|M, V, \Theta_U). \tag{5}$$

We see that the conditional distributions over the V feature vectors and the V mode hyperparameters have exactly the same form.

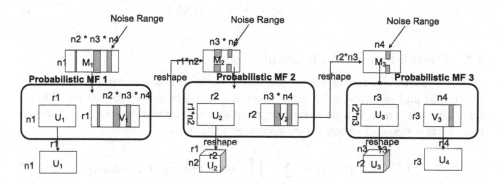

Fig. 4. Illustration of the noise error propagation

Equations 4 and 5, along with Fig. 4, indicate how errors propagate downstream. In particular, in Fig. 4, red columns of the first matricization, M_1, show the columns that are noise polluted. During the decomposition, the corresponding columns of resulting factor matrix V_1 (highlighted also in red) are also tainted with stronger noise than other columns of V_1. This tainting process flows downstream (*subject to matrix re-shape operations*) as shown in Fig. 4. Consequently, for decomposition phase k, the number of samples should be proportional to

$$\sum_{j=k}^{d-1} \prod_{i=j+1}^{d} nd_i, \tag{6}$$

where, $\prod_{i=j+1}^{d} nd_i$ is the noise density of matricization of V_k on mode k. The $\sum_{j=k}^{d-1}$ operation, above, takes into account the accumulation process of the noise on all downstream decomposition steps.

Combining Decomposition and Reconstruction Impacts of Noise.
Assuming that the decomposed tensor will be utilized for an application (such as recommendation) which necessitates reconstruction of the approximate tensor, we need to consider reconstruction and decomposition errors together when assigning the number of Gibbs samples. In other words, for an intermediate matrix, M_k, the number of samples must be allocated proportional to the sum of reconstruction and decomposition errors, i.e, $nd_k + \sum_{j=k}^{d-1} \prod_{i=j+1}^{d} nd_i$. This leads to the following formula for the number $L_{n\_err}(M_k)$ of samples:

$$L_{min}(M_k) + \lceil \gamma_{n\_err} \times (\sum_{j=k}^{d-1} \prod_{i=j+1}^{d} nd_i + nd_k) \rceil \times L_{n\_err(total)},$$

where γ_{n_err} is a scaling parameter such that the sum of all the sample counts is equal to the total number, $L_{n\_err(total)}$, of samples allocated for dealing with *noise errors*:

$$L_{n\_err(total)} = \sum_{k=1}^{d-1} L_{n\_err}(M_k). \tag{7}$$

4.5 Overall Sample Assignment

While considering the error propagation, both internal decomposition error (Sect. 4.3) and external noise error (Sect. 4.4) need to be accounted for. Therefore, the combined sample assignment equation, for matricization, M_k, in the tensor train decomposition process, can be written as

$$\begin{aligned} L(M_k) = &\lceil \gamma_{n\_err} \times (\sum_{j=k}^{d-1} \prod_{i=j+1}^{d} nd_i + nd_k) \rceil \times L_{n\_err(total)} \\ &+ \lceil \gamma_{i\_err} \times (size(U_k) + size(V_k)) \rceil \times L_{i\_err(total)} \\ &+ L_{min}(M_k) \end{aligned} \tag{8}$$

where $L_{min}(M_k)$ is the minimum number of samples a (non-noisy) tensor of the given size would need for accurate decomposition and γ_{n_err} and γ_{i_err} are two scaling parameters, selected such that the total number of samples is equal to the number, L_{total}, of samples allocated for the whole tensor:

$$L_{total} = \sum_{k=1}^{d-1} L(M_k).$$

The parameters, γ_{n_err} and γ_{i_err}, also control the relative impacts of the internal and external noise. In the experiments, they are set such that the number of samples allocated to handle internal and external noise are the same.

5   Experimental Evaluation

In this section, we report experiments that aim to assess the effectiveness of the proposed noise adaptive tensor train decomposition approach.

5.1 Experiment Setup

Key parameters and their values are reported in Table 1.

Table 1. Parameters – default values, used unless otherwise specified, are highlighted

Parameters	Alternative values
Dataset	Ciao; BxCrossing; MovieLens
Noise density	**10%**; 20%; 30%;
Noise intensity (σ)	**1**, 3, 5
Total samples (L_{total})	**90**; 135; 180;
Min. samples ($L_{min}(M_k)$)	$L_{total}/(3 \times (d-1)) = L_{total}/9$

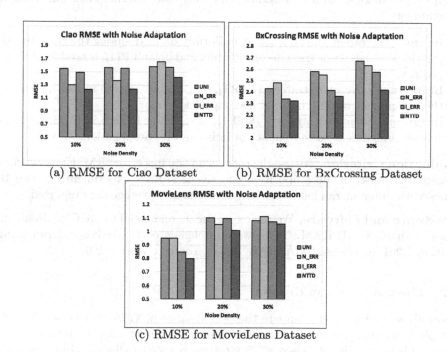

(a) RMSE for Ciao Dataset   (b) RMSE for BxCrossing Dataset

(c) RMSE for MovieLens Dataset

Fig. 5. RMSE with different data sets and noise densities ($L_{total} = 90$)

Data Sets. In these experiments, we used three user-centered datasets: Ciao [43], MovieLens [41] and BxCrossing [42]. Ciao dataset is represented in the form of $143 \times 200 \times 12 \times 4$ (density 5.68E−04) with the schema $\langle user, product, category, helpfullness \rangle$. BxCrossing dataset is represented in the form of $2599 \times 34 \times 16 \times 76$ (density 2.48E−0.5) with the schema $\langle user, book, publishedyear, userage \rangle$. The MovieLens dataset is represented in the form of $247 \times 112 \times 48 \times 21$ (density 8.86E−06) with the schema $\langle user, movie, age, location \rangle$. In Ciao and MovieLens data sets, the tensor cells contain rating values between 1 and 5 or (if the rating does not exist) a special "null" symbol. And for the BxCrossing dataset, the tensor cells contain rating values between 1 and 10.

Noise. To observe the different degrees of noise, we selected a random portion of the non-null cells and randomly perturbed the value. It is a worst-case scenario for NTTD, where the noise is distributed *uniformly on the tensor*, but the experiments show that even in this case, NTTD can take into account the noise density difference across the data modes, implied by the difference in corresponding data densities. Therefore, in the experiments, the noise density for different modes is approximated by the corresponding data density.

Alternative Strategies. We compare the proposed approach against other sampling strategies: uniform, internal-noise only, and external-noise only sample assignment:

- In *uniform* strategy (UNI), L_{total} is uniformly divided among the three matricizations in the tensor train decomposition and default PTTD is used for decomposition.
- In *internal-noise only* strategy (I_ERR), γ_{n_err} is set to zero in Eq. 8, focusing the assignment to only internal decomposition error.
- In *external-noise only* strategy (N_ERR), γ_{i_err} is set to zero in Eq. 8, focusing sample assignment to the impact of noise and its propagation.

Evaluation Criterion. We use the root mean squares error (RMSE) inaccuracy measure to assess the decomposition effectiveness. Each experiment was run 10 times with different random noise distributions and averages are reported.

Hardware and Software. We ran experiments on an eight-core CPU Nehalem Node with 16.00 GB RAM. Codes were implemented in Matlab and run using Matlab R2016b. We used MATLAB Tensor Toolbox Version 2.6.

5.2 Discussion of the Results

Overview. In Fig. 5, we compare the performance of NTTD with noise-adaptive sample assignments against other strategies for different noise densities. As we see in this figure, the proposed NTTD strategy is able to allocate Gibbs samples effectively to significantly reduce RMSE relative to PTTD with uniform sample assignment. Moreover, we also see that internal- and external-only strategies that ignore part of the error can actually hurt the accuracy and perform worse than the uniform strategy. These show that the proposed noise-adaptive strategy is effective in leveraging rough knowledge about external noise distributions and internal decomposition errors to better allocate the Gibbs samples.

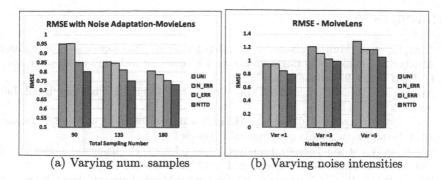

(a) Varying num. samples   (b) Varying noise intensities

Fig. 6. (a) RMSE with different num. of samples; i.e. L_{total} is 90, 135, or 180 (noise density 10%, noise intensity 1; (b) RMSE with different noise intensities; i.e., σ is 1, 3, or 5 (noise density 10%)

Impact of the Total Number of Samples. A key parameter of the NTTD algorithm is the number of total Gibbs samples. As we see in Fig. 6(a), as we would expect, increasing the number of Gibbs samples helps reduce the overall decomposition error. Note that, among the four strategies, NTTD is the one that provides most consistent and quickest drop in error. The figure shows the result for the MovieLens data; the results are similar also for the other data sets.

Impact of the Noise Intensity. In Fig. 6(b), we consider the MovieLens data set with different noise intensities. As we expect, increased noise corresponds to increased RMSE. However, NTTD provides the best results for all noise intensities considered. NTTD is also the best strategy for the other two data sets.

6   Conclusion

Tensor train decomposition is a latent space embedding technique that promises a good trade-off between accuracy and resource requirements – but can be subject to overfitting and be impacted of data noise. Recent research has shown that probabilistic techniques can ease the problem of overfitting caused by noise, especially on sparse data. However, existing techniques ignore potential non-uniformities in the noise distribution. In this paper, we proposed a novel noise-adaptive tensor train decomposition (NTTD) technique that leverages rough information about noise distribution to improve the tensor decomposition performance. NTTD decomposes each intermediate matrix probabilistically through Bayesian factorization. The noise profiles of tensor and their alignments are then leveraged to develop a strategy that considers the internal decomposition error as well as external to obtain a Gibbs sample assignment data noise best suits the noise profile of a given tensor.

References

1. Acar, E., et al.: Multiway analysis of epilepsy tensors. Bioinformatics, 10–18 (2007)
2. Abedjan, Z., et al.: Detecting data errors: where are we and what needs to be done? PVLDB **9**(12), 993–1004 (2016)
3. Andersson, C.A., Bro, R.: The N-way toolbox for MATLAB. Chemom. Intell. Lab. Syst. **52**(1), 1–4 (2000)
4. Bader, B.W., Kolda, T.G., et al.: MATLAB tensor toolbox version 2.5, January 2012. http://www.sandia.gov/~tgkolda/TensorToolbox
5. Bader, B.W., et al.: Temporal analysis of social networks using three-way DEDI-COM. Sandia National Laboratories TR SAND2006-2161 (2006)
6. Balakrishnan, R., Kambhampati, S.: SourceRank: relevance and trust assessment for deep web sources based on inter-source agreement. In: WWW, pp. 1055–1056 (2010)
7. Ballani, J., Grasedyck, L., Kluge, M.: Black box approximation of tensors in hierarchical Tucker format. Linear Algebra Appl. **438**, 639–657 (2013)
8. Ballani, J., Grasedyck, L.: A projection method to solve linear systems in tensor format. Numer. Linear Algebra Appl. **20**, 27–43 (2013)
9. Brin, S., et al.: The anatomy of a large-scale hypertextual web search engine. In: WWW 1998 (1998)
10. Chakrabarti, S.: Dynamic personalized PageRank in entity-relation graphs. In: WWW 2007 (2007)
11. Chen, X., et al.: LWI-SVD: low-rank, windowed, incremental singular value decompositions on time-evolving data sets. In: KDD 2014 (2014)
12. Chi, E.C., Kolda, T.G.: Making tensor factorizations robust to non-Gaussian noise. Technical report. arXiv: 1010.3043v1 (2010)
13. Chu, W., Ghahramani, Z.: Probabilistic models for incomplete multi-dimensional arrays. In: AISTATS 2009 (2009)
14. Chu, X., et al.: RuleMiner: data quality rules discovery. In: CDE 2014 (2014)
15. Cichocki, A., et al.: Tensor networks for dimensionality reduction and large-scale optimization: Part 1 low-rank tensor decompositions Found. Trends Mach. Learn. **9**, 249–429 (2016)
16. Harshman, R.A.: Foundations of the PARAFAC procedure: model and conditions for an explanatory multi-mode factor analysis. In: UCLA Working Papers in Phonetics, pp. 16:1–84 (1970)
17. Hinton, G.E., Van, C.: Keeping the neural networks simple by minimizing the description length of the weights. In: COLT, pp. 5–13 (1993)
18. Holtz, S., et al.: The alternating linear scheme for tensor optimization in the tensor train format. SIAM J. Sci. Comput. **34**, A683–A713 (2012)
19. Houle, M.E., Kashima, H., Nett, M.: Fast similarity computation in factorized tensors. In: Navarro, G., Pestov, V. (eds.) SISAP 2012. LNCS, vol. 7404, pp. 226–239. Springer, Heidelberg (2012). https://doi.org/10.1007/978-3-642-32153-5_16
20. Grasedyck, L., et al.: An introduction to hierarchical (H-) rank and TT-rank of tensors with examples. Comput. Methods Appl. Math **3**, 291–304 (2011)
21. Jeon, I., Papalexakis, E., Kang, U., Faloutsos, C.: HaTen2: billionscale tensor decompositions. In: ICDE 2015 (2015)
22. Jeon, B., et al.: SCouT: scalable coupled matrix-tensor factorization - algorithm and discoveries. In: ICDE 2016 (2016)
23. Jordan, M.I., Ghahramani, Z., Jaakkola, T.S., Saul, L.K.: An introduction to variational methods for graphical models. Mach. Learn. **37**, 183 (1999)

24. Kang, U., et al.: GigaTensor: scaling tensor analysis up by 100 times algorithms and discoveries. In: KDD 2012 (2012)
25. Kroonenberg, P.M., et al.: Principal component analysis of three-mode data by means of alternating least squares algorithms. Psychometrika **45**, 69–97 (1980)
26. Kim, H., Park, H.: Nonnegative matrix factorization based on alternating nonnegativity constrained least squares and active set method. SIAM J. Matrix Anal. Appl. **30**(2), 713–730 (2008)
27. Kim, M., Candan, K.S.: Efficient static and dynamic in-database tensor decompositions on chunk-based array stores. In: CIKM 2014 (2014)
28. Li, X., et al.: Focusing decomposition accuracy by personalizing tensor decomposition (PTD). In: CIKM 2014 (2014)
29. Li, X., et al.: 2PCP: two-phase CP decomposition for billion-scale dense tensors. In: ICDE 2016 (2016)
30. Li, X., et al.: nTD: noise adaptive tensor decomposition. In: WWW 2017 (2017)
31. Li, X., et al.: M2TD: multi-task tensor decomposition for sparse ensemble simulations. In: ICDE 2018 (2018)
32. Oseledets, I.V.: Tensor-train decomposition. SIAM **33**, 2295–2317 (2011)
33. Salakhutdinov, R., et al.: Probabilistic matrix factorization. In: NIPS 2007 (2007)
34. Sun, J., Papadimitriou, S., Yu, P.S.: Window based tensor analysis on high dimensional and multi aspect streams. In: ICDM, pp. 1076–1080 (2006)
35. Tjandra, A., Sakti, S., Nakamura, S.: Compressing recurrent neural network with tensor train. In: NIPS 2017 (2017)
36. Tucker, L.: Some mathematical notes on three-mode factor analysis. Psychometrika **31**, 279–311 (1966)
37. Xiong, L. et al.: Temporal collaborative filtering with Bayesian probabilistic tensor factorization. In: SDM 2010 (2010)
38. Yang, Y., Krompass, D., Tresp, V.: Tensor-train recurrent neural networks for video classification. In: ICML 2017 (2017)
39. Zafarani, R., Liu, H.: Users joining multiple sites: friendship and popularity variations across sites. Inf. Fusion **28**, 83–89 (2016)
40. Neal, R.: Probabilistic inference using Markov chain Monte Carlo methods. Technical Report CRG-TR-93-1, University of Toronto (1993)
41. http://grouplens.org/datasets/movielens/
42. http://www2.informatik.uni-freiburg.de/cziegler/BX/
43. http://www.public.asu.edu/~jtang20/datasetcode/truststudy.htm

ABID: Angle Based Intrinsic Dimensionality

Erik Thordsen$^{(\boxtimes)}$ and Erich Schubert

TU Dortmund University, Dortmund, Germany
{erik.thordsen,erich.schubert}@tu-dortmund.de

Abstract. The intrinsic dimensionality refers to the "true" dimensionality of the data, as opposed to the dimensionality of the data representation. For example, when attributes are highly correlated, the intrinsic dimensionality can be much lower than the number of variables. Local intrinsic dimensionality refers to the observation that this property can vary for different parts of the data set; and intrinsic dimensionality can serve as a proxy for the local difficulty of the data set.

Most popular methods for estimating the local intrinsic dimensionality are based on distances, and the rate at which the distances to the nearest neighbors increase, a concept known as "expansion dimension". In this paper we introduce an orthogonal concept, which does not use any distances: we use the distribution of angles between neighbor points. We derive the theoretical distribution of angles and use this to construct an estimator for intrinsic dimensionality.

Experimentally, we verify that this measure behaves similarly, but complementarily, to existing measures of intrinsic dimensionality. By introducing a new idea of intrinsic dimensionality to the research community, we hope to contribute to a better understanding of intrinsic dimensionality and to spur new research in this direction.

1 Introduction

Intrinsic Dimensionality (ID) estimation is the process of estimating the dimension of a manifold embedding of a given data set either at each point of the data set individually or for the entire data set at large. The dimension of a given algebraic set is a well-understood problem in algebra [21], but lifting these methods to a sample of an unknown function is not trivially possible. Therefore methods that are very different from functional analysis are required to grasp the dimensionality of a discrete data set. Prior work in the field is largely focused on analyzing the differential of point counts in changing volumes [1,10,12,14], as linear algebra gives estimates of these differentials assuming a certain dimensionality. These approaches rely on distances between points and assume the data to be uniformly sampled from their defining space. The resulting ID describes the dimension required to embed a point and its neighborhood in a manifold with small loss of precision. In our novel approach for ID estimation, we derive an estimate based on the cosines between directional vectors of a point to all points

© Springer Nature Switzerland AG 2020
S. Satoh et al. (Eds.): SISAP 2020, LNCS 12440, pp. 218–232, 2020.
https://doi.org/10.1007/978-3-030-60936-8_17

in its neighborhood. The idea is illustrated in Fig. 1: in two-dimensional dense data, we see all directions evenly, whereas in a linear subspace we mostly see similar or opposite directions. Hence we aim at deriving an estimator capable of computing the angles between observed data points. It differs from the distance- and volume-based approaches as it describes the least dimensions required to connect a given point to the rest of the data set. It can, therefore, be understood as a description of the simplicial composition of the data set. Besides describing a different notion of local dimensionality, we provide evidence that our approach is more robust and gives stable estimates on smaller neighborhoods than the volume-based approaches. We hope that in the future the new angle-based interpretation of intrinsic dimensionality will be combined with expansion-rate-based approaches and spur further research in intrinsic dimensionality.

Fig. 1. Motivation of angle-based intrinsic dimensionality: in two-dimensional dense data, we observe all directions evenly, in noised one-dimensional linear data arrows go either in similar or in opposite directions.

2   Related Work

Intrinsic dimensionality has been shown to affect both the speed and accuracy of similarity search problems such as approximate nearest-neighbor search and the algorithms developed for this problem [4,7,18]. Intrinsic dimensionality has also been employed to improve the quality of embeddings [19], to detect anomalies in data sets [15], to determine relevant subspaces [6], and to improve generative adversarial networks (GANs) [5]. Distance-based estimation of intrinsic dimensionality is the "short tail" equivalent of extreme value theory [12,13], and many techniques can be adapted from estimators originally devised for extreme values on the long tail of (censored-) distributions [2], as previously used in disaster control. Important estimators include the Hill estimator [10], the aggregated version of it [16], the Generalized Expansion Dimension [14], method of moments estimators [1], regularly varying functions [1], and probability-weighted moments [1]. ELKI [20] also includes L-moments [11] based adaptations of this and improves the bias of these estimators slightly. A noteworthy recent development is the inclusion of pairwise distances as additional measurements [9] and the idea of also taking virtual mirror images of observed data points into account [3]. We will note interesting parallels between these methods and our new approach.

Angle-based approaches have been successfully used for outlier-detection in high-dimensional data, for example with the method ABOD [17], which considers points with a low variance of the (distance-weighed) angle spectrum to be anomalous, with the assumption that such points are on the "outside" of the data set. Our approach brings ideas from this method to the estimation of intrinsic dimensionality (which in turn has been shown to relate to outlierness [15]).

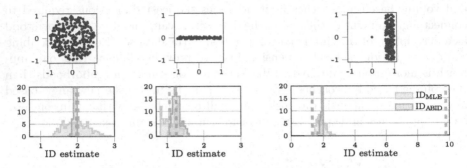

Fig. 2. Three data sets with corresponding ID estimate histograms. The dashed vertical lines correspond to the ID estimate of the point at $(0, 0)$.

3   On the Dimensionality of Functions and Data

The dimensionality of a vector field $\mathbb{R}^d$ is the number of components d; a quantity referred to as *representational dimensionality* because it characterizes the data representation more than the underlying data. Manifolds are embeddings of lower dimensional vector spaces into some higher-dimensional space. These can be linear embeddings, but we can also consider curves such as $(x, \sin x)$, which we consider to be a one-dimensional manifold. This aligns with human intuition, for example when differentiating a circle (the outline) from the corresponding disc (the contained area). The concept of *local intrinsic dimensionality* (LID) [12,13] captures the need for allowing different parts of a data set to have different dimensionality. Nevertheless, the "correct" answer to the question of dimensionality is all but unambiguous, e.g., when considering a point on the *surface* of a ball, or the crossing point of the figure eight.

While existing work focuses on the analysis of distances in enclosing neighborhoods, our novel approach uses the distribution of pairwise angles of neighboring points. The different nature of the resulting ID estimates is showcased in Fig. 2. Both the distance-based ($\mathrm{ID}_{\mathrm{MLE}}$, [10]) and the angle-based ($\mathrm{ID}_{\mathrm{ABID}}$, this article) approach consider the second data set to be dominantly one-dimensional and the other to be mostly two-dimensional. The outlying point $(0, 0)$ in the third data set, however, is judged very differently by both approaches. The distance-based approach $\mathrm{ID}_{\mathrm{MLE}}$ considers its environment to be almost ten-dimensional (≈ 9.78) as all distances are very close, whereas our novel angle-based approach $\mathrm{ID}_{\mathrm{ABID}}$ considers it to be one-dimensional (≈ 1.18) as the observed neighborhood lies in a narrow cone, similar to the second data set. With increased neighborhood size this cone widens and the ID gets closer to two. This effect is similar to visual details of a surface disappearing at a distance.

We will now lay the mathematical foundations for our novel ID estimator. When estimating the ID of a given point from a data set, the general approach is to consider a number of nearby points as an enclosing neighborhood. An assumption shared by all ID estimators is that this neighborhood should be "representative" of an underlying manifold (e.g., uniformly sampled from some parameter space). As angles are independent of vector lengths, assuming a uniform random distribution over the $(d\text{-}1)$-sphere or the d-ball is the same. It is noteworthy that the unit n-sphere contains all unit n'-spheres with $n' \leq n$ as a subset which allows for lower dimensional submanifolds. In the following, we assume the local neighborhood to originate from a unit $(d\text{-}1)$-sphere, and use the distribution of pairwise angles of Cai, Fan, and Jiang [8].

Theorem 1 (Distribution of random angles in a $(d\text{-}1)$-sphere. *The distribution of angles θ between two random points sampled independently and uniformly from a $(d\text{-}1)$-sphere converges, as the number of samples goes to infinity, to*

$$P(\theta) = \frac{\Gamma(\frac{d}{2})}{\Gamma(\frac{1}{2})\Gamma(\frac{d-1}{2})} \cdot \sin(\theta)^{d-2} \tag{1}$$

where Γ is the gamma function and θ is defined on $[0, \pi]$.

Proof. See Cai, Fan, and Jiang [8] for a detailed proof.

Because angles are invariant of the vector lengths, this also holds for points sampled from a d-ball instead of the $(d\text{-}1)$-sphere as well as other rotation invariant distributions such as spherical Gaussians, as long as the origin point is not included in the data (for which the angle is undefined). Note that such points at distance 0 cause problems for most estimators of intrinsic dimensionality and are commonly removed for such estimators, too.

As popularly known from the *curse of dimensionality*, all angles tend to become approximately orthogonal as dimensionality approaches infinity. This causes Eq. (1) to concentrate around $\frac{\pi}{2}$ [8]. The distribution above is unwieldy and expensive to compute (as we need to compute the arcus cosines). We, therefore, prefer to work directly on the cosines. By applying the Legendre duplication formula and doing a change of variables, we obtain the distribution of cosines.

Theorem 2 (Distribution of cosine similarities of points in a $(d\text{-}1)$-sphere). *The distribution of pairwise cosine similarities C between random points sampled independently and uniformly from a $(d\text{-}1)$-sphere is*

$$P(C) = \tfrac{1}{2}B(\tfrac{1+C}{2}; \tfrac{d-1}{2}, \tfrac{d-1}{2}) \tag{2}$$

where $B(x; \alpha, \beta)$ is the beta distribution p.d.f. and C is defined on $[-1, 1]$.

Proof. For this proof, we modify the well known Legendre duplication formula:

$$\Gamma(x)\Gamma(x+\tfrac{1}{2}) = 2^{1-2x}\Gamma(\tfrac{1}{2})\Gamma(2x)$$

$$\frac{\Gamma(x+\tfrac{1}{2})}{\Gamma(x)\Gamma(\tfrac{1}{2})} = \frac{2^{1-2x}\Gamma(2x)}{\Gamma(x)^2} = \frac{1}{B(x,x)} \cdot \tfrac{1}{2}^{2x-1} \tag{3}$$

where $B(\cdot,\cdot)$ is the beta function. By using this in Eq. (1) for $x = \frac{d-1}{2}$, we obtain

$$P(\theta) = \frac{1}{B(\frac{d-1}{2},\frac{d-1}{2})} \cdot \left(\tfrac{1}{2}\sin(\theta)\right)^{d-2}$$

We can now substitute θ with $\arccos(C)$ by a change of variable:

$$
\begin{aligned}
P(C) &= \frac{1}{B(\frac{d-1}{2},\frac{d-1}{2})} \cdot \left(\tfrac{1}{2}\sin(\arccos(C))\right)^{d-2} \cdot \left|\tfrac{\partial}{\partial C}\arccos(C)\right| \\
&= \frac{1}{B(\frac{d-1}{2},\frac{d-1}{2})} \cdot \left(\tfrac{1}{2}\sqrt{1-C^2}\right)^{d-2} \cdot \frac{1}{\sqrt{1-C^2}} \\
&= \frac{1}{B(\frac{d-1}{2},\frac{d-1}{2})} \cdot \left(\frac{(1-C)(1+C)}{2\cdot 2}\right)^{\frac{d-2}{2}} \cdot ((1-C)(1+C))^{-\frac{1}{2}} \\
&= \frac{1}{B(\frac{d-1}{2},\frac{d-1}{2})} \cdot \left(1-\tfrac{1+C}{2}\right)^{\frac{d-1}{2}-1} \cdot \left(\tfrac{1+C}{2}\right)^{\frac{d-1}{2}-1} \cdot \tfrac{1}{2} \\
&= \tfrac{1}{2} B\left(\tfrac{1+C}{2}; \tfrac{d-1}{2}, \tfrac{d-1}{2}\right)
\end{aligned}
$$

which is a beta distribution rescaled to the interval $[-1, 1]$, on which C is defined.

Based on this, we can easily obtain the following helpful corollary:

Corollary 1. *The average cosine similarity of two random points sampled independently and uniformly from a d-ball is given by*

$$\mathbb{E}[C] = 0.$$

The variance and the non-central second moment are given by

$$\mathrm{Var}(C) = \mathbb{E}[C^2] = \tfrac{1}{d}.$$

Proof. This follows immediately from the central moments of beta distributions. By Theorem 2 we have $\frac{1+C}{2} \sim B(\frac{d-1}{2}, \frac{d-1}{2})$. This *symmetric* beta distribution has a mean of $\frac{1}{2}$, and hence $\mathbb{E}[C] = 0$. The variance of this beta distribution given $\alpha = \beta = \frac{d-1}{2}$ is $\mathrm{Var}(\frac{1+C}{2}) = \frac{1}{4d}$, and hence $\mathbb{E}[(\frac{1+C}{2} - \frac{1}{2})^2] = \mathbb{E}[\frac{C^2}{4}] = \frac{1}{4d}$. Because the mean is 0, the variance and the second non-central moment agree trivially.

4   Estimating Intrinsic Dimensionality

Based on this theoretical distribution of cosine similarities in a d-ball, we propose new estimators of intrinsic dimensionality based on the method of moments. Similar to other estimators, we assume the data sample comes from the local neighborhood of a point; usually from a ball. The first moment of Corollary 1 cannot be used for estimation because it does not depend on d. Both the variance and the second non-central moment, however, are suitable for estimating intrinsic dimensionality, as they depend inversely on d. This simple dependency stands in contrast to the expansion-rate based approaches, which generally obtain an exponential relation to the dimensionality, as the volume of a d-ball has d in its exponent. Hence, we hope to obtain a more robust measure even with smaller neighborhood sizes (fewer samples); and as we do not need to compute logarithms it can be computed more efficiently. But we still have two choices: we can either estimate using the variance $\hat{d} = 1/\operatorname{Var}(C)$ or using the non-central second moment $\hat{d} = 1/\mathbb{E}[C^2]$, which only agrees if $\mathbb{E}[C] = 0$ as expected for a uniform ball.

Consider the scenario of many points sampled from a hyperplane, but the point of interest is not on this hyperplane. The local neighborhood will then consist of samples in a circular region on this plane. If we move the point of interest away from the plane, the average cosine tends to 1, and the variance to 0. The variance-based estimate would hence tend to infinity, while the second non-central moment estimate will tend to 1. We argue that this is the more appropriate estimate, as the data concentrates in a single far away area.

Inspired by Amsaleg et al. [3], we investigated the idea of considering the reflections of all points with respect to the point of interest. Such a reflection causes the average cosine in this example to be 0, as every pair of points can be matched to the pair with the second point reflected. In the above example, we would obtain two opposite discs of points and the resulting variance would tend to 1. When adding reflected points, the variance and the non-central second moment become equivalent: Since $c(x_i, -x_j) = -c(x_i, x_j) = c(-x_i, x_j)$, adding reflections yields two positive and two negative copies of each cosine. The resulting average then is 0, and hence $\operatorname{Var}(C') = \mathbb{E}[C'^2] - \mathbb{E}[C']^2 = \mathbb{E}[C'^2] = \mathbb{E}[C^2]$. We, therefore, do not further consider using such reflections of points, besides their implicit presence in the non-central second moment.

Instead of discussing the limit cases of distributions, we will now work with a fixed data sample of k points, centered around a point of interest. For simplicity, we assume that the data has been translated such that the point of interest is always at the origin, and that this point and any duplicates of it have been removed from the sample. We now use *all pairwise* cosine similarities in a $k \times k$ matrix denoted C. The diagonal of this matrix is usually excluded from computations. We use the term C_1 when the ones on the diagonal are to be included. By C^2, we denote the individual squaring of cosines. The next theorem will use both a fixed sample and the matrix C_1 with the diagonal included.

Theorem 3 (Upper bound). *Let $X = \{x_1, \ldots, x_k\} \subset \mathbb{R}^D$ be a sample from a d-dimensional subspace embedded in $\mathbb{R}^D$ for some $d \leq D$. Formally, let X contain at least d linearly independent vectors and let all x_i be linear combinations of a given set of d orthonormal basis vectors. Then the following inequality holds*

$$\mathbb{E}[C_1^2]^{-1} \leq d.$$

Proof. Let $\tilde{X}$ be the $k \times d$ matrix obtained from X by first performing a change of basis to the given orthonormal basis of size d, then normalizing each vector to unit length to produce $\tilde{x}_i$. Neither the change of basis (which is a rotation) nor the posterior normalization affects the cosine similarities, and we hence have

$$c(\tilde{x}_i, \tilde{x}_j) = c(x_i, x_j). \tag{4}$$

It immediately follows that $\tilde{X}$ has a rank of d, as we still have d linearly independent vectors. The matrix $\tilde{C}_1 = \tilde{X}\tilde{X}^T$ then contains entries of the form $\langle \tilde{x}_i, \tilde{x}_j \rangle$. As all $\tilde{x}_i$ are normalized, $\tilde{C}_1$ is equal to the cosine similarities. Per Eq. (4) it then follows that $\tilde{C}_1$ is exactly C_1. Because C_1 is a cosine similarity matrix, the diagonal entries are all 1, and we have $\mathrm{tr}(C_1) = k$. Since $\tilde{X}$ is a $k \times d$ matrix with rank d, we know that the rank of C_1 is d as well. Therefore C_1 has d eigenvalues $\lambda_1, \ldots, \lambda_d$ with $\sum_{i=1}^d \lambda_i = \mathrm{tr}(C_1) = k$. The sum of squared entries $\|C_1\|_2^2$ equals the sum of squared eigenvalues $\sum_{i=1}^d \lambda_i^2$ and is minimized if every eigenvalue equals $\frac{k}{d}$, which means we have the following lower bound:

$$\|C_1\|_2^2 = \sum_{i=1}^d \lambda_i^2 \geq d \cdot \left(\frac{k}{d}\right)^2 = \frac{k^2}{d} \tag{5}$$

and by taking the inverse we obtain the upper bound $\mathbb{E}[C_1^2]^{-1} \leq d$.

This is an upper bound for estimating the intrinsic dimensionality using C_1, and we can use this to also obtain an upper bound for C.

Corollary 2. *Let $X = \{x_1, \ldots, x_k\} \subset \mathbb{R}^D$ be a sample from a d-dimensional subspace embedded in $\mathbb{R}^D$ as defined in Theorem 3. If $k > d$, then*

$$\mathbb{E}[C^2]^{-1} \leq \frac{k-1}{k-d} \cdot d. \tag{6}$$

Proof. As the difference between C and C_1 is the diagonal of ones, Eq. (5) yields

$$\|C\|_2^2 = \|C_1\|_2^2 - k \geq \frac{k^2}{d} - k = \frac{k(k-d)}{d}$$

and hence the average of the remaining $k^2 - k$ cells is

$$\mathbb{E}[C^2] \geq \frac{k-d}{k-1} \cdot \frac{1}{d}$$

which is equivalent to the inequality above. For $k = d$ we obtain a trivial bound.

The difference of including the diagonal or not vanishes for large enough k. One could attempt to regularize $\mathbb{E}[C^2]$ with $\frac{k-1}{k-d}$. The major problem therein is that we do not know d in advance. To control the maximal overestimation of d, a sufficiently large neighborhood can be used to lower the margin of error. For example, to bound $\mathbb{E}[C^2]^{-1} \leq d + c$, at least $k \geq \frac{1}{c}d^2 + (1 - \frac{1}{c})d$ neighbors are required. For the bound $d + 1$ ($c = 1$), this means we require $k \geq d^2$ samples.

To solve this self-referential problem, we can also attempt an iterative refinement. It turns out that the fixed point of this regularization yields exactly the result we obtain by using C_1 instead of C. Because using C_1 corresponds to using a regularized version and because it has a very elegant upper bound, we base our method on this estimate:

Definition 1 (ABID). *Given a data set* $X = \{x_1, \ldots, x_n\} \subset \mathbb{R}^D$, *the regularized angle-based intrinsic dimensionality estimator for a point* x_i *is:*

$$ID_{ABID}(x_i; k) := \mathbb{E}[C_1(B_k(x_i))^2]^{-1}$$

where $B_k(x_i)$ *are the directional vectors from* x_i *to the* k *nearest neighbors of* x_i *and* $C_1(B_k(x_i))$ *are the pairwise cosine similarities within* $B_k(x_i)$.

By choosing the neighborhood of any point in the specified set by the k nearest neighbors, the measure is invariant under scaling. Analogously, one can instead define the neighborhood by a maximum distance to the central point. The sole restriction thereby is that the size of the neighborhood has to be greater or equal to $d+2$ as for any smaller neighborhood, the estimator does not need to be properly regularized. Since the error of the non-regularized estimate is limited for any neighborhood with size quadratic in the intrinsic dimension, we further introduce a non-regularized version for comparative analysis.

Definition 2 (RABID). *Given a data set* $X = \{x_1, \ldots, x_n\} \subset \mathbb{R}^D$, *the raw angle-based intrinsic dimensionality estimator for a point* x_i *is defined as*

$$ID_{RABID}(x_i; k) := \mathbb{E}[C(B_k(x_i))^2]^{-1}$$

where $B_k(x_i)$ *are the directional vectors from* x_i *to the* k *nearest neighbors of* x_i *and* $C(B_k(x_i))$ *are the pairwise cosine similarities of* different *vectors in* $B_k(x_i)$.

Beware that this estimator can cause a division by zero if all k vectors are pairwise orthogonal, and can return values larger than k. In such cases, it is recommended to treat the estimate as k, because the input vectors span a k dimensional subspace. Nevertheless, this estimator is likely unstable for small k, and for large k, it converges to ID_{ABID}.

To interpret the estimates by the new method, it is important to consider the domain they operate on. The angle-based measure is bounded by the spanning dimensionality of the point set. While distributions of angles are usually distorted by non-linear transformations, many transformations such as rotations will retain this bound. Hence the bound may nevertheless apply—at least approximately—for many projections of lower-dimensional manifolds in higher dimensional embeddings. It is easy to see that angle-preserving transformations do not affect our measure, while distance-preserving transformations will not affect distance-based estimators. Our new measure is less affected by local non-linear contractions and expansions such as the decreasing density on the outer parts of Gaussian distributions, but it tends to estimate higher dimensionality than distance-based-approaches when the transformations are non-linear. We do not consider this to be a flaw, just a different design that may or may not have advantages: a common assumption in many methods and applications like manifold learning is to have locally linear transformations that preserve small neighborhoods, which will then affect neither angles nor densities. Our estimator, which can be seen as estimating how many dimensions such a locally linear embedding needs to have, is arguably very close to the idea of such applications.

5   Evaluation

In our comparative evaluation, we consider several ID estimators on many standard evaluation data sets of both artificial and natural origin. As measures of quality, we analyze the estimated ID's consistency both with expected values (for synthetic data) and with each other (for natural data with no true value). We will further inspect the stability of ID estimates for varying neighborhood sizes. Depending on the density of data sets, approaches that require a large neighborhood to stabilize, tend to be inapplicable.

The histograms shown in this section are limited to a region of interest in both x and y direction for interpretability. Outside of the presented range along the x-axis, the distributions always show a smooth drop to zero with no further peaks but may have a long tail.

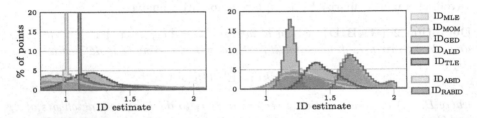

Fig. 3. Histograms of ID estimates of points sampled from a Koch snowflake using the 10 (left) respectively 200 (right) nearest neighbors.

5.1   Reference Estimators

We compare ID_{ABID} and ID_{RABID} to the Hill estimator ID_{MLE} [10], the measure-of-moments-based estimator ID_{MOM} [1], the generalized expansion dimension ID_{GED} [14], the augmented local ID estimator ID_{ALID} [9] and its successor, the tight LID estimator ID_{TLE} [3] using the implementations in the ELKI framework [20]. All of these alternative estimators are based on the expansion rate. The ID_{TLE} is supposed to reduce the necessary sample size in the neighborhood to acquire a good estimate, yet in our experiments tends to give higher estimates than the other distance-based approaches.

5.2   Dimensionality of Fractal Curves

In line with the theoretical foundation of this work and to demonstrate the different semantics of angle-based and distance-based ID estimation, we analyze the estimated ID of a well-known fractal, the Koch snowflake. As seen in Fig. 3, most distance-based approaches estimate a dimensionality roughly around $\frac{\log 4}{\log 3} \approx 1.26$, which is the Hausdorff dimension of the Koch snowflake, when we consider enough neighbors ($k = 200$). This result is not surprising, as the distance-based approaches are conceptually closely related to classical fractal dimensions. Our angle-based estimates, however, estimate a dimensionality of ≈1.6 for larger neighborhoods, which is larger than the fractal dimension, yet smaller than the representation dimension. The difference can be explained by the highly non-linear shape of the snowflake, as two consecutive line segments are overlapping in a singularity. Whilst points sampled from a finite recursion Koch snowflake lie on a curve, they might be locally indistinguishable from samples from $\mathbb{R}^2$. A higher ID estimate may turn out to be more robust for downstream applications such as manifold learning. The results on further fractals were similar.

It is noteworthy that the scale of the neighborhood has a large impact on the estimates. When choosing a neighborhood small enough to mostly stay within a line segment of the fractals (here $k = 10$), the ID estimates approximate 1, as most neighborhoods lie on straight lines. For larger neighborhoods, the estimates approach a proper representation of the manifold space. For too large neighborhoods, however, boundaries of the point set as well as observing points distant on the manifold, yet close in the embedding space, tend to corrupt the estimates.

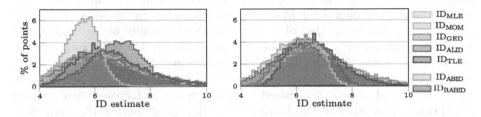

Fig. 4. Histograms of ID estimates of the m6 set with 30 resp. 150 neighbors.

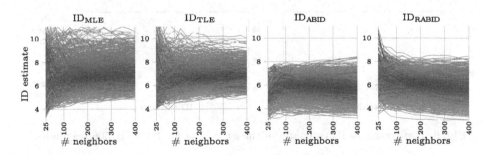

Fig. 5. Trails of estimates of 1000 points for varying neighborhood sizes on the m6 set. Trail colors are assigned in order of ID estimates at 200 neighbors.

5.3 Synthetic Data

Amsaleg et al. [2,3] provide a collection of synthetic and real data sets. The m6 data set consists of points sampled from a 6-dimensional manifold non-linearly embedded in a 36-dimensional space. As can be seen in Fig. 4, for $k = 150$ all estimators agree on the data set to be inherently 6-dimensional at most points. Where distance-based estimators tend to have a long tail towards higher dimensions, the angle-based approaches have an upper bound. Even though this nonlinearity shifts the upper bound beyond 6, the angle-based approaches tend to have a shorter upper tail and drop off faster to zero. When comparing the estimates of each method at different neighborhood sizes between 30 and 300, the angle-based approaches achieve higher scores than the distance-based approaches on both Spearman's and Pearson's correlation of estimates with adjacent neighborhood sizes. In this sense, the angle-based approaches are more stable both in the value as well as the order when varying neighborhood sizes. In more extreme neighborhoods ($<$30 and $>$300), artifacts from having too few samples for a reliable estimate and reaching the boundaries of the data set, respectively, cause results to become less stable. The stability is visualized in Fig. 5 using trails of ID estimates for individual points when varying the neighborhood size. In a perfectly stable result, all lines would be parallel; instability causes lines that cross outside their own color range (which represents the order at $k = 200$) and hence the mixing of the colors. The improved stability of the angle-based estimates is shown by a fairly stable plot from 125 to 300 neighbors, whereas the distance-based estimates deviate much more at $\leq$150 and $\geq$250 neighbors respectively already. Additionally, we can see in this plot that the average (the purple region) of the distance-based estimates tends to increase with growing neighborhood size whereas the distribution of ID_{ABID} appears stable upwards of 100 neighbors. We can observe the upper bound property of ID_{ABID} compared to ID_{RABID}. The higher stability means that smaller neighborhoods suffice for estimation and that the neighborhood parameter is easier to choose.

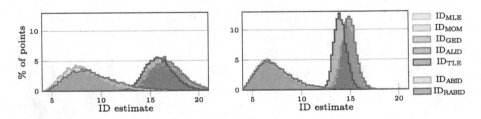

Fig. 6. Histograms of ID estimates of the m10c data for 100 resp. 500 neighbors.

The data set of Amsaleg et al. [3] with the highest intrinsic dimension, m10c, is a 24-dimensional hypercube embedded in 25 dimensions. Here, we observed a larger discrepancy between the approaches, shown in Fig. 6. However, m10c consists of only 10000 points, which is the number of corners of a $\log_2(10000) \approx 13$

dimensional hypercube. Hence, we doubt that this small sample can reliably represent a full 24-dimensional manifold, but the data likely is of much lower dimensionality. The estimates of the distance-based approaches move towards this value as the neighborhood size increases. Each of these 10000 points then is essentially the corner of a 13-dimensional hypercube; and will see the other data points as forming a hypercone, producing smaller angles than if the data would evenly surround the point. We believe it is because of this effect (essentially a variant of the curse of dimensionality) that the angle-based approaches estimate a far lower ID.

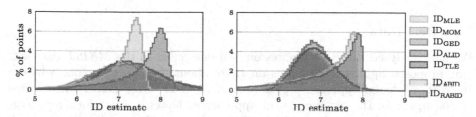

Fig. 7. Histograms of ID estimates of points on an 8-dimensional noisy lattice using 100 respectively 500 nearest neighbors.

To support this theory, we created a data set consisting of one point sampled from each cell an 8-dimensional grid of $4^8 = 65536$ points. We obtain a data set that is more evenly distributed than uniform random sampling and truly spans an 8-dimensional space. On this data set, only the angle-based approaches were able to estimate the correct dimension for most points as can be seen in Fig. 7. The points where ID_{ABID} and ID_{RABID} estimate lower values are likely the *many* points at the corners, edges, and sides of this lattice.

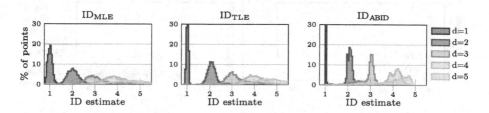

Fig. 8. Histograms of ID estimates of nested hypercubes with a neighborhood size of 100 colored by the hypercube from which they were sampled.

To test the reliability of estimators in a mixture of manifolds, we created instances of 1- through 5-dimensional hypercubes linearly projected into the same 5-hypercube. The projection was chosen such that every d_i-dimensional hypercube intersects every d_j-dimensional hypercube in a $\min(d_i, d_j)$-dimensional subspace. For every hypercube, we sampled 5000 points uniformly at random and computed ID estimates using a neighborhood of different sizes. In all experiments, the angle-based approaches were visibly more capable of differentiating between the different dimensional subspaces, which can be seen from the sharper spikes in Fig. 8. Being capable of separating lower-dimensional subspaces is an important feat, as noise in the embedding space can be considered a high-dimensional manifold containing the manifold of interest, and we believe this new ID estimate may help subspace discovery approaches that, based on intrinsic dimensionality (e.g, [6]). We observe that the angle-based approaches are more robust against noise and in the presence of overlapping subspaces.

5.4 Real Data

We also analyzed the ID estimates on real data such as the *MNIST* data set. Our proposed approach estimates an ID of about 6 for most points, whereas the distance-based approaches peak around 10 to 11. From neighborhood sizes of 100 upwards, the distance-based approaches, however, start forming a second peak at the same ID as the angle-based approaches, visible in Fig. 9. A possible explanation could be that the *MNIST* data set is not uniformly random on the manifold, whereby small environments are too noisy for distance-based approaches. On the 5000-dimensional variant *Gisette* of this data set, the added noise harshly increased the estimates of the distance-based approaches. The angle-based approaches, however, estimate a slightly lower ID of about 4 for most points. We consider the smaller change of the angle-based estimates as more plausible, even though the proposed estimates for the *Gisette* data set might be slightly too low as the high-dimensional noise might have sparsened the local neighborhoods too much. Nevertheless, we observe that the angle-based approach can be more robust against noise in such a semi-real scenario.

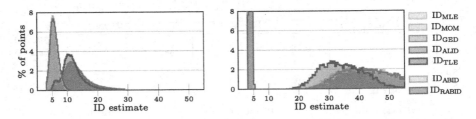

Fig. 9. Histograms of ID estimates for *MNIST* and *Gisette* with 300 neighbors.

5.5 Estimator Interactions

These experiments can also give some insight into the differences and interactions of the different estimators. As expected from theory, ID_{ABID} and ID_{RABID} converge towards the same value for sufficiently large neighborhood sizes.

Because it is trivial to compute both estimators at once, we can use the difference of the estimates to assess the quality; if they differ much we may need larger sample sizes, if they are close the sample size should be sufficient for this dimensionality. Because the angle-based estimators appear to require fewer samples than the distance-based approaches, this may also help to choose neighborhood sizes for these methods. Secondly, if the angle-based estimates are much smaller than the distance-based estimates, the data set might not be sufficiently densely sampled for this dimensionality; if the angle-based estimates are much larger than the distance-based estimates, the embedding may be highly non-linear (as in the Koch snowflake example), or may not preserve local density.

6   Conclusions

In this paper, we propose a novel approach to estimate local intrinsic dimensionality, along with two estimators, ID_{ABID} and ID_{RABID}. Instead of analyzing the expansion rate, as previous distance-based approaches do, the novel approach focuses on the geometry characterized by pairwise angles. We have given an a priori derivation of the novel estimators derived from integral geometry. Our experimental evaluation suggests that the novel approach may be more robust against noise, computes a bit stabler estimates, and gives estimates as reasonable as distance-based estimators, albeit of a different nature. We have further discussed how the difference between estimates can hint at particular effects in the data. The presented approach does not yet fully utilize all interactions of the pairwise angles within a neighborhood, which could lead to an improved ID estimation in future work by incorporating ideas of [3]. Future work may also investigate using higher-order moments, as well as robust estimation techniques for the second moment, such as the median average deviation or L-moments [11]. As $\mathbb{E}[C] \gg 0$ indicates points remote from their neighbors, this can be interesting to integrate into an outlier detection method based on intrinsic dimensionality, which would yield a hybrid of ABOD [17] and LID outlier detection [15].

References

1. Amsaleg, L., et al.: Estimating local intrinsic dimensionality. In: KDD, pp. 29–38 (2015). https://doi.org/10.1145/2783258.2783405
2. Amsaleg, L., et al.: Extreme-value-theoretic estimation of local intrinsic dimensionality. Data Min. Knowl. Discov. **32**(6), 1768–1805 (2018). https://doi.org/10.1007/s10618-018-0578-6
3. Amsaleg, L., et al.: Intrinsic dimensionality estimation within tight localities. In: SDM, pp. 181–189 (2019). https://doi.org/10.1137/1.9781611975673.21
4. Aumüller, M., Ceccarello, M.: The role of local intrinsic dimensionality in benchmarking nearest neighbor search. In: Amato, G., Gennaro, C., Oria, V., Radovanović, M. (eds.) SISAP 2019. LNCS, vol. 11807, pp. 113–127. Springer, Cham (2019). https://doi.org/10.1007/978-3-030-32047-8_11
5. Barua, S., Ma, X., Erfani, S.M., Houle, M.E., Bailey, J.: Quality evaluation of GANs using cross local intrinsic dimensionality. CoRR abs/1905.00643 (2019)

6. Becker, R., Hafnaoui, I., Houle, M.E., Li, P., Zimek, A.: Subspace determination through local intrinsic dimensional decomposition. In: Amato, G., Gennaro, C., Oria, V., Radovanović, M. (eds.) SISAP 2019. LNCS, vol. 11807, pp. 281–289. Springer, Cham (2019). https://doi.org/10.1007/978-3-030-32047-8_25
7. Bratic, B., Houle, M.E., Kurbalija, V., Oria, V., Radovanovic, M.: The influence of hubness on NN-descent. Int. J. Artif. Intell. Tools **28**(6), 1960002:1–1960002:23 (2019). https://doi.org/10.1142/S0218213019600029
8. Cai, T.T., Fan, J., Jiang, T.: Distributions of angles in random packing on spheres. J. Mach. Learn. Res. **14**(1), 1837–1864 (2013)
9. Chelly, O., Houle, M.E., Kawarabayashi, K.: Enhanced estimation of local intrinsic dimensionality using auxiliary distances. Technical report NII-2016-007E, National Institute of Informatics (2016)
10. Hill, B.M.: A simple general approach to inference about the tail of a distribution. Ann. Stat. **3**(5), 1163–1174 (1975). https://doi.org/10.1214/aos/1176343247
11. Hosking, J.R.M.: L-moments: analysis and estimation of distributions using linear combinations of order statistics. J. Royal Stat. Soc. B **52**(1), 105–124 (1990). https://doi.org/10.2307/2345653
12. Houle, M.E.: Local intrinsic dimensionality I: an extreme-value-theoretic foundation for similarity applications. In: Beecks, C., Borutta, F., Kröger, P., Seidl, T. (eds.) SISAP 2017SISAP 2017. LNCS, vol. 10609, pp. 64–79. Springer, Cham (2017). https://doi.org/10.1007/978-3-319-68474-1_5
13. Houle, M.E.: Local intrinsic dimensionality II: multivariate analysis and distributional support. In: Beecks, C., Borutta, F., Kröger, P., Seidl, T. (eds.) SISAP 2017. LNCS, vol. 10609, pp. 80–95. Springer, Cham (2017). https://doi.org/10.1007/978-3-319-68474-1_6
14. Houle, M.E., Kashima, H., Nett, M.: Generalized expansion dimension. In: ICDM Workshops, pp. 587–594 (2012). https://doi.org/10.1109/ICDMW.2012.94
15. Houle, M.E., Schubert, E., Zimek, A.: On the correlation between local intrinsic dimensionality and outlierness. In: Marchand-Maillet, S., Silva, Y.N., Chávez, E. (eds.) SISAP 2018. LNCS, vol. 11223, pp. 177–191. Springer, Cham (2018). https://doi.org/10.1007/978-3-030-02224-2_14
16. Huisman, R., Koedijk, K.G., Kool, C.J.M., Palm, F.: Tail-index estimates in small samples. J. Bus. Econ. Stat. **19**(2), 208–216 (2001). https://doi.org/10.1198/073500101316970421
17. Kriegel, H., Schubert, M., Zimek, A.: Angle-based outlier detection in high-dimensional data. In: KDD, pp. 444–452 (2008). https://doi.org/10.1145/1401890.1401946
18. Radovanovic, M., Nanopoulos, A., Ivanovic, M.: Hubs in space: popular nearest neighbors in high-dimensional data. J. Mach. Learn. Res. **11**, 2487–2531 (2010)
19. Schubert, E., Gertz, M.: Intrinsic t-stochastic neighbor embedding for visualization and outlier detection - a remedy against the curse of dimensionality? In: Beecks, C., Borutta, F., Kröger, P., Seidl, T. (eds.) SISAP 2017. LNCS, vol. 10609, pp. 188–203. Springer, Cham (2017). https://doi.org/10.1007/978-3-319-68474-1_13
20. Schubert, E., Zimek, A.: ELKI: a large open-source library for data analysis - ELKI release 0.7.5, Heidelberg. CoRR abs/1902.03616 (2019)
21. Watanabe, S.: Algebraic Geometry and Statistical Learning Theory, vol. 25. Cambridge University Press, Cambridge (2009). https://doi.org/10.1017/CBO9780511800474

Sampled Angles in High-Dimensional Spaces

Richard Connor[✉] and Alan Dearle

University of St Andrews, Jack Cole Building, North Haugh, St Andrews,
Fife KY16 9SX, Scotland, UK
{rchc,al}@st-andrews.ac.uk

Abstract. Similarity search using metric indexing techniques is largely a solved problem in *low-dimensional* spaces. However techniques based only on the triangle inequality property start to fail as dimensionality increases.

Since proper metric spaces allow a finite projection of any three objects into a 2D Euclidean space, the notion of *angle* can be validly applied among any three (but no more) objects. High dimensionality is known to have interesting effects on angles in vector spaces, but to our knowledge this has not been studied in more general metric spaces. Here, we consider the use of angles among objects in combination with distances.

As dimensionality becomes higher, we show that the variance in sampled angles reduces. Furthermore, sampled angles also become correlated with inter-object distances, giving different distributions between query solutions and non-solutions. We show the theoretical underpinnings of this observation in unbounded high-dimensional Euclidean spaces, and then examine how the pure property is reflected in some real-world high dimensional spaces. Our experiments on both generated and real world datasets demonstrate that these observations can have an important impact on the tractability of search as dimensionality increases.

Keywords: Metric search · High dimensional space

1 Introduction

The context of interest is searching a (large) finite set of objects S which is a subset of an infinite set U, where (U, d) is a metric space: that is, a pair (U, d), where U is a domain of objects and d is a total distance function $d : U \times U \to \mathbb{R}$, satisfying postulates of non-negativity, identity, symmetry, and triangle inequality. The general requirement is to efficiently find members of S which are similar to an arbitrary member of U given as a query, where the distance function d gives the only way by which any two objects may be compared. There are many important practical examples captured by this general mathematical framework, see for example [4,14]. There are two main types of query: *range* and

© Springer Nature Switzerland AG 2020
S. Satoh et al. (Eds.): SISAP 2020, LNCS 12440, pp. 233–247, 2020.
https://doi.org/10.1007/978-3-030-60936-8_18

nearest-neighbour search. The *range search* for some query $q \in U$ and threshold $t \in \mathbb{R}$ is defined as having the solution set $R = \{s \in S \mid d(q,s) \leq t\}$. More practical in many contexts is the *nearest-neighbour (kNN)* search where the solution set comprises the k closest objects to a query.

The essence of metric search is to spend time pre-processing the finite set S so that solutions to queries can be efficiently calculated. In all cases distances among members of S and selected *reference* or *pivot* objects are calculated during pre-processing. At query time the relative distances between the query and the same pivot objects can be used to make deductions about which data values may, or may not, be candidate solutions to the query. Such deductions are based upon the *triangle inequality* property of the metric.

1.1 Distances and Angles

In this paper we consider not just the measured distances among objects, but also the angles implied by these distances. In any metric space, the triangle inequality property also implies a finite 3-embedding in 2D Euclidean space [5], and so it is valid to discuss the *angles* of a triangle constructed according to the distances among any three objects selected from the space. It is important to stress that, in this paper, this is the only notion of angle that we use; thus our discussion is valid with respect to any proper metric space, not just vector spaces.

In the context of metric search, we are interested in the distribution of angles $\angle pqs_i$ where objects p and q are fixed, and s_i is sampled within a relatively small bounded distance from q. This is typical of a situation where p represents some reference object, q represents a query object, and s_i is sampled from the solution objects of the query. Note that the situation described generalises to both range and nearest-neighbour queries.

In general, we compare this distribution of angles with the alternative distribution of $\angle pqx_i$ where the same p and q are selected, but where x_i is sampled from the entire metric space without constraint. We find that in many cases, especially in high-dimensional spaces, these distributions differ significantly. This information can be used to effect within existing metric access methods, and furthermore gives a geometric explanation to phenomena that have been previously observed, but not previously explained, in approximate indexing techniques.

The main observation of this paper is that the following usually hold in high-dimensional metric spaces:

- if three values p, q, x are randomly sampled, then the mean of the angle $\angle pqx$ is 60°, and the variance in this measurement decreases as dimensionality increases;
- however if values p, q are fixed, and then s_i is sampled from a sufficiently small fixed distance bound of q, then the mean of the angle $\angle pqs_i$ is greater than 60°, and again the variance decreases as dimensionality increases.

We show how these observations can be used to effect in approximate search techniques.

1.2   Contributions

It is generally known that, as the dimension of a vector space increases, the probability of two independently selected vectors being close to orthogonal increases. In Sect. 3 we show that in an unbounded Euclidean space, for any values of a and b, and c sampled within a distance bound t of b, the mean angle $\angle abc$ is $90°$, and the variance decreases according to the dimension of the space. This corollary allows the angles to be calculated using only the distances among three objects, rather than from the values of two vectors, and thus allows the possibility of extension into general metric spaces.

We show that this effect can be used to determine an upper bound on the probability of a randomly selected point lying in the intersection of hyperspheres centred on a and b, and how this probability may be used to construct approximate search mechanisms. We show that this probability is related to the function $\sin^n \theta$ where n is the Euclidean dimension. This can lead to very low probabilities in some cases, and thus highly accurate approximations.

In Sect. 4 we show by experiment that this theory holds perfectly in a bounded uniform Euclidean space, as long as the position of b and the distance d are fixed to ensure that the hypersphere described by $<b, d>$ is fully enclosed in the space. However, in many high-dimensional search spaces, this does not hold. This is because the hypersphere $<q, t>$, where q is a query object and t is a distance bound which includes elements of the finite search space, may include a significant region that lies outside the boundaries of the space. Nonetheless for such spaces there still exists a predictable distribution of sampled angles which is different from randomly sampled angles. We introduce an observed correlation between *outlierness* and the distribution of angles in these spaces.

Finally in Sect. 5 we study some "real-life" high-dimensional metric search spaces to check if the theoretical observations still hold. We find that, while compromised from the pure model, there is still a useful distinction between the angle distributions of query solutions and non-solutions, and furthermore this is observable in non-Euclidean metric spaces as well as Euclidean spaces. We show some experiments which use a variant of LAESA to demonstrate a practical application of our observations. For all of the datasets used, a relaxation on the exclusion condition based on the angle-enhanced analysis allows substantially more exclusion while still maintaining almost perfect accuracy.

2   Related Work

The distribution of vectors within high-dimensional vector spaces is discussed in a book chapter by Hopcroft and Kannan [2]. This introduces the notion of the $90°$ angle norm in discussion of an "annulus" within the hypersphere. However, the subtleties of the hypersphere being only partially embedded within the data space are overlooked.

In [3], the authors state that it is "a matter of folklore" that "all high dimensional random vectors are almost always nearly orthogonal to each other". They quantify this with a probability density function, directly proportional to

$\sin^{(n-2)}\theta$, for the angle between two vectors randomly sampled from a uniform distribution on the surface of an n-dimensional hypersphere. This is consistent with our distribution, which is proportional to $\sin^n\theta$, considering points uniformly distributed within a segment of the hypersphere. They also observe the close relationship between the function $\sin^n\theta$ and a related normal distribution, as we use in Sect. 3.

Pramanik et al. [12] give an expression for the volume of a hypersphere in an n-dimensional Euclidean space, and imply a derived PDF for angles which is proportional to $\sin^n\theta$. They do not give a derivation of their volume formula. Perhaps as a result of this, their implication of the relationship between volume and probability is incorrect, and in fact the density function should be proportional to $\sin^{(n-2)}\theta$ as above.

They go on to use an angle-based relaxation of a ball partitioning mechanism to improve performance of a single-pivot mechanism, the *AB Tree*, which they show to be effective in terms of increased performance versus a small loss in accuracy. There are however a number of issues with their presentation which we clarify in our work. First, they overlook the fact that as dimensions increase the theoretical distribution becomes ever less true due to the inability of a bounded data space to contain the query hypersphere. They present graphs showing a perfect distribution over a search space which we have been unable to reproduce. Most importantly, the conceptual basis of their optimisation depends on the angle $\angle qp_i s_j$, where q is the query, p_i is the centre of a ball partition, and s_j is a potential solution to the query. This implies that the radius of the search space around the query object is larger than the radius of the data which is being pivoted by p_i, which could not occur in a high-dimensional space.

We have been able to synthesise large numbers of uniformly distributed values within very low-volume hyperspheres thanks to a technique shown by Voelker et al. [13]. This has been extremely valuable as, in high dimensions, it is effectively impossible to otherwise find uniformly distributed points within a given hypersphere.

Finally, there are many papers that give results based on relaxing the strict condition of triangle inequality in ball partitioning, increasing the efficiency of the mechanisms at cost of giving approximate results. We believe that our work here goes a long way to explain the effectiveness of such mechanisms in high-dimensional spaces.

3   Unbounded Euclidean Spaces

The orthogonality of vectors in high dimensional spaces, once quantified, can give useful insights with respect to single pivot exclusion, in particular towards assessing the probability of an object lying within the intersection of two hyperspheres. Traditional metric search techniques allow only that this is zero when the sum of the radii is greater than the distance between the centres; we are interested in quantifying the probability of an object from the finite space lying within an intersection of the infinite space. However given a high probability of restricted angles in high dimensional spaces, many overlapping hyperspheres will have a very low probability of the geometric intersection containing any elements of a given finite space.

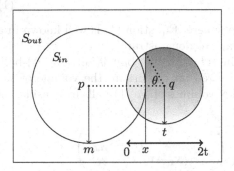 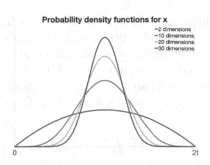

Fig. 1. On the left hand side, the figure shows how the intersection of the hyperspheres is contained in the segment of the hypersphere around q defined by θ. The graph on the right hand side shows the probability of a randomly selected point from within the hypersphere also being in this segment, as x in the left hand figure varies between 0 and $2t$. This therefore gives an upper bound on the probability of such an object in the intersection.

The left hand side of Fig. 1 shows how a restriction of angles is useful in metric search. p and q represent hypersphere centres, where a finite metric space S has been divided during pre-processing into S_{out} and S_{in} according to a distance m from a pivot p. q represents a query, to which solutions are being sought within the threshold distance t. Since the hyperspheres intersect, according to the distance $d(p, q)$ calculated at query time, S_{in} cannot be excluded using the metric properties alone.

With respect to the hypersphere centred around q, consider the segment defined by the angle θ. If, for all elements $s_i \in S$ such that $d(q, s_i) \leq t$, the angle $\angle pqs_i$ is greater than θ, then the finite intersection is empty and the set S_{in} can be excluded from the search. If there is a high probability of vectors pq and qs_i being close to orthogonal, there will be a correspondingly high probability of the intersection being empty.

In a general Euclidean space of course this can never be guaranteed; however as we will show, as the dimension of the space increases, the probability of an individual point from a uniform distribution being within the intersection may become very small. The right hand side of Fig. 1 gives probability density functions (PDFs), in various dimensions of Euclidean space, for the displacement x for a randomly selected point within the solution space.

It can be seen that, for higher dimensions, the probability of a point lying with the intersection is very low. We will proceed to give a quantification of an upper bound which is easily calculated. In the remainder of this section, we derive a PDF and quantify the examples shown in the figure.

3.1   Volume of a Hypersphere

The volume of a hypersphere of radius r can be expressed in terms of the volume of the unit hypersphere (i.e. $r = 1$) as

$$V_n(r) = v_n r^n \tag{1}$$

where v_n is the volume of the unit hypersphere. Equation (1) is well known in a more general context, and straightforward to demonstrate[1].

The intersection of a hyperplane in $\mathbb{R}^n$ with a n-ball is an $(n-1)$-ball. Considering a unit $(n-1)$-ball b_{n-1} centred on the origin, the volume of the unit n-ball can be written as an integral of volumes of $(n-1)$-balls by considering hyperplanes orthogonal to the X_1-axis:

$$v_n = \int_b dx_1 \dots dx_n = \int_{-1}^1 \left(\int_{b_{n-1} \cap \{X_1 = z\}} dx_2 \dots dx_n \right) dz \tag{2}$$

As depicted in the left-hand side of Fig. 2, the intersection $b_n \cap \{X_1 = x\}$ is an $(n-1)$-ball of radius $r = \sqrt{1-x^2}$, thus its volume is $V_{n-1}(\sqrt{1-x^2})$ and Eq. (2) can be rewritten as

$$v_n = \int_{-1}^1 V_{n-1}(\sqrt{1-x^2}) dx$$

which then, according to Eq. 1 gives

$$v_n = v_{n-1} \int_{-1}^1 \left(\sqrt{1-x^2} \right)^{n-1} dx$$

as v_{n-1} may be removed from the integral as it is a constant.

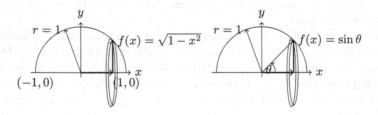

Fig. 2. Volume of a (3D) unit sphere: $\int_{-1}^1 \pi \left(\sqrt{1-z^2} \right)^2 dz = \int_0^\pi \pi \sin^3 \theta \, d\theta$

Finally, by considering, as shown on the right hand side of Fig. 2, that x can be written as $\cos \theta$ and then $f(x) - \sin \theta$, integrating by substitution we have

$$\int_{-1}^1 \left(\sqrt{1-x^2} \right)^{n-1} dx = \int_\pi^0 \left(\sqrt{(1 - \cos^2 \theta)} \right)^{n-1} (-\sin \theta) d\theta = \int_0^\pi (\sin \theta)^n d\theta$$

---

[1] $V_n(R) = \int_{B_n(R)} 1 \, dx_1 \dots dx_n = \int_{B_n(1)} R^n \, dy_1 \dots dy_n = R^n V_n(1)$, where we integrate by substitution with $x_i = R y_i$ for all i.

So finally putting all the pieces together we have an expression for the volume of a hypersphere of radius r in n dimensions:

$$V_n(r) = r^n k \int_0^\pi sin^n x dx \qquad (3)$$

for a constant k.

3.2   Derivation of the PDF

To construct a PDF, we note that Eq. 3 derives from a Riemann integral of infinitesimal hyperspheres, in $n - 1$ dimensions, each orthogonal to a diameter through the centre of the n-dimensional hypersphere. Considering the left-hand side of Fig. 1, the integration may notionally be performed along the axis pq within any $(n - 1)$-dimensional hyperplane containing p and q. Then the angle θ in the figure corresponds to the integral variable x in Eq. 3. Thus, the volume in the green-shaded area of the figure is given by the definite integral $t^n k \int_0^\theta sin^n x dx$. Within a uniformly populated space, the PDF of a point being within the defined segment, with respect to the angle θ, is therefore directly proportional to $h(x) = sin^n x, x \in [0, \pi]$.

The PDFs shown in the right hand side of Fig. 1 are produced by applying this function to $cos \frac{x}{t}$, where x is the distance from q along the line pq, in order to convert the angular dependence to a distance along the pq axis. The outcome is then divided by the volume of the hypersphere around q to normalise the area under the curves.

Quantifying this PDF is non-trivial. However, for high values of n, the function $sin^n x$ becomes almost indistinguishable from a related Gaussian, and in turn the related PDF becomes almost indistinguishable from that of a normal distribution, and thus readily available. For large n, for example $n > 15$, the PDF function is almost indistinguishable from that of a normal distribution with $\mu = \frac{\pi}{2}$ and $\sigma = \frac{1}{\sqrt{n}}$. This observation has also been made by [3], and is discussed in [1].

3.3   Examples of Overlap in Unbounded Euclidean Spaces

Table 1 gives probability calculations, for Euclidean spaces of various dimensions, for the situation shown in Fig. 1, where $d(p, q) = m + \frac{t}{2}$. These figures correspond with the probability density functions shown in the Figure. Two points are notable: first, how small the probabilities become as dimensions increase, even with this significant amount of overlap; secondly, how the normal distribution estimate gives an increasingly small error as the dimension increases.

Table 1. Probability of Inclusion

Dimension	Probability	Normal estimate
2	0.195	0.228
10	0.0405	0.0479
20	0.0074	0.0093
30	0.0015	0.0020

Table 2. Proportion of queries within unit cube

Metric	Inside cube	Outside cube
Euc10	87.57%	12.43%
Euc20	51.15%	48.85%
Euc30	35.88%	64.12%

4   Experiments in Generated Euclidean Spaces

In the following experiments we use a number of different generated Euclidian spaces with individual coordinates drawn from a Gaussian distribution. Data points in these spaces have 10, 20 and 30 coordinates and are referred to as EUC10, EUC20 and EUC30.

In the following experiments we examine the mean and variance of angles abc within various spaces. In the first experiment a, b and c are sampled uniformly from within the space. The results of this experiment are shown in the brown (left hand) distributions in Fig. 3. In all cases the average angle is close to 60° with standard deviations of 16.5°, 11.25°, 9.11° for Euc10, 20 and 30 respectively. As can be observed from the figure the standard deviation drops as the dimensionality of the data-set increases.

In the second experiment a and b are sampled uniformly from the space but the third point c is constrained to be both within a threshold of the query point b and within the unit cube. For each experiment the radius of the hypersphere is calibrated to return one-millionth of the data-set. For EUC10, EUC20 and EUC30 these are: 0.229, 0.602 and, 0.727 respectively.[2] The results of this experiment are shown in the blue (right hand) distributions in Fig. 3. As can be seen the angle is close to 90° and like the earlier experiment the standard distribution of angles reduces with increasing dimensionality of the data-set.

4.1   Query Regions Lying Outwith the Unit Cube

In the above experiment we constrained the third points c to be within the sampled space. To determine the proportion of the query ball that lies outwith

---

[2] If the radius of the hypersphere is constrained to be within the unit cube rather than at the defined radii the results are identical.

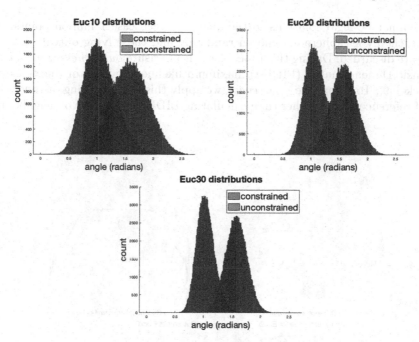

Fig. 3. Distance distributions for constrained and unconstrained triples

the unit cube we performed the following experiment on each of the Euclidean spaces. We randomly sampled one thousand points from within the space. For each point we uniformly sampled a further thousand points from within the hypercube of radius set to be the standard thresholds (described above) and measured if the point is within the unit cube or not. The results of running the experiment is shown in Table 2. As can be seen, the proportion of the hypersphere outwith the unit cube increases dramatically as the dimensionality of the data-set increases.

4.2 Prediction of the Angle Distribution

To understand the effect of the relationship between where queries are in the space and the resultant angles we conducted the following experiment. We sweep a hyper sphere up the diagonal of the unit hyper cube (in some dimension) from the origin to the opposite corner $(1, .., 1)$ in intervals of 0.01. The radius of the hypersphere is set using the standard thresholds used above. We examine the mean and variance of angles $\angle abc$ as follows. a is a fixed viewpoint which is always the centre of the unit cube[3], b is set to be a point along the diagonal of the cube $(0, 0, ..., 0)$ to $(1, 1, ..., 1)$, and for each instance of b, c is sampled from within a fixed hypersphere centred around b as before. As before, we discard any points that are not within the $(0, .., 0) - (1, .., 1)$ hypercube – i.e. those points

---

[3] We separately established that the viewpoint does not affect the measured angles.

that cannot be legal solutions to the query. In each case 1 million points from within the hyper sphere are chosen randomly and those lying outwith the unit cube are discarded. During this process we also measure *outlierness* using a Local Intrinsic Dimensionality (LIDIM) maximum likelihood estimator due to Levina & Bickel [9]. To determine *outlierness*, we apply this formula using distances to a set of reference points rather than calculating LIDIM using a set of neighbouring points.

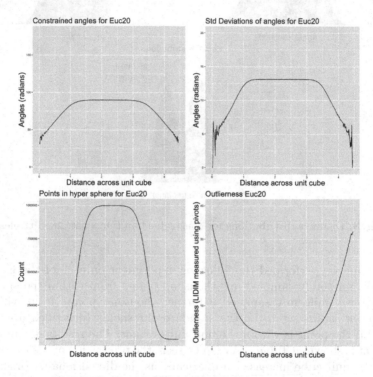

Fig. 4. Constrained Angles in the EUC20 Dataset

The results of these experiments are shown in Fig. 4. These four plots demonstrate a number of different interesting facets of the query solutions. Firstly, it can be seen that angles ∠*abc* are far from uniform. At the edge of the unit cube they rise and fall rapidly as the hyper sphere approaches the vertices of the unit cube. When the hypersphere approaches the centre of the cube the angles tend towards 90°. Secondly the distribution of angles are not constant. As the sphere approaches the centre of the cube they rise to a relatively constant variation of approximately 13.1°. Thirdly, the number of points inside the cube vary greatly. Close to the vertices of the cube the number of legal points tend towards zero whereas in the centre all of the 1 million sampled points are within the cube. Lastly, a good approximation to how much of an outlier a point is can be made by using the LIDIM formula with a fixed set of reference points as described above.

5   Experiments in Other Spaces

In this section we test the value of the angle analysis by examining pivot exclusion, enhanced by angle analysis, in a LAESA-like framework. We should stress that these experiments are to test the proof of the concept only; we believe that more sophisticated mechanisms can take advantage of the angle information to much better effect.

Table 3. Data sets

Name	Dimensions	Derivation	Preparation	Metric
MfAlex	4096	MirFlickr	fc6 layer AlexNet, no RELU	Euclidean
DeCaf	4096	Profiset	fc7 layer AlexNet, post-RELU	Euclidean
AnnSift	128	MirFlickr	ℓ_2 normalised	Euclidean
MfGist	480	MirFlickr	ℓ_1 normalised	Jensen-Shannon

We used four different high-dimensional data sets with different properties, as summarised in Table 3. *MfAlex* is derived from the application of the AlexNet [8] convolutional neural network on the MirFlickr image collection[4]. The data used is extracted from the first fully connected layer (fc6). *DeCaf* descriptors [7] are extracted from the *Profiset* image collection[5] using AlexNet, from which the fc7 *post-Relu* layer is extracted. *AnnSift* descriptors [10] are taken from the *ANN_SIFT1M* dataset[6]. Although queried with the ℓ_2 distance, these vectors are ℓ_2 normalised and thus this metric acts as a proxy for Cosine distance. *MfGist* is derived using GIST [11] image descriptors over the MirFlickr 1M image collection. These descriptors are queried using the Jensen-Shannon distance, which has been shown to be the best metric for near-duplicate detection [6].

Of the four data sets, only the first therefore represents a true Euclidean space where each dimension contains a range of positive and negative values. We have deliberately chosen this range of data sets to examine whether the angular properties which are clear in unbounded Euclidean spaces follow in more general metric spaces. The four spaces all contain one million objects, and in each case a ground truth is known for one thousand queries, each of which has 100 known nearest neighbours.

The queries are divided into two equal sets, the first of which is used to perform analysis over the space, and the second of which is used to test a search mechanism using that analysis.

5.1   Correlation of Outlierness and Angle

Our hypothesis is that for any high-dimensional space, the distribution of angles $\angle p_i q s_j$, where p_i is selected from a set of reference points, q is a query and

---

[4] https://press.liacs.nl/mirflickr.
[5] http://disa.fi.muni.cz/profiset/.
[6] http://corpus-texmex.irisa.fr/.

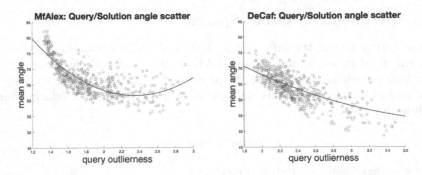

Fig. 5. Correlation between query outlierness and mean angle $\angle p_i q s_j$, where s_j is a solution to query q. The lines show the best-fit quadratics, used in the experiments in Sect. 5.

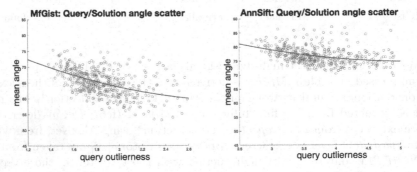

Fig. 6. Correlation between query outlierness and mean angle $\angle p_i q s_j$ for the non-Euclidean spaces.

s_j is selected from solutions to that query, will be constrained in comparison with randomly sampled angles from the space, and will be correlated with the *outlierness* of the query (Figs. 5, 6).

A randomly selected set of 256 objects was selected from the dataset to act as reference objects. For each query q, for each p_i in the reference set, and for each s_j in the known solution set, the angle $\angle p_i q s_j$ was measured and the mean and variance recorded.

For each q, an approximation to the outlierness was calculated based on the distances $d(p_i, q)$ for each reference object using the maximum likelihood estimator as described above [9]. The scatter plots in Fig. 6 show a clear relationship between query outlierness and the mean angle. It is clear in all cases that the majority of angles are greater 60° and are thus distinct from the angles within a randomly selected triplet. The angles depicted on the scatters are averaged over the query's 100 nearest neighbours, and in all cases the standard deviation is quite low - almost always less than 10°. The implication, in terms of the analysis shown in Sect. 3, is that exclusion may safely occur in many situation where its safety cannot be guaranteed by triangle inequality alone. We quantify this in the next section.

5.2   Use in Querying

In this section, a simple search mechanism is applied in order to give experimental validation of the principles outlined[7].

The search mechanism used is a variant of LAESA. From the data set S, a randomly selected subset P of 256 reference objects[8] is removed. The pre-processing phase comprises the calculation of a distance table between each reference object $p_i \in P$ and each remaining member s_j of S. At query time, the distance between the query and p_i is calculated. The possibility of exclusion of each object s_j is determined by scanning the appropriate row of the pre-calculated distance table. In the normal LAESA algorithm, exclusion may occur if and only if $|d(p_i, s_j) - d(p_i, q)| > t$, where t is the threshold distance for that query. In that case, it is impossible for the hypersphere of radius t centred on q to contain s_j, and the distance $d(q, s_j)$ does not require to be calculated.

The pure LAESA mechanism is adapted to perform exclusion even in some cases where $|d(p_i, s_j) - d(p_i, q)| \leq t$, as depicted in Fig. 1. For each query q, the set of distances $d(q, p_i)$ is calculated as usual. These distances are first used to measure an estimate of outlierness, as described in Sect. 5.1, and thus to determine an estimate γ of the mean angle $\angle pqs_i$ in cases where $d(q, s_i)$ is small. In our experiments, a fixed amount of variance τ is allowed, with the intent that, for all solutions, the angle $\angle pqs_i$ is highly likely to lie within the bounds $\gamma \pm \tau$. Now, for all values s_j from the finite set, and for each p_i, the angle θ is calculated[9] from the values $d(p_i, s_j), d(p_i, q)$ and t. If the angle θ lies outside the range $\gamma \pm \tau$, then s_j is excluded without performing the calculation $d(q, s_j)$.

In the experiments over all the data sets we report outcomes using a range of fixed tolerances between 0.3 and 0.65 rad. Finally, each experiment is repeated with a tolerance of $\frac{\pi}{2}$ radians, which effectively makes exclusion impossible other than when allowed by the pure LAESA mechanism.

Results for the four data sets are shown in Fig. 7. The left hand graph shows the cost per query for the different tolerances; as expected, a smaller tolerance, resulting in a larger cutoff angle θ, gives a lower cost. The right hand graph shows recall for the same tolerances. As noted, this mechanism always gives perfect precision.

The results fully justify our derived model. For all data sets, a tolerance value of around 0.6 rad gives almost perfect recall. This value implies that all query solutions, in these spaces, do indeed lie within an arc of 1.2 rad in the

---

[7] Source code is available from https://github.com/aldearle/SISAP2020_angles or from the authors.

[8] The same number is used across all sets to allow fair comparison, even although for example the cost of performing an exclusion for *AnnSift* may be greater than the cost of measuring the distance directly.

[9] Unless $|d(p_i, s_j) - d(p_i, q)| > t$, when exclusion can occur in any case.

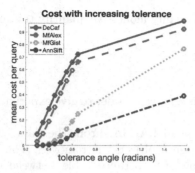
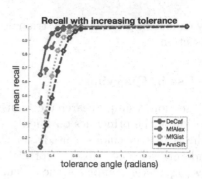

Fig. 7. Test results with LAESA with increasing tolerance. The left figure shows cost, as the number of distance calculations performed divided by the size of the data. The right hand figure shows recall. The values at the right hand side of each graph equate to those for an unmodified LAESA mechanism.

2D projection, implying that over half of the angular space is empty. It is also noteworthy that the data sets with higher costs for the unmodified LAESA give perfect recall with lower tolerance levels. It is reasonable to assume that this is a consequence of a higher inherent dimensionality leading to a tighter clustering of the angles within the 2D projection.

6   Conclusions and Future Work

We have taken an observation from high-dimensional vector spaces and applied it to general metric spaces by way of a derived approximate search paradigm. We have shown an underlying mathematical model which explains a related effect in unbounded, uniform Euclidean spaces, and demonstrated it experimentally. We have shown that, unfortunately, the effect does not hold perfectly in bounded high dimensional search spaces. This is because the radius required to capture query solutions in a finite space far exceeds the boundaries of the space.

We have demonstrated nonetheless an interesting restriction in the distribution of angles in metric spaces, and in particular that the angles from a reference point, via a query, to a query solution are significantly different from angles randomly sampled from the space.

We have outlined how this may be used to conduct a probabilistic search, and a trade-off is shown between query efficiency and accuracy in spaces which are otherwise intractable for exact search. We believe that this topic has much further promise; our present analysis of the spaces is based on a relatively crude measure of query outlierness, and we believe a more sophisticated analysis of the space may result in a finer-grained understanding of the angle distribution, as well as further query mechanisms based on it. In particular, we have not yet examined the effect of the restricted angles on hyperplane exclusion mechanisms, nor in conjunction with the four-point property of supermetric spaces.

References

1. Large powers of sine appear gaussian — why? https://math.stackexchange.com/questions/2293330. Accessed 22 June 2020
2. Blum, A., Hopcroft, J., Kannan, R.: High-Dimensional Space, p. 4–28. Cambridge University Press (2020). https://doi.org/10.1017/9781108755528.002
3. Cai, T., Fan, J., Jiang, T.: Distributions of angles in random packing on spheres. J. Mach. Learn. Res. **14**(21), 1837–1864 (2013)
4. Chávez, E., Navarro, G.: Metric databases. In: Encyclopedia of Database Technologies and Applications. Idea Group (2005)
5. Connor, R., Cardillo, F.A., Vadicamo, L., Rabitti, F.: Hilbert exclusion: improved metric search through finite isometric embeddings. ACM Trans. Inf. Syst. **35**(3), 17:1–17:27, December 2016. https://doi.org/10.1145/3001583
6. Connor, R.C.H., Cardillo, F.A.: Quantifying the specificity of near-duplicate image classification functions. In: Proceedings of the 11th Joint Conference on Computer Vision, Imaging and Computer Graphics Theory and Applications (VISIGRAPP 2016) - Volume 4: VISAPP, Rome, Italy, 27–29 February 2016, pp. 647–654 (2016)
7. Donahue, J., Jia, Y., Vinyals, O., Hoffman, J., Zhang, N., Tzeng, E., Darrell, T.: DeCAF: a deep convolutional activation feature for generic visual recognition. In: ICML 2014, Beijing, China, pp. 647–655 (2014)
8. Krizhevsky, A., Sutskever, I., Hinton, G.E.: ImageNet classification with deep convolutional neural networks. In: Advances in Neural Information Processing Systems 25, pp. 1097–1105. Curran Associates, Inc. (2012)
9. Levina, E., Bickel, P.J.: Maximum likelihood estimation of intrinsic dimension. In: Advances in Neural Information Processing Systems, pp. 777–784. MIT Press (2005)
10. Lowe, D.G.: Object recognition from local scale-invariant features. In: Proceedings of the International Conference on Computer Vision, Kerkyra, Corfu, Greece, 20–25 September 1999, pp. 1150–1157 (1999)
11. Oliva, A., Torralba, A.: Modeling the shape of the scene: a holistic representation of the spatial envelope. Int. J. Comput. Vis. **42**, 145–175 (2001)
12. Pramanik, S.K., Li, J., Ruan, J., Bhattacharjee, S.K.: Efficient search scheme for very large image databases. In: Internet Imaging, vol. 3964 (1999)
13. Voelker, A.R., Gosmann, J., Stewart, T.C.: Efficiently sampling vectors and coordinates from the n-sphere and n-ball. Technical report, Centre for Theoretical Neuroscience, Waterloo, ON, January 2017
14. Zezula, P., Amato, G., Dohnal, V., Batko, M.: Similarity Search: The Metric Space Approach, vol. 32. Springer, Heidelberg (2006). https://doi.org/10.1007/0-387-29151-2

Local Intrinsic Dimensionality III: Density and Similarity

Michael E. Houle[✉]

National Institute of Informatics,
2-1-2 Hitotsubashi, Chiyoda-ku, Tokyo 101-8430, Japan
meh@nii.ac.jp

Abstract. In artificial intelligence, machine learning, and other areas in which statistical estimation and modeling is common, distributions are typically assumed to admit a representation in terms of a probability density function (pdf). However, in many situations, such as mixture modeling and subspace methods, the distributions in question are not always describable in terms of a single pdf. In this paper, we present a theoretical foundation for the modeling of density ratios in terms of the local intrinsic dimensionality (LID) model, in a way that avoids the use of traditional probability density functions. These formulations provide greater flexibility when modeling data under the assumption of local variation in intrinsic dimensionality, in that no explicit dependence on a fixed-dimensional data representation is required.

1 Introduction

1.1 Probability Density and Dimensionality

Modelers in computer science disciplines such as artificial intelligence, machine learning, and pattern recognition typically base their analyses on a distributional view of their data sources, either in terms of parametric approaches involving standard distributions, or through nonparametric approaches.

For applications in which the data is continuous in nature, and in the supporting domains of continuous statistical modeling and probability theory, the concept of probability density is ubiquitous. In continuous domains, the absolute probability associated with any single event is typically zero, due to the infinite number of possible outcomes. For this reason, theoreticians and practitioners alike have been concerned with the relative likelihood of generation of one given sample value compared to another, as taken over infinitesimally-small volumes of positive probability measure that contain the values of interest. These density values, when integrated over the domain, account for its full probability measure, which is 1 by definition.

Not all distributions of interest admit a probability density function (pdf). Perhaps the most famous and important example is that of mixture distributions derived from a collection of random variables. Sampling from a mixture distribution involves a two-stage process, in which a distribution is first selected from

S. Satoh et al. (Eds.): SISAP 2020, LNCS 12440, pp. 248–260, 2020.
https://doi.org/10.1007/978-3-030-60936-8_19

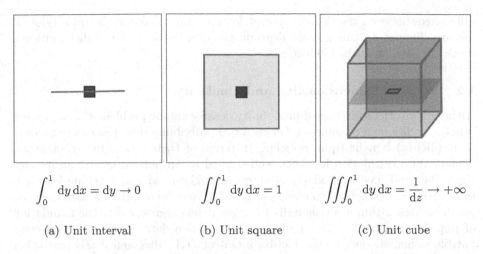

$$\int_0^1 \mathrm{d}y\,\mathrm{d}x = \mathrm{d}y \to 0 \qquad \iint_0^1 \mathrm{d}y\,\mathrm{d}x = 1 \qquad \iiint_0^1 \mathrm{d}y\,\mathrm{d}x = \frac{1}{\mathrm{d}z} \to +\infty$$

(a) Unit interval   (b) Unit square   (c) Unit cube

Fig. 1. Application of a 2-D pdf in uniform distributions in 1-D, 2-D, and 3-D domains. A 2-D pdf expresses probability density as a ratio between probability mass and area. Since the 1-D interval has area 0, and since the 3-D cube has infinite area, integration of the 2-D pdf over the interval and the cube (in terms of the infinitesimal product $\mathrm{d}y\,\mathrm{d}x$) would produce the values 0 and $+\infty$, respectively.

the mixture according to some probability associated with it, and then a value is generated from the selected distribution. Although each of the constituent distributions of the mixture can be associated with a probability density function, in general the mixture distribution as a whole may not. As an illustrative example, consider the situation in Fig. 1 in which with equal probability a sample can be selected from one of three uniform distributions: over a 1-dimensional unit interval, over a 2-dimensional unit square, or over a 3-dimensional unit cube. Each of the three pdfs of the mixture, when integrated over its own domain (interval, square, or cube), would return a value of 1. However, taking the 2-D pdf of the square domain and integrating it over the 1-D interval domain would produce a result of 0, whereas its integration over the 3-D cube would diverge to $+\infty$. Accordingly, it is not possible to devise a single probability density function capable of representing this mixture distribution as a whole.

Standard techniques for the direct estimation of probability density (such as multivariate kernel density estimation [35]) can be problematic when the data dimensionality is assumed to be large and fixed. As the example in Fig. 1 shows, standard formulations of probability density as an integrand over the distribution domain explicitly depend on the dimensionalities of the infinitesimal volumes over which the integration is performed. In settings for which the numbers of features is very large, data analysis models that rely on forms of density estimation can exhibit bias in terms of the number of features that are relevant to individual localities within the domain, and the degree of correlation and other interactions among these features. Variations in the numbers of relevant features from locality to locality within the domain, together with vast differences between the (local) relevant feature dimension and the total (global) feature dimension, can greatly degrade the effectiveness of the distributional model in data analysis applications.

1.2 Volume, Dimensionality, and Similarity

Although direct estimation of probability density can be problematic, data analysis that relies on a comparison between two probability densities can (explicitly or implicitly) benefit from modeling in terms of their ratios. In recent years, density ratio estimation has been well-studied as an important area in its own right [34], and has been widely adopted throughout AI and machine learning. In addition to being used to compare the relative concentration of data between two locations within a common distribution, it has also served as the foundation of popular measures of the similarity between two data distributions on compatible domains, such as the Kullback-Leibler (KL) divergence [27] and other f-divergences [28]. Although direct estimation of ratios has been shown to be inherently biased for small sample sizes [9,33], division of one probability density by another has the advantage of producing a unitless quantity independent of local dimensional assumptions. Nevertheless, it should be noted that KL-divergence and other distributional similarity measures typically integrate these ratios of probability density over the entire domain (of one of the two distributions), thereby reintroducing fixed-dimensional infinitesimal volumes into the data model.

The domains of continuous distributions are typically equipped with a similarity (or dissimilarity) measure through which the interrelationships among data are modeled and assessed. In many situations in machine learning/and data mining, modelers and practitioners tend to view similarity-based density estimation as an acceptable proxy for volume-based density estimation. However, in general this is not the case: volume tends to increase not linearly with distance radius, but with that radius raised to the power of the data dimension.

In this paper, we explicitly acknowledge this disparity between radius and volume by investigating the issue of density ratios from the perspective of the recent Local Intrinsic Dimensionality (LID) [15,16] model. LID is a distributional form of intrinsic dimensional modeling in which the volume of a ball of radius r is taken to be the probability measure associated with its interior, denoted by $F(r)$. The function F can be regarded as the cumulative distribution function (cdf) of an underlying distribution of distances.

Theoretical properties of LID in multivariate analysis have been studied recently [17]. LID and related expansion-based measures of intrinsic dimensionality [18] have also seen applications in such areas as similarity search [3,8,19–22], dependency analysis [31], feature selection and ranking [4,13,23], outlier detection and its analysis [11,24,32], and deep learning [29,30]. Practical estimators of LID have been developed based on interpoint distances within neighborhood samples [1,2], which trace their roots to the estimation of the generalized pareto distribution (GPD) shape parameter [14] developed within the statistical discipline of extreme value theory (EVT).

1.3   Overview

The goal of the paper is to introduce a theoretical foundation that can serve for modeling density ratios in terms of LID, without recourse to the traditional formulation of probability density. After first briefly surveying of the LID model in the next section, in Sect. 3 we present several results relating the limits of functions with properties appropriate to density ratio modeling: namely, the functions are assumed to be smooth (continuously differentiable), positive, vanishing at zero, and of the form of a cdf of an induced distance distribution relative to a reference location in a distribution of interest. In each theoretical statement, the local intrinsic dimensionality is shown to influence the density ratio model in a natural way. The paper is then concluded in Sect. 4.

2   Local Intrinsic Dimensionality

In this section, we give a brief overview of the extreme-value-theoretic LID model of intrinsic dimensionality first introduced in [15]. For more information on the model and its connections to the statistical theory of extreme values, please see [10, 16, 17].

The LID model falls into the expansion family of intrinsic dimensional estimation [18, 26]. Like earlier expansion models, LID draws its motivation from the relationship between volume and radius in an expanding ball around points of interest. Unlike these models, the LID interprets volume as a function of the same form as a univariate cumulative distribution function (cdf), representing the probability measure $F(r)$ captured by a ball of radius r. Although motivated by applications involving the distribution of distance values, the model formulation (as stated in [16]) generalizes this notion even further, to any smooth real-valued function that is non-zero in the vicinity of $r \neq 0$.

Definition 1 ([16]). *Let F be a real-valued function that is non-zero over some open interval containing $r \in \mathbb{R}$, $r \neq 0$. The intrinsic dimensionality of F at r is defined as follows whenever the limit exists:*

$$\mathrm{IntrDim}_F(r) \triangleq \lim_{\epsilon \to 0} \frac{\ln\left(F((1+\epsilon)r)/F(r)\right)}{\ln\left((1+\epsilon)r/r\right)} = \lim_{\epsilon \to 0} \frac{\ln\left(F((1+\epsilon)r)/F(r)\right)}{\ln(1+\epsilon)}.$$

Under the same assumptions on F, when F can be interpreted as the cdf of a distance distribution, the definition of LID can be regarded as equivalent to a notion of indiscriminability. Intuitively, if an underlying distance measure is indiscriminative at a given distance r, then expanding the distance by some small factor should incur a relatively large increase in probability measure as a proportion of the current value, $F(r)$. Accordingly, the indiscriminability of the distance variable is defined as the limit of the ratio of two quantities: the proportional rate of increase of $F(r)$, and the proportional rate of increase in r.

Definition 2 ([16]). *Let F be a real-valued function that is non-zero over some open interval containing $r \in \mathbb{R}$, $r \neq 0$. The* indiscriminability of F at r *is defined as follows whenever the limit exists:*

$$\mathrm{InDiscr}_F(r) \triangleq \lim_{\epsilon \to 0} \left[\frac{F((1+\epsilon)r) - F(r)}{F(r)} \middle/ \frac{(1+\epsilon)r - r}{r} \right] = \lim_{\epsilon \to 0} \frac{F((1+\epsilon)r) - F(r)}{\epsilon \cdot F(r)}.$$

When F satisfies certain smoothness conditions in the vicinity of r, its intrinsic dimensionality and indiscriminability have been shown to be identical:

Theorem 1 ([16]). *Let F be a real-valued function that is non-zero over some open interval containing $r \in \mathbb{R}$, $r \neq 0$. If F is continuously differentiable at r, then*

$$\mathrm{ID}_F(r) \triangleq \frac{r \cdot F'(r)}{F(r)} = \mathrm{IntrDim}_F(r) = \mathrm{InDiscr}_F(r).$$

Let $\mathbf{x}$ be any reference location within a data domain $\mathcal{S}$ equipped with a distance measure d. To any point $\mathbf{y} \in \mathcal{D}$ we can associate the distance $r = d(\mathbf{x}, \mathbf{y})$; in this way, any global data distribution over $\mathcal{D}$ induces a local distance distribution with respect to $\mathbf{x}$. Motivated by a need to characterize the local intrinsic dimensionality in the vicinity of individual reference points, we are interested in the limit of $\mathrm{ID}_F(r)$ as the distance r tends to 0. For convenience, for non-zero distances r we refer to $\mathrm{ID}_F(r)$ as the *indiscriminability of F at r*, and to $\mathrm{ID}_F^* \triangleq \lim_{r \to 0} \mathrm{ID}_F(r)$ as the *local intrinsic dimension* (or *LID*) of F.

In the ideal case where the data in the vicinity of $\mathbf{x}$ is distributed uniformly within a submanifold in $\mathcal{D}$, ID_F^* would equal the dimension of the submanifold; however, in general these distributions are not ideal, the manifold model of data does not perfectly apply, and ID_F^* is not necessarily an integer. Nevertheless, the local intrinsic dimensionality would give an indication of the dimension of the submanifold containing $\mathbf{x}$ that would best fit the data distribution in the vicinity of $\mathbf{x}$.

The indiscriminability function ID_F can be seen to fully characterize its associated function F.

Theorem 2 (LID Representation Theorem [16]**).** *Let $F : \mathbb{R} \to \mathbb{R}$ be a real-valued function, and assume that ID_F^* exists. Let x and w be values for which x/w and $F(x)/F(w)$ are both positive. If F is non-zero and continuously differentiable everywhere in the interval $[\min\{x, w\}, \max\{x, w\}]$, then*

$$\frac{F(x)}{F(w)} = \left(\frac{x}{w}\right)^{\mathrm{ID}_F^*} \cdot G_F(x, w), \text{ where}$$

$$G_F(x, w) \triangleq \exp\left(\int_x^w \frac{\mathrm{ID}_F^* - \mathrm{ID}_F(t)}{t} \, dt\right),$$

whenever the integral exists.

In [16], conditions on x and w are provided for which the $G_F(x, w)$ can be seen to tend to 1 as $x, w \to 0$. The function $G_F(x, w)$ is related to the slowly-varying functions of long-standing interest within the EVT research community [10, 25], as it governs the rate of convergence of F to its ideal (asymptotic) form. Next, we revisit this issue so as to prove statements useful for the analysis of limits of density ratios, that do not explicitly rely on formulations involving G_F.

3 LID-Aware Density Ratios

In this section, we present statements concerning the limits of ratios of two functions, with properties that allow them to be applied to situations in which density ratios are to be modeled. Instead of considering the ratio of pdfs over an infinitesimal volume whose dimensions depend on knowledge of a full-dimensional coordinate system (as in the usual sense of density ratio estimation), we instead consider as the underlying volume spheres with infinitesimal radii. Local density ratios can be then be described as ratio of two cdf functions of distance distributions, each representing the probability measure captured by a sphere, whose radii (the function arguments) both tend to zero in some fashion.

Although this work is motivated by an interest in modeling limits of density ratios, the theoretical statements will be presented in a more generally applicable way. Here, we assume only that the functions are smooth (continuously differentiable), positive, and vanish at zero—all properties that are assumed, either explicitly or implicitly, in traditional pdf-based density ratio estimation. For each of the statements, we show how the existence of ratio limits relates to the underlying local intrinsic dimensionalities of the two functions involved, in a natural way.

3.1 Limits of Ratios of Different Functions

Intuitively speaking, the first result to be presented shows that for a density ratio limit to be greater than zero, the LID values of the numerator function and denominator function must be the same—any difference in the LID values would result in the limit either vanishing or diverging.

Lemma 1. *Let* $\alpha, \beta : \mathbb{R}^{\geq 0} \to \mathbb{R}^{\geq 0}$ *be functions such that* $\alpha(0) = \beta(0) = 0$, *and for some value of* $r > 0$, *their restrictions to the interval* $(0, r)$ *are continuously differentiable and positive. Let us also assume that* ID_α^* *and* ID_β^* *both exist and are positive. Further, let* $\Delta = \mathrm{ID}_\beta^* - \mathrm{ID}_\alpha^*$, *and let* β_0 *be the function obtained by decomposing* β *into* $\beta(u) = u^\Delta \cdot \beta_0(u)$. *Consider the limits of ratios*

$$\lambda \triangleq \lim_{u \to 0} \frac{\beta(u)}{\alpha(u)} \quad and \quad \lambda_0 \triangleq \lim_{u \to 0} \frac{\beta_0(u)}{\alpha(u)}.$$

If λ *exists and is positive, then* $\mathrm{ID}_\alpha^* = \mathrm{ID}_\beta^*$. *Alternatively, if* λ_0 *exists and is positive, then the following statements hold:*

1. $\lambda = 0$ *if and only if* $\mathrm{ID}_\alpha^* < \mathrm{ID}_\beta^*$ *(in which case* $\Delta > 0$*);*
2. $\lambda > 0$ *if and only if* $\mathrm{ID}_\alpha^* = \mathrm{ID}_\beta^*$ *(in which case* $\Delta = 0$ *and* $\lambda = \lambda_0$*);*
3. λ *diverges to* $+\infty$ *if and only if* $\mathrm{ID}_\alpha^* > \mathrm{ID}_\beta^*$ *(in which case* $\Delta < 0$*).*

Proof. For the limit λ_0 to exist, $\alpha(0) = 0$ implies that $\beta_0(0) = 0$ as well. Note also that β_0 must be continuously differentiable and positive over the full range $(0, r)$. L'Hôpital's rule can therefore be applied together with Theorem 1, yielding

$$\lambda_0 = \lim_{u \to 0} \frac{\beta_0(u)}{\alpha(u)} = \lim_{u \to 0} \frac{\beta_0'(u)}{\alpha'(u)} = \lim_{u \to 0} \frac{\mathrm{ID}_{\beta_0}(u) \cdot \beta_0(u)}{\mathrm{ID}_{\alpha}(u) \cdot \alpha(u)}.$$

The right-hand limit can be separated to produce

$$\lambda_0 = \lim_{u \to 0} \frac{\beta_0(u)}{\alpha(u)} = \frac{\mathrm{ID}_{\beta_0}^*}{\mathrm{ID}_{\alpha}^*} \lim_{u \to 0} \frac{\beta_0(u)}{\alpha(u)} = \frac{\mathrm{ID}_{\beta_0}^*}{\mathrm{ID}_{\alpha}^*} \cdot \lambda_0.$$

These equalities imply that whenever $\lambda_0 > 0$, we have that $\mathrm{ID}_{\alpha}^* = \mathrm{ID}_{\beta_0}^*$, and also that $\mathrm{ID}_{\beta}^* = \mathrm{ID}_{\beta_0}^* + \Delta$. Similar arguments also show that $\mathrm{ID}_{\alpha}^* = \mathrm{ID}_{\beta}^*$ whenever λ exists and is positive.

The existence of limit λ_0 has implications for the existence of λ, which can be expressed as

$$\lambda = \lim_{u \to 0} \frac{\beta(u)}{\alpha(u)} = \lim_{u \to 0} \frac{u^{\Delta} \cdot \beta_0(u)}{\alpha(u)} = \lambda_0 \lim_{u \to 0} u^{\Delta}.$$

Whenever λ_0 exists and is positive, we have that λ behaves as the limit of u^{Δ} as $u \to 0$, in that $\lambda = 0$ if and only if $\Delta > 0$, and λ diverges to $+\infty$ if and only if $\Delta < 0$. Otherwise, $\lambda > 0$ if and only if $\Delta = 0$, in which case $\lambda = \lambda_0$.   □

Lemma 1 tells us that in an asymptotic sense, density ratios are only meaningful when the local intrinsic dimensionalities of the two functions are equal. This accords well with what is already known of standard density ratios as defined using probability densities. However, the LID model has a distinct advantage in that it is implicitly defined for any locality that admits a 'local' distance distribution, and that when modeling a 'global' distribution, no arbitrary universally-fixed local intrinsic dimensionality need be imposed.

3.2 Parameterized Limits Involving a Single Function

Next, we turn our attention to the effect of applying a common function to the numerator and denominator of an existing ratio. Equivalently, this situation can be regarded as the effect on an existing ratio of two parameterized values of the same function, as the common parameter tends to zero. Here, we state and prove results showing that the LID Representation Theorem applies with simplified conditions, in terms of the existence of a limit involving the parameterization itself.

Lemma 2. *Let $F : \mathbb{R}^{\geq 0} \to [0, 1]$ be a non-decreasing function, and assume that ID_F^* exists. Let $\alpha, \beta : \mathbb{R}^{\geq 0} \to \mathbb{R}^{\geq 0}$ be functions such that $\alpha(0) = \beta(0) = 0$, and for some value of $r > 0$, their restrictions to the interval $[0, r)$ are continuously differentiable and strictly monotonically increasing. For any constant $c \neq 0$,*

$$\lim_{u \to 0} \left(\frac{\beta(u)}{\alpha(u)} \right)^c \cdot G_F(\alpha(u), \beta(u)) = \lim_{u \to 0} \left(\frac{\beta(u)}{\alpha(u)} \right)^c = \lambda^c \qquad (1)$$

whenever the limit $\lambda = \lim_{u \to 0} \frac{\beta(u)}{\alpha(u)}$ *exists and is positive. If instead* $c > 0$ *and* λ *diverges to* $+\infty$, *or if* $c < 0$ *and* $\lambda = 0$, *then the limits in Eq. 1 both diverge to* $+\infty$. *Otherwise, if* $c > 0$ *and* $\lambda = 0$, *or if* $c < 0$ *and* λ *diverges to* $+\infty$, *then the limits in Eq. 1 are both zero.*

Proof. Since ID_F^* is assumed to exist, for any real value $\epsilon \in (0, r)$ there must exist a value $0 < \delta < \epsilon < r$ such that $t < \delta$ implies that $|\mathrm{ID}_F(t) - \mathrm{ID}_F^*| < \epsilon$. Therefore, when $0 < \alpha(u) < \delta$ and $0 < \beta(u) < \delta$,

$$\ln G_F\left(\alpha(u), \beta(u)\right) = \int_{\alpha(u)}^{\beta(u)} \frac{\mathrm{ID}_F^* - \mathrm{ID}_F(t)}{t} \, dt$$

$$|\ln G_F\left(\alpha(u), \beta(u)\right)| \leq \epsilon \cdot \left| \int_{\alpha(u)}^{\beta(u)} \frac{1}{t} \, dt \right| = \epsilon \cdot \left| \ln \frac{\beta(u)}{\alpha(u)} \right|.$$

Exponentiating, and multiplying through by $(\beta(u)/\alpha(u))^c$, we obtain

$$\left(\frac{\beta(u)}{\alpha(u)} \right)^{c - \epsilon_0} \leq \left(\frac{\beta(u)}{\alpha(u)} \right)^c \cdot G_F\left(\alpha(u), \beta(u)\right) \leq \left(\frac{\beta(u)}{\alpha(u)} \right)^{c + \epsilon_0},$$

where $\epsilon_0 = \epsilon$ if $\beta(u) \geq \alpha(u)$, and $\epsilon_0 = -\epsilon$ otherwise.

Let us assume that the limit $\lambda = \lim_{u \to 0} \beta(u)/\alpha(u)$ exists. If $\lambda > 0$, then $\lim_{u \to 0} (\beta(u)/\alpha(u))^c$ also exists, and equals λ^c. In this case, as ϵ and δ tend to 0, the monotonicity of $\alpha(u)$ and $\beta(u)$ implies that u is driven to 0 as well. Thus, we have that

$$\lim_{u \to 0} \left(\frac{\beta(u)}{\alpha(u)} \right)^c \cdot G_F\left(\alpha(u), \beta(u)\right) = \lambda^c = \lim_{u \to 0} \left(\frac{\beta(u)}{\alpha(u)} \right)^c.$$

However, if $\lambda = 0$ or diverges to $+\infty$, then similar arguments show that the limits in Eq. 1 both diverge to $+\infty$ (if $c > 0$ and λ diverges, or $c < 0$ and $\lambda = 0$) or both converge to 0 (if $c > 0$ and $\lambda = 0$, or $c < 0$ and λ diverges). $\qquad \square$

Lemma 2 states conditions for which the LID representation function G_F can be ignored. Using the lemma leads to the following simplified restatement of the LID Representation Theorem itself.

Theorem 3. *Let* $F : \mathbb{R}^{\geq 0} \to [0, 1]$ *be a non-decreasing function, and assume that* ID_F^* *exists and is positive. Let* $\alpha, \beta : \mathbb{R}^{\geq 0} \to \mathbb{R}^{\geq 0}$ *be functions such that* $\alpha(0) = \beta(0) = 0$, *and for some value of* $r > 0$, *their restrictions to the interval* $[0, r)$ *are continuously differentiable and strictly monotonically increasing. Then*

$$\lim_{u \to 0} \frac{F(\beta(u))}{F(\alpha(u))} = \lim_{u \to 0} \left(\frac{\beta(u)}{\alpha(u)} \right)^{\mathrm{ID}_F^*} = \lambda^{\mathrm{ID}_F^*} \tag{2}$$

whenever the limit $\lambda = \lim_{u \to 0} \frac{\beta(u)}{\alpha(u)}$ *exists. If instead* λ *diverges to* $+\infty$, *then the limits in Eq. 2 both diverge to* $+\infty$.

Proof. Applying Theorem 2 to the first limit in Eq. 2 yields

$$\lim_{u \to 0} \frac{F(\beta(u))}{F(\alpha(u))} = \lim_{u \to 0} \left(\frac{\beta(u)}{\alpha(u)} \right)^{\mathrm{ID}_F^*} \cdot G_F\left(\alpha(u), \beta(u)\right).$$

The result then follows from Lemma 2, with $c = \mathrm{ID}_F^*$. $\square$

3.3   Limits of Ratios of Inverse Functions

We next show the effect of taking the limit of ratios of the inverse of functions, in terms of the original functions as well as their local intrinsic dimensionalities. Here, we find that the limit of the ratio of the functions equals that of the ratio of their inverses, raised to the power of the local intrinsic dimension of either function.

When the original function limits are interpreted as the probability measure captured within neighborhoods whose radii tend to zero, the limits of their inverse functions can be regarded as the radii associated with neighborhoods whose captured probability measure tends to zero (as the neighborhoods shrink). Assessing the limits of ratios involving inverse functions thus gives modelers greater flexibility in designing estimators for density ratios.

Theorem 4. *Let $\alpha, \beta : \mathbb{R}^{\geq 0} \to \mathbb{R}^{\geq 0}$ be functions such that $\alpha(0) = \beta(0) = 0$, and for some value of $r > 0$, their restrictions to the interval $[0, r)$ are continuously differentiable and strictly monotonically increasing. Let us also assume that ID_α^* and ID_β^* both exist and are positive. If the limit ratio $\lambda \triangleq \lim_{u \to 0} \frac{\beta(u)}{\alpha(u)}$ exists, then*

$$\lambda = \lim_{p \to 0} \left(\frac{\alpha^{-1}(p)}{\beta^{-1}(p)} \right)^{\mathrm{ID}_\alpha^*} = \lim_{p \to 0} \left(\frac{\alpha^{-1}(p)}{\beta^{-1}(p)} \right)^{\mathrm{ID}_\beta^*}. \tag{3}$$

In addition, the limits in Eq. 3 diverge to $+\infty$ whenever λ diverges to $+\infty$.

Proof. Since $\alpha(0) = \beta(0) = 0$, the strict monotonicity of α and β over the range $[0, r)$ imply that their inverse functions $\alpha^{-1}(p)$ and $\beta^{-1}(p)$ both exist for all $p \in [0, s)$, where $s \triangleq \min\{\alpha(r), \beta(r)\}$. Moreover, the inverse function theorem implies that α^{-1} and β^{-1} are themselves continuously differentiable over $[0, s)$.

Consequently, using Theorem 3, the limit ratio λ can be expanded as follows:

$$\lambda = \lim_{u \to 0} \frac{\beta(u)}{\alpha(u)} = \lim_{u \to 0} \frac{\alpha(\alpha^{-1}(\beta(u)))}{\alpha(\beta^{-1}(\beta(u)))}$$

$$= \lim_{p \to 0} \left(\frac{\alpha^{-1}(p)}{\beta^{-1}(p)} \right)^{\mathrm{ID}_\alpha^*} = \lim_{p \to 0} \left(\frac{\alpha^{-1}(p)}{\beta^{-1}(p)} \right)^{\mathrm{ID}_\beta^*},$$

after the substitution of $p = \beta(u)$. Hence, Theorem 3 guarantees that all these limits exist and equal λ if the limit λ exists; otherwise, if λ diverges to $+\infty$, then the limits all diverge. Note that the last equality holds due to Lemma 1, which states that $\mathrm{ID}_\alpha^* = \mathrm{ID}_\beta^*$ for the case when $\lambda > 0$. $\square$

It is worth noting here that, as shown by Lemma 1, these limits are only positive when the LID values of the two original functions are identical; otherwise, the limits either both vanish or both diverge.

4   Conclusion

In this paper, we presented theoretical statements that can serve as a foundation for modeling density ratios in terms of local intrinsic dimensionality. These formulations give greater flexibility when modeling data under the assumption of local variation in intrinsic dimensionality, in that no explicit dependence on a fixed-dimensional data representation is required. In particular, the result of Theorem 4 on the existence and nature of the ratio of inverse functions allows modelers to move flexibly and interchangeably between a distance-based model of local probability density on the one hand, and a probability-based model of neighborhood radius ratios on the other—all taking into proper account the effect of local intrinsic dimensionality.

As mentioned earlier, distributional modeling of density ratios is already a well-established strategy within the machine learning community [34]. The density ratio formulation considered in this paper can (in principle) be substituted for conventional pdf-based density ratios in models involving smooth (continuously differentiable) distributions—the ratio of pdfs thereby being replaced by the limits of ratios of cdfs of distance distributions. In data mining and other settings where the data is not explicitly modeled in terms of smooth distributions, the connection to LID-aware density ratios can still be made, albeit less directly. A data set consisting of n points can be regarded as a sample drawn from some unknown global distribution that, from any given point of interest $\mathbf{x}$, induces a distribution of distances. Letting F denote the cdf of the distribution of distances to $\mathbf{x}$, the expected number of data points lying within distance r of $\mathbf{x}$ is simply $n \cdot F(r)$. If r is the radius of the k-nearest neighborhood of $\mathbf{x}$, then k/n can serve as a (crude) estimate of $F(r)$. In this sense, within the underlying continuous model, formulations involving k, n and r can be viewed as sample-based estimators of formulations involving F and ID_F. Continuous limit-based and LID-aware forms such as those considered in this paper can consequently be applied to yield explanations of asymptotic behavior, as the sample size n tends to infinity. Moreover, an LID-aware asymptotic analysis could conceivably suggest alternative heuristics with improved performance characteristics.

In the data mining setting of outlier detection [7], density ratio modeling has already assumed a place of central importance. The classic (and still state-of-the-art) LOF family of outlier detection methods [5] assess the outlierness of a test data point through a ratio of densities, one at the test point, and the other with respect to an aggregation of points in the vicinity of the test point. LOF formulations essentially estimate density using neighborhood-based criteria, in terms of the radii within which a predetermined number of points are captured. However, for those settings where the dimensional characteristics are assumed to vary from locality to locality, Theorem 4 implies that any use of the ratio of distances to estimate density ratios should also take the local intrinsic dimensionality into account. The theoretical results of this paper underscore the need for the development of LID-aware local outlier detection techniques; work in this direction is already underway.

LID-aware density ratio modeling may also have useful applications in density-based clustering [6] and other non-parametric unsupervised learning settings that exploit similarity information. One such possibility involves the well-known DBSCAN family of clustering methods [12], which relies on absolute density thresholding to determine clusters, where density is (typically) estimated in terms of the numbers of points enclosed within neighborhoods of fixed radius. Extension of the DBSCAN strategy to account for LID-aware density would potentially allow for the formation of clusters whose densities are not necessarily high in an absolute sense, but instead high relative to the density within some background distribution. This has the advantage that locally-dense configurations of data points within sparse regions may still be discoverable as clusters, even in the presence of other clusters in regions that are much more dense.

In the larger sense, with its unification of the notions of distance, probability, and local dimensionality, LID-aware modeling presents an opportunity to reconcile established heuristics in the area of data mining with the distributional framework that serves as the foundation of machine learning and other areas in AI. By breaking the dependence of models on a global dimension parameter, it also offers better support in subspace-based modeling and other contexts in which sparse-featured solutions are sought. The theoretical results presented in this paper on LID-aware density ratios will hopefully encourage and support further research in this direction.

Acknowledgments. Michael E. Houle acknowledges the financial support of JSPS Kakenhi Kiban (B) Research Grant 18H03296.

References

1. Amsaleg, L., et al.: Extreme-value-theoretic estimation of local intrinsic dimensionality. Data Min. Knowl. Discov. **32**(6), 1768–1805 (2018)
2. Amsaleg, L., Chelly, O., Houle, M.E., Kawarabayashi, K., Radovanović, M., Treeratanajaru, W.: Intrinsic dimensionality within tight localities. In: Proceedings of the 19th SIAM International Conference on Data Mining (SDM), Calgary, AB, Canada, pp. 181–189 (2019)
3. Aumüller, M., Ceccarello, M.: The role of local intrinsic dimensionality in benchmarking nearest neighbor search. In: Amato, G., Gennaro, C., Oria, V., Radovanović, M. (eds.) SISAP 2019. LNCS, vol. 11807, pp. 113–127. Springer, Cham (2019). https://doi.org/10.1007/978-3-030-32047-8_11
4. Becker, R., Hafnaoui, I., Houle, M.E., Li, P., Zimek, A.: Subspace determination through local intrinsic dimensional decomposition. In: Amato, G., Gennaro, C., Oria, V., Radovanović, M. (eds.) SISAP 2019. LNCS, vol. 11807, pp. 281–289. Springer, Cham (2019). https://doi.org/10.1007/978-3-030-32047-8_25
5. Breunig, M.M., Kriegel, H.P., Ng, R., Sander, J.: LOF: identifying density-based local outliers. In: Proceedings of the ACM International Conference on Management of Data (SIGMOD), Dallas, TX, pp. 93–104 (2000)

6. Campello, R.J.G.B., Kröger, P., Sander, J., Zimek, A.: Density-based clustering. Wiley Interdiscip. Rev. Data Min. Knowl. Discov. **10**(2) (2020)
7. Campos, G.O., et al.: On the evaluation of unsupervised outlier detection: measures, datasets, and an empirical study. Data Min. Knowl. Discov. **30**(4), 891–927 (2016)
8. Casanova, G., et al.: Dimensional testing for reverse k-nearest neighbor search. Proc. VLDB Endow. **10**(7), 769–780 (2017)
9. Cochran, W.G.: Sampling Techniques. Wiley, Hoboken (1977)
10. Coles, S.: An Introduction to Statistical Modeling of Extreme Values. Springer, London (2001). https://doi.org/10.1007/978-1-4471-3675-0
11. de Vries, T., Chawla, S., Houle, M.E.: Density-preserving projections for large-scale local anomaly detection. Knowl. Inf. Syst. (KAIS) **32**(1), 25–52 (2012)
12. Ester, M., Kriegel, H.P., Sander, J., Xu, X.: A density-based algorithm for discovering clusters in large spatial databases with noise. In: Proceedings of the 2nd ACM International Conference on Knowledge Discovery and Data Mining (KDD), Portland, OR, pp. 226–231 (1996)
13. Hashem, T., Rashidi, L., Bailey, J., Kulik, L.: Characteristics of local intrinsic dimensionality (LID) in subspaces: local neighbourhood analysis. In: Amato, G., Gennaro, C., Oria, V., Radovanović, M. (eds.) SISAP 2019. LNCS, vol. 11807, pp. 247–264. Springer, Cham (2019). https://doi.org/10.1007/978-3-030-32047-8_22
14. Hill, B.M.: A simple general approach to inference about the tail of a distribution. Ann. Stat. **3**(5), 1163–1174 (1975)
15. Houle, M.E.: Dimensionality, discriminability, density and distance distributions. In: Proceedings of the 13th IEEE International Conference on Data Mining Workshops, ICDM Workshops, Dallas, TX, pp. 468–473 (2013)
16. Houle, M.E.: Local intrinsic dimensionality I: an extreme-value-theoretic foundation for similarity applications. In: Beecks, C., Borutta, F., Kröger, P., Seidl, T. (eds.) SISAP 2017. LNCS, vol. 10609, pp. 64–79. Springer, Cham (2017). https://doi.org/10.1007/978-3-319-68474-1_5
17. Houle, M.E.: Local intrinsic dimensionality II: multivariate analysis and distributional support. In: Beecks, C., Borutta, F., Kröger, P., Seidl, T. (eds.) SISAP 2017. LNCS, vol. 10609, pp. 80–95. Springer, Cham (2017). https://doi.org/10.1007/978-3-319-68474-1_6
18. Houle, M.E., Kashima, H., Nett, M.: Generalized expansion dimension. In: ICDM Workshop Practical Theories for Exploratory Data Mining (PTDM), pp. 587–594 (2012)
19. Houle, M.E., Ma, X., Nett, M., Oria, V.: Dimensional testing for multi-step similarity search. In: Proceedings of the 12th IEEE International Conference on Data Mining (ICDM), Brussels, Belgium, pp. 299–308 (2012)
20. Houle, M.E., Ma, X., Oria, V.: Effective and efficient algorithms for flexible aggregate similarity search in high dimensional spaces. IEEE Trans. Knowl. Data Eng. **27**(12), 3258–3273 (2015)
21. Houle, M.E., Ma, X., Oria, V., Sun, J.: Efficient algorithms for similarity search in axis-aligned subspaces. In: Traina, A.J.M., Traina, C., Cordeiro, R.L.F. (eds.) SISAP 2014. LNCS, vol. 8821, pp. 1–12. Springer, Cham (2014). https://doi.org/10.1007/978-3-319-11988-5_1
22. Houle, M.E., Nett, M.: Rank-based similarity search: reducing the dimensional dependence. IEEE Trans. Pattern Anal. Mach. Intell. **37**(1), 136–150 (2015)

23. Houle, M.E., Oria, V., Wali, A.M.: Improving k-NN graph accuracy using local intrinsic dimensionality. In: Beecks, C., Borutta, F., Kröger, P., Seidl, T. (eds.) SISAP 2017. LNCS, vol. 10609, pp. 110–124. Springer, Cham (2017). https://doi.org/10.1007/978-3-319-68474-1_8

24. Houle, M.E., Schubert, E., Zimek, A.: On the correlation between local intrinsic dimensionality and outlierness. In: Marchand-Maillet, S., Silva, Y.N., Chávez, E. (eds.) SISAP 2018. LNCS, vol. 11223, pp. 177–191. Springer, Cham (2018). https://doi.org/10.1007/978-3-030-02224-2_14

25. Karamata, J.: Sur un mode de croissance réguliere, théorèmes fondamentaux. Bull. Soc. Math. France **61**, 55–62 (1933)

26. Karger, D.R., Ruhl, M.: Finding nearest neighbors in growth-restricted metrics. In: Proceedings of the 34th Annual ACM Symposium on Theory of Computing (STOC), Montreal, QC, Canada, pp. 741–750 (2002)

27. Kullback, S.: Information Theory and Statistics. Wiley, Hoboken (1959)

28. Liese, F., Vajda, I.: On divergences and informations in statistics and information theory. IEEE Trans. Inf. Theory **52**(10), 4394–4412 (2006)

29. Ma, X., et al.: Characterizing adversarial subspaces using local intrinsic dimensionality, pp. 1–15 (2018)

30. Ma, X., et al.: Dimensionality-driven learning with noisy labels, pp. 3361–3370 (2018)

31. Romano, S., Chelly, O., Nguyen, V., Bailey, J., Houle, M.E.: Measuring dependency via intrinsic dimensionality. In: ICPR16, pp. 1207–1212, December 2016

32. Schubert, E., Gertz, M.: Intrinsic t-stochastic neighbor embedding for visualization and outlier detection. In: Beecks, C., Borutta, F., Kröger, P., Seidl, T. (eds.) SISAP 2017. LNCS, vol. 10609, pp. 188–203. Springer, Cham (2017). https://doi.org/10.1007/978-3-319-68474-1_13

33. Scott, A.J., Wu, C.: On the asymptotic distribution of ratio and regression estimators. J. Am. Stat. Assoc. **76**, 98–102 (2006)

34. Sugiyama, M., Suzuki, T., Kanamori, T.: Density Ratio Estimation in Machine Learning. Cambridge University Press, Cambridge (2012)

35. Terrell, D.G., Scott, D.W.: Variable kernel density estimation. Ann. Stat. **20**(3), 1236–1265 (1992)

Analysing Indexability of Intrinsically High-Dimensional Data Using TriGen

David Bernhauer[1,2] and Tomáš Skopal[1]

[1] SIRET RG, Faculty of Mathematics and Physics, Charles University,
Prague, Czech Republic
bernhdav@fit.cvut.cz, skopal@ksi.mff.cuni.cz
[2] Faculty of Information Technology, Czech Technical University in Prague,
Prague, Czech Republic

Abstract. The TriGen algorithm is a general approach to transform distance spaces in order to provide both exact and approximate similarity search in metric and non-metric spaces. This paper focuses on the reduction of intrinsic dimensionality using TriGen. Besides the well-known intrinsic dimensionality based on distance distribution, we inspect properties of triangles used in metric indexing (the triangularity) as well as properties of quadrilaterals used in ptolemaic indexing (the ptolemaicity). We also show how LAESA with triangle and ptolemaic filtering behaves on several datasets with respect to the proposed indicators.

1 Introduction

The real-world datasets for similarity search often exhibit high intrinsic dimensionality manifested by distance distribution with low variance and high mean [5]. The reason could be the high complexity of the similarity model within a given domain (lot of independent features), but often this is just a consequence of automated feature extraction processes, e.g., the inference of deep features [6]. Intrinsically high-dimensional data cannot be used for efficient exact search but, luckily, there have been developed many approximate methods [9] to tackle this problem for the price of a lower retrieval precision. Some of these methods elegantly avoid the direct problem of high intrinsic dimensionality by not indexing actual distances, but just permutations of pivots [4,7]. These methods enabled competitive application of similarity search in real-world domains where maximal retrieval precision is not as critical as the performance. However, we must keep in mind these methods are limited in tuning the precision at runtime (from query to query) as well as they are restricted to pivot-based indexing schemes.

The TriGen algorithm [11] was proposed as a universal method for fast exact and approximate search in metric and non-metric spaces. So far, it was not analyzed as a method for (intrinsic) dimensionality reduction. In this paper

This research has been supported by Czech Science Foundation (GAČR) project Nr. 19-01641S.

S. Satoh et al. (Eds.): SISAP 2020, LNCS 12440, pp. 261–269, 2020.
https://doi.org/10.1007/978-3-030-60936-8_20

we empirically analyze this missing aspect. We also investigate the impact of TriGen modifications on the potential of ptolemaic indexing [8] that achieves better performance than metric indexing (though limited to ptolemaic metrics).

2   Background

When indexing data for fast similarity search, we face two fundamental concepts – the data indexability and the indexing model.

2.1   Indexability

The indexability generally refers to an ability to search efficiently a dataset $\mathbb{S} \subset \mathbb{U}$ under a similarity model $(\mathbb{U}, d)$, regardless the indexing method used. The key is the distribution of data or, specifically, in case of similarity search it is the distribution of distances $d(x, y)$ among data objects $x, y \in \mathbb{S}$. The classic index-ability indicator for a metric space model $(\mathbb{U}, d)$ is the intrinsic dimensionality [5], defined as the ratio of squared mean and doubled variance of the distance distribution; $\text{iDim}(\mathbb{S}, d) = \frac{\mu^2}{2\sigma^2}$. The lower iDim, the better indexability.

Alternatively, the ball overlap factor (BOF) [11] describes the ability to par-tition the dataset into non-overlapping ball-shaped regions. The BOF counts for how many object pairs will constitute overlapping balls (each ball radius is the distance to the ball center's kth nearest neighbor).

2.2   TriGen Transformation

The TriGen algorithm [11] transforms the input distance space $(\mathbb{U}, d)$ by use of triangle-generating or -violating modifiers and a dataset sample $\mathbb{S}^* \subseteq \mathbb{S} \subset \mathbb{U}$ into a target space $(\mathbb{U}, f(d))$. A modifier $f : R \rightarrow R_0^+$ must be an increasing function with $f(0) = 0$ to preserve the ordering of distances[1] and thus search results with respect to sequential scan. The triangle-generating (concave) modifiers "inflate" all the triangles in the space to become more equilateral; then the dataset is less indexable as the intrinsic dimensionality increases. The triangle-violating (con-vex) modifiers have the opposite effect – "squeezing" the triangles and lowering the intrinsic dimensionality. The idea behind the triangle-violating modifiers is that they lower the intrinsic dimensionality (more efficient search) for the price of a retrieval error (some triangles break which shows in incorrect filtering by querying). The indexability indicators, like the intrinsic dimensionality or BOF, together with the T-error measuring the ratio of broken triangles, guide TriGen to determine the right modifier.

Unlike other methods that map the source distance space into the Euclidean space, the TriGen model is based solely on transformation of distances, hence there is no need for an expensive and static embedding of metric objects into vectors. In consequence, once a modifier is computed for a particular problem, its

---

[1] Ranking of objects $x_i \in \mathbb{U}$ based on $d(q, x_i)$ is the same as based on $f(d(q, x_i))$.

change (e.g., a precision guarantee) can be easily recomputed and the already created index just updated (no change in descriptors). This allows to switch between several TriGen modifiers at query time, providing thus flexible exact-to-approximate search (e.g., the NM-tree [10]). Other TriGen follow-ups include extensions to non-symmetric distances [3], and the genetic TriGen variants [1, 2].

In this paper, we inspect the TriGen in the role of dimensionality reduction method. In high-dimensional datasets (as measured by intrinsic dimensionality), all of the non-trivial triangles tend to be almost-equilateral. Then application of TriGen with triangle-violating modifiers could act as a lossless dimensionality reduction method by squeezing the triangles without the violation of triangle inequality (breaking the triangles by squeezing them too much). Our hypothesis is, the higher the (intrinsic) dimensionality of data is, the more almost-equilateral the triangles are, and so the more aggressive modifier could be applied while still keeping the triangles unbroken. Simply said, we analyze the question if TriGen could "cancel" the curse of dimensionality (to some extent) in similarity search.

2.3   Metric and Ptolemaic Indexing

The metric access methods (metric indexes) [5] use some construction of lower bounds using the triangle inequality. In the simplest case of pivot tables (aka LAESA), the three objects in the triangle are the query object q, a dataset object x, and a pivot p (i.e., $LB_\triangle(q, x) = |d(q, p) - d(p, x)|$). If the triangle is equilateral, $LB_\triangle(q, x) = 0$ and so the dataset object x cannot be filtered by the lower bound. On the other hand, if the triangle is (squeezed to) a line segment, the lower bound gets maximal (i.e., $LB_\triangle(q, x, p) = d(q, x)$) and so it is "super-effective" for filtering.

Similarly, ptolemaic access methods (ptolemaic indexes) [8] use some construction of lower bounds using the Ptolemy's inequality that operates on quadrilaterals (quadruplets, respectively). In the simplest (LAESA) case there are four objects in the quadrilaterals: the query object q, a dataset object x, and two pivots p_1, p_2, while a lower bound can be derived as

$$LB_{pt}(q, x, p_1, p_2) = \frac{|d(q, p_1) \cdot d(x, p_2) - d(q, p_2) \cdot d(x, p_1)|}{d(p_1, p_2)} \tag{1}$$

As the quadrilaterals are more complex than triangles, there is not a single best or worst quadrilateral example for the lower bound construction. Also the inflating and squeezing effect of TriGen modifiers is not clear in case of quadrilaterals, and so for ptolemaic indexing.

3   Triangle and Quadrilateral Distribution

The intrinsic dimensionality, as an indexability indicator, considers only distances themselves but does not consider that some distance combinations cannot be present in triangles at the same time, which is important for the filtering by metric access methods. The BOF compensates this issue, but it cannot be easily generalized for Ptolemaic inequality or non-metric cases. Therefore, we define the *triangularity* to quantify the shape of triangle on a real-value scale from equilateral triangle, through line segment to broken triangle. Similarly, we define *Ptolemaicity* to quantify the shape of quadrilateral on a scale from tetrahedron, through line segment to broken equilateral.

Hence, we need to aggregate three distances forming a triangle into one number, with extremes for equilateral triangles and line segments. We could adopt the TriGen criteria (presented in [5]) used for determining the number of triangles that do not satisfy the triangle inequality. The *triangularity* is defined for a triangle $a = d(x, y), b = d(y, z), c = d(x, z)$ by Eq. 2 – this ratio determines how "equilateralish" (or "inflated") a triangle is. The triangularity is 1 for equilateral triangle, $1/2$ means the triangle forms line segment ("squeezed"), and for values below $1/2$ the triangle is broken (does not satisfy the triangle inequality).

$$\text{Triangularity}(a, b, c) = \frac{a + b}{2c}, \text{ where } a \leq b \leq c \qquad (2)$$

After TriGen preprocessing, we expect the distribution will be shifted to line segments ("squeezed") instead of almost-equilateral triangles. Knowledge of this common property makes the *triangularity* a good indicator of datasets with statistically high probability to exhibit bad indexability.

Moreover, we try TriGen for Ptolemaic indexing, though the TriGen modifiers were originally proposed for indexing using the lower bounds based on triangle inequality and not the Ptolemy's inequality (Eq. 3). We would like to find out how the Ptolemy's inequality holds in comparison with the triangle inequality. We define *ptolemaicity* of a quadrilateral as Eq. 4, where $d(w, x)d(y, z), d(w, y)d(x, z) \leq d(w, z)d(x, y)$. The greatest ptolemaicity value is 1, which represents regular tetrahedron and results in bad indexability. ptolemaicity $1/2$ represents a line segment and for values below $1/2$ the equilateral is broken (does not satisfy Ptolemy's inequality).

$$(\forall w, x, y, z \in \mathbb{U}) \; d(w, x)d(y, z) + d(w, y)d(x, z) \geq d(w, z)d(x, y) \qquad (3)$$

$$\text{Ptolemaicity}(w, x, y, z) = \frac{d(w, x)d(y, z) + d(w, y)d(x, z)}{2d(w, z)d(x, y)} \qquad (4)$$

4   Analysis of High-Dimensional Data

We have analyzed several datasets and looked at the intrinsic dimensionality and the retrieval efficiency (using the LAESA algorithm). Two low-dimensional datasets are from SISAP datasets: the 20-dimensional NASA dataset, and the 112-dimensional Colors dataset. As high-dimensional datasets we used a sample of the 2048-dimensional AlexNet image (V3C1) dataset, and several artificial datasets of dimensionality 2 to 2048 (randomly generated vectors). For all datasets we have used the Euclidean space, which is both metric and ptolemaic.

Table 1. Datasets statistics (iDim, distance computations with metric LAESA).

Dataset (dim)	without TriGen		with TriGen (zero error)	
	iDim	Dist. Comp.	iDim	Dist. Comp.
NASA (20)	5.184 ± 0.007	2.12%	4.593 ± 0.007	1.15%
Colors (112)	2.742 ± 0.003	2.63%	2.553 ± 0.003	2.08%
Random (128)	181.328 ± 0.304	100%	28.663 ± 0.022	95.78%
Random (2048)	1967.66 ± 184.295	100%	37.035 ± 0.175	99.3%
V3C1 (2048)	30 ± 0.050	86.65%	9.215 ± 0.012	45.39%

In Table 1 on the left, we present intrinsic dimensionality comparison and efficiency improvement of the metric LAESA (with randomly chosen 50 pivots) against sequential search. The iDim of Colors dataset is lower than iDim NASA dataset, however, LAESA performs better on NASA. Note the embedding dimensionality and iDim are dramatically different in case of V3C1 and Colors. Figure 1 shows distance distribution histograms for all datasets.

In Fig. 2a (dashed), we present triangularity distribution. As we expected, the distribution is shifted to the right side for high-dimensional datasets. This is the main assumption for transforming metric space using the TriGen into a more indexable one. Similarly, we have visualized the ptolemaicity distribution in Fig. 2b (dashed), which displays the same properties.

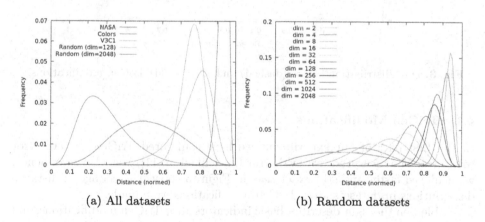

(a) All datasets   (b) Random datasets

Fig. 1. Distance distribution comparison

Both triangularity and ptolemaicity distributions are similar, which means TriGen could be used for modification of Ptolemaic space, too. If the TriGen transforms both spaces consistently then, based on figures, Ptolemy's inequality is violated earlier, because there is a higher number of line segments.

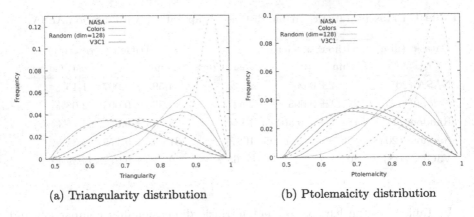

(a) Triangularity distribution   (b) Ptolemaicity distribution

Fig. 2. Distribution of triangularity or ptolemaicity in datasets before (dashed) and after (solid) TriGen modifications.

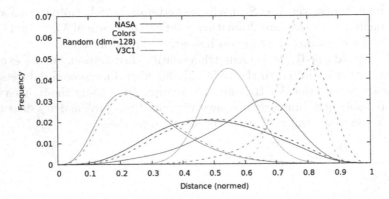

Fig. 3. Dist. distribution before (dashed) and after (solid) TriGen modifications.

4.1   TriGen Modifications

In the first part of our experiment, we have configured TriGen to zero error tolerance. The measured retrieval error (as defined in [11]) was also zero, hence, we achieved faster and still exact search. Figure 3 shows the change of distance distributions in datasets after TriGen modifications were made.

Table 1 on the right describes basic indicators after TriGen modifications, and we observe that triangle-violating modifications reduced the intrinsic dimensionality. The retrieval efficiency improved for all datasets (for some only slightly, but two times for NASA and V3C1). It indicates the presence of an inner structure beyond all conventional indicators, except for Random (2048) that is not indexable for exact search. However, TriGen can still transform a seemingly not-indexable dataset (V3C1, Random(128)) into partially indexable even for exact search.

Both triangularity (Fig. 2a) and ptolemaicity (Fig. 2b) distributions are flatter and shifted to the left as we expected. The ptolemaicity distribution is flatter than triangularity distribution, which means that Ptolemy's inequality is more prone to a violation when used with TriGen.

4.2   Comparison of Real Performance

The TriGen algorithm controls the ratio of triangles satisfying the triangle inequality (so-called T-error tolerance) by a weight parameter that determines the convexity/concavity of the modifier. In the previous experiments we set T-error tolerance = 0 that (empirically) guarantees zero retrieval error. In Fig. 4a, we can see the dependence of distance computations and retrieval error on the weight (V3C1 dataset). We used just the triangle-violating (squeezing) modifications where −10 weight is heavy squeezing and −0.1 weight is almost no squeezing. We used LAESA with 50 randomly chosen pivots utilizing metric filtering, ptolemaic filtering, or both, and compared it with the sequential search.

The important observation is the ptolemaic filtering[2] has a similar pattern as the metric filtering. The general difference is in the shift of the ptolemaic curves to the right. The combination of triangle and ptolemaic filtering utilizes the benefits of both approaches. Triangle filtering deals with retrieval error caused by the Ptolemy's inequality violation and the Ptolemy's filtering deals with better efficiency, because of its ability to create better lower bounds.

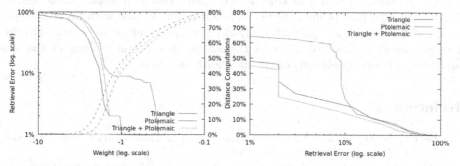

(a) Efficiency (dashed) and retrieval error (solid) based on TriGen weight.

(b) Efficiency per error

Fig. 4. Efficiency and retrieval error (LAESA with 50 pivots on V3C1 dataset).

Another point of view is presented in Fig. 4a, where pairs of efficiency and retrieval error values from Fig. 4b are aggregated into single efficiency per error value. So, we get rid of the TriGen weight parameter and only observe how the real efficiency is dependent on real retrieval error, obtaining more readable results than when depicted individually.

---

[2] We used simple random selection of pivot pairs in ptolemaic filtering instead the better but slower Balanced heuristic [8].

4.3 Discussion

The intrinsic dimensionality is not always sufficient to predict the real efficiency of an indexing algorithm. First, because of some inner structure that can hardly be described by a single number. Second, the high number of low distances, triangularities, or ptolemaicities does not imply better indexability. A good example can be randomly generated vectors with one outlier, which will shift the whole histogram to the left.

The TriGen can be used for both precise and approximate search. The combination of both filtering inequalities improves not only efficiency but also lowers the retrieval error. There is a possibility in the future to try other kinds of inequalities and their ability to scale with TriGen.

5 Conclusions

We have introduced structure-sensitive empirical measures for the analysis of metric and Ptolemaic spaces and defined the triangularity and the ptolemaicity as the quantifiers of triangle and quadrilateral shapes. Analysis of high-dimensional data shows that it is possible to use TriGen as dimensionality reduction method that improves the efficiency of similarity search.

Although the TriGen was designed for transforming non-metric spaces into metric ones, we have shown that the inverse application on high-dimensional data is possible as well and efficient for both exact and approximate search. Moreover, experiments indicate that TriGen could be used with different types of filtering inequalities (like Ptolemy's). The combination of several filtering inequalities synergically deals with the advantages (better efficiency) and disadvantages (worse precision) of the individual methods.

References

1. Bernhauer, D., Skopal, T.: Approximate search in dissimilarity spaces using GA. In: GECCO, pp. 279–280. ACM (2019)
2. Bernhauer, D., Skopal, T.: Non-metric similarity search using genetic TriGen. In: Amato, G., Gennaro, C., Oria, V., Radovanović, M. (eds.) SISAP 2019. LNCS, vol. 11807, pp. 86–93. Springer, Cham (2019). https://doi.org/10.1007/978-3-030-32047-8_8
3. Boytsov, L., Nyberg, E.: Pruning algorithms for low-dimensional non-metric k-NN search: a case study. In: Amato, G., Gennaro, C., Oria, V., Radovanović, M. (eds.) SISAP 2019. LNCS, vol. 11807, pp. 72–85. Springer, Cham (2019). https://doi.org/10.1007/978-3-030-32047-8_7
4. Chavez, E., Figueroa, K., Navarro, G.: Effective proximity retrieval by ordering permutations. IEEE TPAMI 30(9), 1647–1658 (2008)
5. Chávez, E., Navarro, G., Baeza-Yates, R., Marroquín, J.L.: Searching in metric spaces. ACM Comput. Surv. 33(3), 273–321 (2001)
6. Donahue, J., et al.: DeCAF: a deep convolutional activation feature for generic visual recognition. In: ICML, pp. I-647–I-655. JMLR.org (2014)

7. Esuli, A.: Use of permutation prefixes for efficient and scalable approximate similarity search. Inf. Process. Manag. **48**(5), 889–902 (2012)
8. Hetland, M.L., Skopal, T., Lokoč, J., Beecks, C.: Ptolemaic access methods: challenging the reign of the metric space model. Inf. Syst. **38**(7), 989–1006 (2013)
9. Patella, M., Ciaccia, P.: Approximate similarity search: a multi-faceted problem. J. Discret. Algorithms **7**(1), 36–48 (2009)
10. Skopal, T., Lokoč, J.: NM-tree: flexible approximate similarity search in metric and non-metric spaces. In: Bhowmick, S.S., Küng, J., Wagner, R. (eds.) DEXA 2008. LNCS, vol. 5181, pp. 312–325. Springer, Heidelberg (2008). https://doi.org/10.1007/978-3-540-85654-2_30
11. Skopal, T.: Unified framework for fast exact and approximate search in dissimilarity spaces. ACM Trans. Database Syst. **32**(4), 29-es (2007)

Reverse k-Nearest Neighbors Centrality Measures and Local Intrinsic Dimension

Oscar Pedreira[1(✉)], Stephane Marchand-Maillet[2], and Edgar Chávez[3]

[1] Universidade A Coruña, CITIC, Elviña, 15071 A Coruña, Spain
oscar.pedreira@udc.es
[2] Viper Group - University of Geneva, Geneva, Switzerland
Stephane.Marchand-Maillet@unige.ch
[3] CICESE Research Center, Ensenada, Mexico
elchavez@cicese.mx

Abstract. The estimation of local intrinsic dimensionality has applications ranging from adversarial attack disclosure to clustering and outlier detection, indexing, and data fingerprinting. In this paper, we analyze measures of network centrality in the kNN graph and their relation to LID measures. Our method ranks the dataset by its centrality, measured as the number of reverse or mutual kNN of each object. The computation of these measures involves only kNN queries, allowing a speedup in its computation using probabilistic indexing. A property of independent interest is the rank being independent of k for a wide range of k values, leading to parameter-free density estimation and applications.

Keywords: Similarity search · Local intrinsic dimensionality

1 Introduction

Metric access methods use the properties of metric spaces and precomputed information to avoid the sequential scan of the collection in searches. Search performance depends directly on the collection's *intrinsic dimensionality* [6,14,18], which therefore measures its *indexability*. The performance can be close to that of a sequential scan when the intrinsic dimensionality is high. The literature on similarity search usually refers to this as the *curse of dimensionality* [19]. The intrinsic dimensionality depends on the nature of the data and the distance function. When the distances between any two objects tend to concentrate around the mean of the distance distribution, the *discriminative* capacity of the distance function is lower and the intrinsic dimensionality is higher.

The intrinsic dimensionality has been studied as a property of the entire collection. However, Houle [2,9] proposed the concept of *local intrinsic dimensionality* (LID), which involves only the distances in the neighbourhood of a

Partially funded by: BIZDEVOPS-Global (RTI2018-098309-B-C32), MINECO & FEDER; Centros singulares de investigación de Galicia (ED431G/01), Grupo de Referencia Competitiva (ED431C 2017/58), and ConectaPEME GEMA (IN852A 2018/14), Xunta de Galicia & FEDER.

S. Satoh et al. (Eds.): SISAP 2020, LNCS 12440, pp. 270–278, 2020.
https://doi.org/10.1007/978-3-030-60936-8_21

specific object. Amsaleg et al. [2] proposed different ways of estimating the local intrinsic dimensionality based on the distribution of the distances. The concept of local intrinsic dimensionality has many applications. For example, Houle et al. [12] used it in the proposal of a fingerprinting method for secure search. Hoyos et al. [13] showed that LID can be used for partitioning of a collection based on the indexability of its objects, and showed that objects in the collection can be ranked according to their centrality in the half-space proximal graph (HSP) [5].

In this paper, we focus on two centrality measures of objects in the general kNN graph: the number of mutual nearest neighbours, and the number of reverse nearest neighbours. These two measures give us information about the density of different regions in the space. Regions of the space with high values for the number of reverse near neighbours are regions with a high density. This may have important applications for random sampling algorithms, for example, to detect bias in datasets. We show that there is a relationship between these measures and the LID. However, the purpose of these measures is not estimating the LID, but to rank the objects of the collection according to their indexability, so different subsets can be indexed in different ways depending on their complexity.

The paper is structured as follows: Sect. 2 presents related work. Section 3 defines two measures to rank objects according to their indexability, and Sect. 4 presents experimental results on different datasets. Section 5 presents the conclusions of the paper.

2 Background and Related Work

The intrinsic dimensionality is a recurrent topic in similarity search. The intrinsic dimensionality of a collection is determined by the nature of the data and the distance function. In high-dimensional spaces, the distances tend to be close to the mean of the distance distribution, so the distance function has low discriminative power. These scenarios deteriorate the performance of indexing methods since their pruning criteria tend to fail when all distances are too similar. However, the problem of high-dimensional spaces is not just that of performance, since distance functions with low discriminative power can affect the quality of the results of the application where that distance is being used [3].

The intrinsic dimensionality has been studied as a property of the entire collection, typically estimated using the mean and standard deviation of the distance distribution [6]. In this case, its estimation can give us no more information than the collection's indexability. Houle defined the local intrinsic dimensionality as a measure of the dimensionality in the neighbourhood of an object [9–11], and showed it is not necessarily uniform in the collection. Amsaleg et al. [2] proposed LID estimators based in the distance distribution. Let X be a distance random variable in $[0, w)$, and $x_1 \leq \cdots \leq x_m$ observations of X. The maximum likelihood estimator (MLE) of LID is defined in [2] as: $\widehat{\mathrm{ID}}_X = - \left(\frac{1}{m} \sum_{i=1}^{m} \ln \frac{x_i}{w} \right)^{-1}$.

Since dimensionality is not necessarily uniform, the indexability can also be different in different regions. This assumption opens the possibility of indexing different parts of a collection differently, depending on their complexity.

For example, Brisaboa et al. [4] showed empirically the existence of *nested metric spaces*, and that indexing them independently can improve the search performance. Hoyos et al. [13] proposed to partition the collection in terms of its indexability by ranking the objects in the collection depending on their centrality in the half-space proximal graph, and showed that the centrality measure could be used for identifying harder objects in the dataset.

3 Collection Partitioning Based on Centrality Measures

In this paper, we analyze the use of two centrality measures in the general kNN graph to rank the objects by their indexability. Let (U, d) be a metric space, and $S \subseteq U$ a finite collection of $n = |S|$ objects. The kNN-graph on a metric dataset connects each object $u \in S$ to its k nearest neighbors, kNN(u).

- *Number of reverse neighbors* (REVERSE): this centrality measure counts for each object $u \in S$ the number of objects $v \in S$ for which $u \in k$NN(v). The number of reverse neighbours (RkNN) has already been used, for example, for identifying boundary objects in multidimensional datasets [7].
- *Number of mutual neighbors* (MUTUAL): this measure counts for each object $u \in S$ the number of objects $v \in S$ for which $u \in k$NN(v) $\wedge$ $v \in k$NN(u).

These two measures give us information about the density of objects in the neighbourhood of an object. We hypothesise that, since the values of these measures depend on the density of the neighbourhood of an object, they can be used to rank the objects in the collection by their indexability.

Both measures can be obtained directly on a kNN graph on S. Building the graph can be expensive. However, to rank objects by their indexability, we build an approximation of the kNN graph by indexing the collection with HNSW [15] and then running approximate kNN queries for each $u \in S$. The values the measures are conditioned by the value of k. A small k makes its computation cheap but would give too few values to discriminate.

4 Experimental Evaluation

Experimental Setup. We used collections of feature vectors from images with L_2:

- COLORS: included in the metric space library [8], it contains $112,544$ vectors of 112 features extracted from the color histograms of the images.
- SIFT: ANN_SIFT1B contains 1 billion images represented by 128-dimensional feature vectors [16]. We used $100,000$ vectors from the base set.
- DeepFeatures: contains a subset of $100,000$ vectors of dimension $4,096$ from the DeepFeatures collection [1], where each vector contains features extracted using a convolutional neural network trained on images and places.

For each collection, we used $100,000$ vectors, where 90% of them were used for indexing and the remaining 10% were used as queries (kNN, $k = 10$). The kNN graph was built on each collection using values of k of 10, 512, and $1,024$. Then, the objects in each collection were ranked in ascending and descending order on the value of MUTUAL and REVERSE. We also estimated the LID in the neighbourhood of each object using MLE and ranked them on this value. In this case, we used the same settings of [2], computing MLE with the distances from each object to its $1,000$ kNNs in a sample of $10,000$ objects.

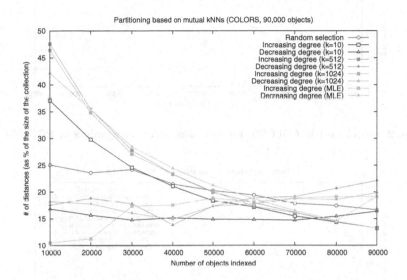

Fig. 1. Search cost in COLORS for increasing/decreasing values of MUTUAL.

We created subsets of each collection with sizes ranging from $10,000$ to $90,000$ by selecting objects at random, and in increasing and decreasing orders of MLE, MUTUAL, and REVERSE. Each subset was indexed with the *Spatial Approximation Tree* (SAT) [17], an efficient structure that recursively partitions the collection by proximity to the root of each subtree, and has no parameters.

Experimental Results. Figures 1 and 2 show the search cost (expressed as the % of the DB explored) for COLORS. For sizes $10,000$–$30,000$, the objects with a low number of mutual neighbours showed a higher search cost. For subsets of size $40,000$ and higher, the differences are not so significant. The results are similar for the subsets created attending to the value of MLE, which suggests that there exists a relationship between LID and the centrality measures. Notice that for size $90,000$, the results are different for each configuration, since the objects are processed in a different order that can affect the index.

Figures 3 and 4 show the results for SIFT. SIFT is harder than COLORS, although they have a similar dimension (128 and 112 respectively). The search

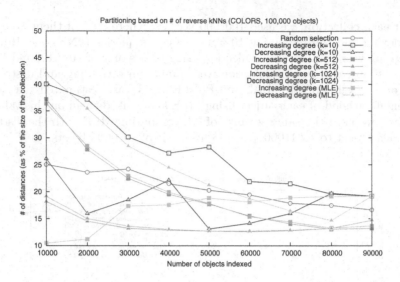

Fig. 2. Search cost in COLORS for increasing/decreasing values of REVERSE.

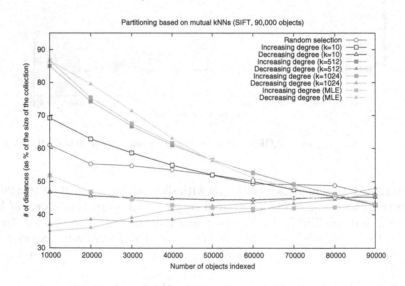

Fig. 3. Search cost in SIFT for increasing/decreasing values of MUTUAL.

cost was around 20%–25% for COLORS and between 50%–60% for SIFT. The results regarding MUTUAL and REVERSE are very similar to those obtained in COLORS. Both measures can distinguish subsets with higher and lower search costs. Another difference with COLORS is that in SIFT the results do not depend that much on the value of k in REVERSE. The results obtained with both measures have a relationship with those obtained with MLE. However, notice that

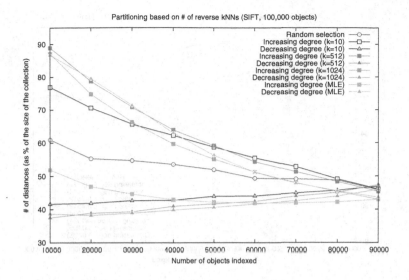

Fig. 4. Search cost in SIFT for increasing/decreasing values of REVERSE.

for sizes 10,000 to 40,000 the number of mutual and reverse neighbours present better results than MLE in identifying easily indexable objects.

Figures 5 and 6 show the results for DeepFeatures. This collection is the hardest in our setup, with a search cost around 95%–99%. In this collection, the differences between a random selection and a selection guided by MUTUAL and

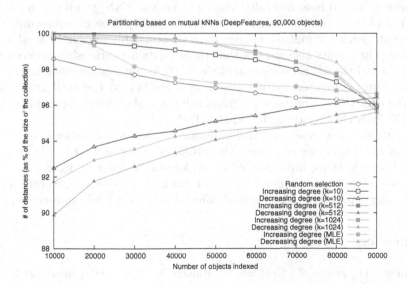

Fig. 5. Search cost in DeepFeatures for increasing/decreasing values of MUTUAL.

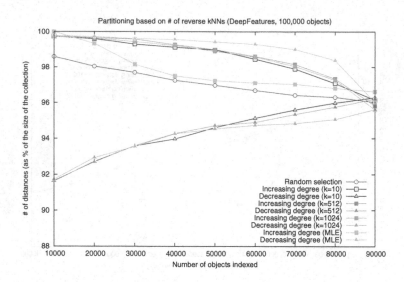

Fig. 6. Search cost in DeepFeatures for increasing/decreasing values of REVERSE.

REVERSE are not so significant as in COLORS and SIFT. However, we can see that both measures can identify the most and least costly objects.

5   Conclusions and Future Work

We have analyzed how centrality measures on the kNN graph can be used to rank the objects in a collection by their indexability. The experimental results show that number of mutual and reverse nearest neighbours can be used for this purpose. The results also suggest that there exists a relationship between these measures and the MLE estimation of the LID. An important difference of these two measures compared to that based on centrality on the HSP graph [13] is that the construction of the kNN graph using a probabilistic approximate index like HNSW is cheaper than that of the HSP.

We are working on an extended evaluation with more collections and indexes. Also, in this paper, we computed the MLE estimator of LID in the same way for all the collections, but it should be explored how other configurations would affect its value in each collection. Future work also includes exploring how we could index separately the subsets of different complexity within each collection.

References

1. Amato, G., Falchi, F., Gennaro, C., Rabitti, F.: YFCC100M hybridnet fc6 deep features for content-based image retrieval. In: Proceedings of ACM Workshop on Multimedia COMMONS (MMCommons 2016), pp. 11–18. ACM (2016)

2. Amsaleg, L., et al.: Estimating local intrinsic dimensionality. In: Proceedings of ACM International Conference on Knowledge Discovery and Data Mining (KDD 2015), pp. 29–38. ACM (2015)
3. Beyer, K., Goldstein, J., Ramakrishnan, R., Shaft, U.: When is "nearest neighbor" meaningful? In: Beeri, C., Buneman, P. (eds.) ICDT 1999. LNCS, vol. 1540, pp. 217–235. Springer, Heidelberg (1999). https://doi.org/10.1007/3-540-49257-7_15
4. Brisaboa, N.R., Luaces, M.R., Pedreira, O., Places, Á.S., Seco, D.: Indexing dense nested metric spaces for efficient similarity search. In: Pnueli, A., Virbitskaite, I., Voronkov, A. (eds.) Perspectives of Systems Informatics. PSI 2009. Lecture Notes in Computer Science, vol. 5947, pp. 98–109. Springer, Heidelberg (2010). https://doi.org/10.1007/978-3-642-11486-1_9
5. Chavez, E., et al.: Half-space proximal: a new local test for extracting a bounded dilation spanner of a unit disk graph. In: Anderson, J.H., Prencipe, G., Wattenhofer, R. (eds.) Principles of Distributed Systems. OPODIS 2005. Lecture Notes in Computer Science, vol. 3974, pp. 235–245. Springer, Heidelberg (2006). https://doi.org/10.1007/11795490_19
6. Chávez, E., Navarro, G., Baeza-Yates, R., Marroquín, J.L.: Searching in metric spaces. ACM Comput. Surv. **33**(3), 273–321 (2001)
7. Chenyi, X., Hsu, W., Lee, M.L., Ooi, B.C.: Border: efficient computation of boundary points. IEEE Trans. Knowl. Data Eng. **18**(3), 289–303 (2006)
8. Figueroa, K., Navarro, G., Chávez, E.: Metric spaces library (2007). http://www.sisap.org/Metric_Space_Library.html
9. Houle, M.E.: Dimensionality, discriminability, density and distance distributions. In: Proceedings of 13th International Conference on Data Mining Workshops, pp. 468–473. IEEE (2013)
10. Houle, M.E.: Local intrinsic dimensionality I: an extreme-value-theoretic foundation for similarity applications. In: Beecks, C., Borutta, F., Kröger, P., Seidl, T. (eds.) Similarity Search and Applications. SISAP 2017. Lecture Notes in Computer Science, vol. 10609, pp. 64–79. Springer, Cham (2017). https://doi.org/10.1007/978-3-319-68474-1_5
11. Houle, M.E.: Local intrinsic dimensionality II: multivariate analysis and distributional support. In: Beecks, C., Borutta, F., Kröger, P., Seidl, T. (eds.) Similarity Search and Applications. SISAP 2017. Lecture Notes in Computer Science, vol. 10609, pp. 80–95. Springer, Cham (2017). https://doi.org/10.1007/978-3-319-68474-1_6
12. Houle, M.E., Oria, V., Rohloff, K.R., Wali, A.M.: LID-fingerprint: a local intrinsic dimensionality-based fingerprinting method. In: Marchand-Maillet, S., Silva, Y.N., Chávez, E. (eds.) SISAP 2018. LNCS, vol. 11223, pp. 134–147. Springer, Cham (2018). https://doi.org/10.1007/978-3-030-02224-2_11
13. Hoyos, A., Ruiz, U., Marchand-Maillet, S., Chávez, E.: Indexability-based dataset partitioning. In: Amato, G., Gennaro, C., Oria, V., Radovanović, M. (eds.) SISAP 2019. LNCS, vol. 11807, pp. 143–150. Springer, Cham (2019). https://doi.org/10.1007/978-3-030-32047-8_13
14. Indyk, P., Motwani, R.: Approximate nearest neighbors: towards removing the curse of dimensionality. In: Proceedings of ACM Symposium on Theory of computing (STOC 1998), pp. 604–613. ACM (1998)
15. Jégou, H., Douze, M., Schmid, C.: Product quantization for nearest neighbor search. IEEE Trans. Pattern Anal. Mach. Intell. **33**(1), 117–128 (2011)
16. Jégou, H., Tavenard, R., Douze, M., Amsaleg, L.: Searching in one billion vectors: re-rank with source coding. In: Proceedings of International Conference on Acoustics, Speech, and Signal Processing (ICASSP 2011), pp. 861–864 (2011)

17. Navarro, G.: Searching in metric spaces by spatial approximation. VLDB J. **11**(1), 28–46 (2002)
18. Shaft, U., Ramakrishnan, R.: Theory of nearest neighbors indexability. ACM Trans. Database Syst. **31**, 814–838 (2006)
19. Volnyansky, I., Pestov, V.: Curse of dimensionality in pivot based indexes. In: Proceedings of International Workshop on Similarity Search and Applications (SISAP 2009), pp. 39–46. IEEE (2009)

Clustering

BETULA: Numerically Stable CF-Trees for BIRCH Clustering

Andreas Lang[ID] and Erich Schubert[(✉)][ID]

TU Dortmund University, Dortmund, Germany
{andreas.lang,erich.schubert}@tu-dortmund.de

Abstract. BIRCH clustering is a widely known approach for clustering, that has influenced much subsequent research and commercial products. The key contribution of BIRCH is the Clustering Feature tree (CF-Tree), which is a compressed representation of the input data. As new data arrives, the tree is eventually rebuilt to increase the compression. Afterward, the leaves of the tree are used for clustering. Because of the data compression, this method is very scalable. The idea has been adopted for example for k-means, data stream, and density-based clustering.

Clustering features used by BIRCH are simple summary statistics that can easily be updated with new data: the number of points, the linear sums, and the sum of squared values. Unfortunately, how the sum of squares is then used in BIRCH is prone to catastrophic cancellation.

We introduce a replacement cluster feature that does not have this numeric problem, that is not much more expensive to maintain, and which makes many computations simpler and hence more efficient. These cluster features can also easily be used in other work derived from BIRCH, such as algorithms for streaming data. In the experiments, we demonstrate the numerical problem and compare the performance of the original algorithm compared to the improved cluster features.

1 Introduction

The BIRCH algorithm [23–25] is a widely known cluster analysis approach, that won the 2006 SIGMOD Test of Time Award. It scales well to big data even with limited resources because it processes the data as a stream and aggregates it into a compact summary of the data. BIRCH has inspired many subsequent works, such as two-step clustering [10], data bubbles [7], and stream clustering methods such as CluStream [1] and DenStream [9]. Clustering is the unsupervised learning task aimed at discovering potential structure in a data set when no labeled data or pattern examples are available. It is inherently underspecified and subjective [5,12] and, unfortunately, also very difficult to evaluate. Instead, it is best approached as explorative data analysis, generating hypotheses about

Part of the work on this paper has been supported by Deutsche Forschungsgemeinschaft (DFG) within the Collaborative Research Center SFB 876 "Providing Information by Resource-Constrained Analysis", project A2. https://sfb876.tu-dortmund.de/.

© Springer Nature Switzerland AG 2020
S. Satoh et al. (Eds.): SISAP 2020, LNCS 12440, pp. 281–296, 2020.
https://doi.org/10.1007/978-3-030-60936-8_22

potential structures in the data, that afterward need to be verified by some other procedure, which is domain-specific and may require a domain expert to inspect the results. Many clustering algorithms and evaluation measures have been proposed with unclear advantages of one over another. Because many of the underlying problems (e.g., k-means clustering) are NP-hard, we often use approximation techniques and great concern is directed at the scalability.

Scalability is where the BIRCH algorithm shines. It is a multi-step procedure for numerical data that first aggregates the data into a tree-based data structure much smaller than the original data. This condensed representation is then fed into a clustering method, which now is faster because of the reduced size. The main contribution of BIRCH is a flexible logic for aggregating the data so that an informative representation is retained even when the size is reduced substantially.

When studying BIRCH closely, we noticed that it is susceptible to a numerical problem known as "catastrophic cancellation". This arises when two large and similar floating-point values are subtracted: many bits of the significand (mantissa) cancel out, and only few bits of valid result remain. In this paper, we show how to avoid this numerical problem and demonstrate that it can arise in real data even at low dimensionality. We propose a replacement cluster feature tree (BETULA) that does not suffer from this numeric problem while retaining all functionality. Furthermore, it is often even easier to use. This structure can easily be integrated into most (if not all) derived methods, in particular also for data streams.

2   Related Work

The BIRCH algorithm was presented at the SIGMOD conference [24], then expanded in a journal version [25]. Still, both versions omit integral details of the algorithm (e.g., Eqs. 15 to 17 below to compute distances using cluster features), which are found only in their technical report [23] or their source code. Nevertheless, the intriguing ideas of the clustering features and the CF-Tree inspired a plethora of subsequent work. Bradley et al. [6] use the same "clustering features" as BIRCH, but call them "sub-cluster sufficient statistics". The CF-Tree has also been used for kernel density estimation [26], with a threshold set on the variance to guarantee approximation quality. In two-step clustering [10], BIRCH is extended to mixed data, by adding histograms over the categorical variables.

Because BIRCH is sequentially inserting data points into the CF-tree, the tree construction can be suspended at any time. The leaves can then be processed with a clustering algorithm; when new data arrives the tree construction is continued and we trivially obtain a stream clustering algorithm [15]. CluStream [1] extends this idea with pyramidal time frames to enable the clustering of parts of the data stream by integrating temporal information. HPStream [2] extends CluStream to projected/subspace clustering. DenStream [9] uses clustering features for density-based stream clustering to detect clusters of arbitrary shape (in contrast to earlier methods that focus on k-means-style clustering). Breunig et al. [8] adopt clustering features to perform hierarchical density-based

OPTICS clustering [3] on large data. The ClusTree [17] combines R-trees with BIRCH clustering features to process data streams. BICO [13] aims at improving the theoretical foundations (and hence, performance guarantees) of BIRCH by combining it with the concept of coresets. For this, it is necessary to add reference points to the clustering features and use a strict radius threshold.

3   BIRCH and BETULA

In this section, we will describe the basic BIRCH tree building algorithm, and introduce the changes made for BETULA to become numerically more reliable.

3.1   BIRCH Clustering Features

The central concept of BIRCH is a summary data structure known as Clustering Features $\mathrm{CF^{BIRCH}} = (\boldsymbol{LS}, SS, N)$. Each clustering feature represents N data points, summarized using the linear sum vector $\boldsymbol{LS} \in \mathbb{R}^d$ (with $LS_i = \sum_x x_i$), the sum of squares $SS \in \mathbb{R}$ (originally not a vector, but a scalar $SS = \sum_i \sum_x x_i^2$) and the count $N \in \mathbb{N}$. The center of a clustering feature can be trivially computed as LS/N. By the algebraic identity $\mathrm{Var}(X) = E[X^2] - E[X]^2$, BIRCH computes the variance of a clustering feature as $\mathrm{Var}(X) = \frac{1}{N}SS - (\frac{1}{N}\sum_i LS_i)^2$. We will discuss the numerical problems with this approach in Sect. 3.5.

A new data sample x can be easily integrated into the clustering feature using $\mathrm{CF^{BIRCH}} + x = (\boldsymbol{LS} + x, SS + \sum_i x_i^2, N + 1)$. Because all of these are sums, two clustering features can also easily be combined (c.f., additivity theorem in [24]) $\mathrm{CF}_A^{\mathrm{BIRCH}} + \mathrm{CF}_B^{\mathrm{BIRCH}} = (\boldsymbol{LS}_A + \boldsymbol{LS}_B, SS_A + SS_B, N_A + N_B)$. A single data point x can hence be interpreted as the clustering feature containing $(x, \sum_i x_i^2, 1)$.

3.2   Clustering Feature Tree (CF-Tree)

The cluster features are organized in a depth-balanced tree called CF-Tree. A leaf stores a set of clustering features (each representing one or many data points), while the inner nodes store the aggregated clustering features of each of its children. The tree is built by sequential insertion of data points (or, at a rebuild, the insertion of earlier clustering features). The insertion leaf is found by choosing the "nearest" clustering feature at each level (five different definitions of closeness will be discussed in Sect. 3.4). Within the leaf node, the data point is added to the best clustering feature if it is within the merging "threshold", otherwise a new clustering feature is added to the leaf. Leaves that exceed a maximum capacity are split, which can propagate to higher levels of the tree and cause the tree to grow when the root node overflows. If the tree exceeds the memory limit, a new tree is built with an increased merging threshold by reinserting the existing clustering features of the leaf level. After modifying a node, the aggregated clustering features along the path to the root are updated.

The discussion of BIRCH in textbooks ends with the CF-Tree, although we do not yet have clusters. This is because the outstanding idea of BIRCH is that of data aggregation into clustering features, and we can run different clustering algorithms afterward. The BIRCH authors mention hierarchical clustering, k-means, and CLARANS [20]. For best results, we would want to use an algorithm that not only uses the mean of the clustering feature, but that also uses the weight and variance. The weight can be fairly easily used in many algorithms, but the variance is less obvious to integrate. In Sect. 3.7 we will propose how to perform Gaussian Mixture Modeling and use the variance information.

3.3   BETULA Cluster Features

The way variance is computed from BIRCH cluster features using the popular equation $\text{Var}(X) = E[X^2] - E[X]^2$ is prone to the numerical problem known as "catastrophic cancellation". This equation can return zero for non-constant data, and because of rounding even negative values (and hence, undefined standard deviation). In the context of BIRCH, we cannot resort to the numerically more reliable textbook definition for variance, $\text{Var}(X) := \frac{1}{N} \sum (x - \mu)^2$, because this requires two passes over the data set (one to find μ, then one for Var). But we also cannot just ignore the problem, because not all clustering features will be close to 0, where the numerical accuracy is not a problem. Schubert and Gertz [21] discuss methods to compute variance and covariance for weighted data, which forms the base for our approach. For this, they collect three running statistics, very similar to the three components of BIRCH clustering features: (i) the sum of weights, (ii) the weighted mean (centroid vector), and (iii) the weighted sum of squared deviations from the mean. Clearly (i) corresponds to N in the clustering feature, (ii) is equivalent to LS/N, but (iii) is $S := \sum_x n_x \|x - \mu\|^2$ (where n_x is the weight of the data point, often simply 1). Hence, we propose the following replacement cluster feature for BETULA:

$$\text{CF}^{\text{BETULA}} := (n, \mu, S) \tag{1}$$

where n is the aggregated weight of all data points (BETULA also allows for weighted data samples), μ denotes the current mean vector, and S is the sum of squared deviations from the mean. The last component can either be a scalar value as in BIRCH (the sum over all components) or a vector of squared deviations. For our experiments, we chose the latter option; a similar modification to BIRCH can be found in various publications (e.g., [1,2,9,16,17]). A single data point of weight n_x is equivalent to a cluster feature $\text{CF}^{\text{BETULA}}_x = (n_x, x, 0)$ (because it has zero deviation from the mean). Similar to the additivity theorem of BIRCH, we can efficiently combine two BETULA cluster features into one:

$$n_{AB} = n_A + n_B \tag{2}$$

$$\mu_{AB} = \mu_A + \frac{n_B}{n_{AB}}(\mu_B - \mu_A) \tag{3}$$

$$S_{AB} = S_A + S_B + n_B(\mu_A - \mu_B)(\mu_{AB} - \mu_B) \tag{4}$$

The derivation of these equations follows directly from the update equations for the weighted (co-) variance of [21]. Because their experiments indicate that using the sum of squared deviations, $S := \sum_x n_x(x - \mu)^2$ has slight computational advantages, we follow suit. We could also have stored $\text{Var} = S/n$ instead (we did not measure a noticeable performance difference between these two options).

3.4   Distance and Absorption Measures

The BIRCH algorithm uses two different measures during tree construction. The first is a distance between two clustering features, which is used to find the closest leaf in the tree. The second is an absorption criterion, used together with a threshold to decide when to add the new data to an existing or as a new clustering feature. Both measures can be defined on the original data, but also in terms of the clustering feature values to compute them efficiently.

Distance Measures: BIRCH proposes five different distance measures enumerated as D0 to D4. The first two correspond simply to the Euclidean distance of the centers (D0) and the Manhattan distance of the centers (D1). The third, average inter-cluster distance (D2), is based on the quadratic mean distance between points of different clusters, while the average intra-cluster distance (D3) uses the quadratic mean distance within the combined cluster. Variance-increase distance (D4) is the variance of the resulting cluster minus the variance of the separated clusters. Similar ideas can be found in hierarchical clustering: centroid linkage (D0, D1), average linkage (D2, D3), and Ward linkage (D4).

$$D0(A, B) = \|\mu_A - \mu_B\| = \sqrt{\sum_i (\mu_{A,i} - \mu_{B,i})^2} \tag{5}$$

$$D1(A, B) = \|\mu_A - \mu_B\|_1 = \sum_i |\mu_{A,i} - \mu_{B,i}| \tag{6}$$

$$D2(A, B) = \sqrt{\frac{1}{n_A n_B} \sum_{x \in A} \sum_{y \in B} \|x - y\|^2} \tag{7}$$

$$D3(A, B) = \sqrt{\frac{1}{n_{AB}(n_{AB}-1)} \sum_{x,y \in AB} \|x - y\|^2} \tag{8}$$

$$D4(A, B) = \sqrt{\sum_{x \in AB} \|x - \mu_{AB}\|^2 - \sum_{x \in A} \|x - \mu_A\|^2 - \sum_{x \in B} \|x - \mu_B\|^2} \tag{9}$$

Absorption Criteria: Absorption in BIRCH is based on a second criterion and a threshold. Conceptually, the threshold can be seen as a maximum radius of a cluster feature; if adding a point would increase the radius beyond the allowed maximum, a new cluster feature is created instead of merging. Intuitively, the radius should be defined as $\max_x \|x - \mu\|$; but this value cannot be efficiently computed from the summary statistics. Instead, the "radius" can be approximated using different criteria. In BIRCH, these criteria were defined on a single clustering feature AB; they are computed by virtually merging two clustering features and evaluating the criteria on the result. We can easily remove the distinction between distance and absorption criteria, but one may nevertheless want to choose them differently (e.g., choosing the nearest leaf by Euclidean distance, but thresholding on minimum variance), as they serve a different purpose. The first criterion proposed in BIRCH is called "radius" R (Eq. 10), the second is the "diameter" D (Eq. 11). Both this "radius" and "diameter" are not maximum values, but averages: the average distance to the center is (R), and the average distance of any two points is (D), which happens to be the same as $D3(A, B)$. Many implementation attempts (such as sklearn's) of BIRCH simply use the distance between the two cluster centers instead (E) – this cannot be defined in the original BIRCH architecture but is easy to add.

$$R(AB) = \quad R(A, B) = \sqrt{\frac{1}{n_{AB}} \sum_{x \in AB} \|x - \mu_{AB}\|^2} \qquad (10)$$

$$D(AB) = \quad D3(A, B) = \sqrt{\frac{1}{n_{AB}(n_{AB}-1)} \sum_{x,y \in AB} \|x - y\|^2} \qquad (11)$$

$$E(A, B) = \quad D0(A, B) = \|\mu_A - \mu_B\| \qquad (12)$$

The values of D and R are almost identical (use Eq. 13 below): they differ only by a factor of $\frac{2n}{n-1}$; similar to the regular radius and diameter. Because of the way they are used in BIRCH, we cannot expect them to perform very differently.

3.5   Catastrophic Cancellation in BIRCH

The numerical problem in BIRCH arises from the "textbook" equation for variance, $\mathrm{Var}(X) = E[X^2] - E[X]^2$. This equation—while mathematically correct—is prone to catastrophic cancellation when used with floating-point arithmetic, unless $E[X]^2 \ll E[X^2]$ holds [21]. In clustering, we cannot assume that all clusters are close to the origin, and the ideal leaves have a small variance and represent the data by their differences in the mean. Because of this, it may not be sufficient to center the data globally. Furthermore, we do not know the center beforehand, and in BIRCH we only want to do a single pass over the data for performance.

Unfortunately, both of the original absorption criteria R and D, as well as distance measures D2–D4 are computed using the above "textbook" equality

$$\mathrm{Var}(X) = \frac{1}{2n^2} \sum_{x,y \in X} \|x - y\|^2 = \frac{1}{n} \sum_{x \in X} \|x - \mu_X\|^2 \qquad (13)$$

which yields the following equalities for BIRCH (equivalent to $n \cdot \mathrm{Var}(X) = S$)

$$S = \sum_{x \in X} \|x - \mu_X\|^2 = \frac{1}{2n} \sum_{x \in X} \|x - y\|^2 = SS - \frac{1}{n} \|LS\|^2. \qquad (14)$$

The BIRCH authors hence proposed to compute these measures (we omit D0, D1, and E as they do not involve squares) based on clustering features as:

$$D2(A, B) = \sqrt{\tfrac{1}{N_A N_B}(N_B SS_A + N_A SS_B \overset{\triangle}{-} 2LS_A^T LS_B)} \tag{15}$$

$$D3(A, B) = \sqrt{\tfrac{2}{N_A+N_B-1}(SS_A + SS_B \overset{\triangle}{-} \tfrac{1}{N_A+N_B} \|LS_A + LS_B\|^2)} \tag{16}$$

$$D4(A, B) = \sqrt{\tfrac{1}{N_A}\|LS_A\|^2 + \tfrac{1}{N_B}\|LS_B\|^2 \overset{\triangle}{-} \tfrac{1}{N_A+N_B}\|LS_A + LS_B\|^2} \tag{17}$$

$$R(AB) = \sqrt{\tfrac{1}{N_{AB}}(SS_{AB} \overset{\triangle}{-} \tfrac{1}{N_{AB}} \|LS_{AB}\|^2)} \tag{18}$$

$$D(AB) = \sqrt{\tfrac{2}{N_{AB}-1}(SS_{AB} \overset{\triangle}{-} \tfrac{1}{N_{AB}} \|LS_{AB}\|^2)} \tag{19}$$

The subtractions flagged with a warning symbol $\triangle$ can suffer from catastrophic cancellation and hence numerical problems. It may come unexpected that in the "variance increase" equation (D4) all SS terms cancel out, and we only get the vector product of the linear sums, but this is the König-Huygens theorem.

The effect of the catastrophic cancellation usually leads to an underestimation of the actual variance, and hence of the distances. Because of this, data points may be assigned to the wrong branch or node. While the result will not be completely off, it is easy to avoid these problems in the first place. More severe problems arise when using the resulting variance in the subsequent steps, such as in clustering. Because most implementations of BIRCH only use the centers of the leaf entries for clustering (e.g., scikit-learn does not even use the weight, and only supports Euclidean distance D0), this has not been observed frequently.

Much of the later work based on BIRCH is prone to the same problem in one way or another. In CF-kernel density estimation [26], the variance is bounded to guarantee approximation quality – underestimating the variance invalidates this guarantee. The (diagonal) Mahalanobis distance used in [6] divides by the standard deviation, which can become 0 due to instabilities; the division tends to amplify the errors. CluStream [1] uses the standard deviation of the arrival times, estimated with the unstable equation. HPStream [2] relies on per attribute standard deviations for subspace clustering. DenStream [9] uses the radius R to estimate density, while data bubbles [8] rely on the standard deviation to estimate the extent. ClusTree [17] estimates the variance in this unstable way. All of these methods can easily be modified to use BETULA cluster features.

Using the improved BETULA cluster features introduced in Sect. 3.3, which we will simply denote by CF, we can easily avoid these numerical problems, because these features directly aggregate the squared errors instead of the sum of squares, as previously used for online estimation of variance [21].

3.6   Improved Distance Computations

In BETULA cluster features, we use the mean μ instead of the linear sum because this makes the subsequent operations more efficient (and elegant). The update equations for merging CFs also involve the mean (c.f. Eq. 4), and we can now compute the BIRCH distances in a more numerically stable way. Using BETULA cluster features CF $= (n, \mu, S)$, and Eq. (14), we can compute the distances and absorption criteria now as follows (the derivation is included in the preprint [19]):

$$D0(A, B) = \|\mu_A - \mu_B\| \tag{20}$$

$$D1(A, B) = \|\mu_A - \mu_B\|_1 \tag{21}$$

$$D2(A, B) = \sqrt{\tfrac{1}{n_A}S_A + \tfrac{1}{n_B}S_B + \|\mu_A - \mu_B\|^2} \tag{22}$$

$$D3(A, B) = \sqrt{\tfrac{2}{n_{AB}(n_{AB}-1)}\left(n_{AB}(S_A + S_B) + n_A n_B \|\mu_A - \mu_B\|^2\right)} \tag{23}$$

$$D4(A, B) = \sqrt{\tfrac{n_A n_B}{n_{AB}} \|\mu_A - \mu_B\|^2} \tag{24}$$

$$R_{AB} = \sqrt{\tfrac{1}{n_{AB}}S_{AB}} = \sqrt{\tfrac{1}{n_{AB}}(S_A + S_B + \tfrac{n_A \cdot n_B}{n_{AB}} \|\mu_A - \mu_B\|^2)} \tag{25}$$

$$D_{AB} = \sqrt{\tfrac{2}{(n_{AB}-1)}S_{AB}} = D3(A, B) \tag{26}$$

With these numerically more stable equations, we can build a CF-Tree using BETULA cluster features instead of the original BIRCH clustering features.

3.7   Gaussian Mixture Modeling with BETULA Cluster Features

Gaussian Mixture Modeling (GMM) with the EM algorithm [11] is a popular, but fairly expensive clustering algorithm. Every iteration, the probability density functions (pdfs) of each Gaussian are evaluated at every data point, then the distribution parameters are updated based on all points weighted by their probabilities. Because this is a soft clustering, a tolerance threshold or an iteration limit are used for convergence. Formally, the method is linear in the number of data points, but in practice, it is fairly expensive because of the many pdfs to compute and the number of iterations. To scale this algorithm to large data sets (large n) as well as many clusters k, it is beneficial to use a data summarization technique such as BIRCH or BETULA. Several variations of GMM exist: we can restrict cluster shapes and we can have independent or shared model parameters. MAP estimation can be employed to improve the robustness [14], because there are other numerical pitfalls that can lead to degenerate clusters. We only consider some of the more popular variants in this work: the spherical model with varying weight and identical volume in each dimension (IGMM), the diagonal model with varying weight and different volume in each dimension (DGMM), and the fully variable model that models covariance (CGMM). If we only have a scalar for SS, then this is well-suited for the simplest model: A spherical model, in which the direction of variance does not matter. When using a vector for S, we can incorporate this per-axis information into the cluster models. For the arbitrarily oriented model, we would need to use a covariance in each cluster feature. This is possible using the corresponding equations for the covariance of [21], but the memory requirement increases to $1 + d + \binom{d}{2} = 1 + \frac{d(d+1)}{2}$ values per cluster feature. Because of this, we do not include this in the experiments.

For clustering, the main tree structure is usually discarded, and only the cluster features within the leaf nodes are kept. For the initialization of the algorithm, we apply the kmeans++ [4] initialization on the leaf entries. Afterward, the Gaussian Mixture Modeling algorithm is executed.

In classic GMM, we usually process a single data sample at a time. When processing cluster features, these represent multiple objects. To improve the quality of the clustering, rather than just using the cluster mean to represent a Cluster Feature, we use the Gaussian distribution of the data in the CF, which we assume is better (at least for GMM). To estimate the responsibilities of each cluster for each clustering we then use $\int_x \mathcal{N}(x|\mu_1, \sigma_1^2)\mathcal{N}(x|\mu_2, \sigma_2^2)dx = \mathcal{N}(\mu_1|\mu_2, \sigma_1^2 + \sigma_2^2)$. Using the law of total probability, these values are normalized to sum to 1, exactly as in the usual EM procedure. When updating the cluster models, the weight of the cluster features is trivially usable as additional weight, and we can update the model variance using Eq. (4).

By utilizing BETULA cluster features and EM-GMM it is possible to cluster big data sets with limited memory and high numerical stability as shown in Section 4. It is also possible to distribute this procedure into a cluster by partitioning the data and aggregating the models of all nodes (c.f. [21]).

4   Evaluation

We compare the following alternative implementations of GMM:

Textbook   Standard EM [11] using the equation $E[X^2] - E[X]^2$

Stable   Numerically stable EM implementation (from ELKI [21, 22])

BIRCH   EM-style using the original BIRCH clustering features

BETULA EM-style using our new BETULA cluster features

The evaluation of clustering algorithms is inherently difficult because they are used in an unsupervised context, where no labeled data is available. Real data is usually dirty and contains undesirable artifacts (such as anomalies, duplicate values, and discretization effects) that can cause problems for methods that assume continuous data. GMM is no exception: e.g., constant attributes will break many implementations. In these experiments, we do not aim at showing the superiority of Gaussian Mixture Modeling over other approaches. The limitations of it are well understood, in particular when data has non-convex clusters.

Instead, we focus on the following research questions:

RQ1 How numerically (un-)stable is BIRCH and does BETULA help?
RQ2 Is the quality of BETULA comparable with BIRCH and regular GMM?
RQ3 How does BETULA scale with data set size (and compare to BIRCH)?
RQ4 Are the results applicable to real data?

4.1 Experimental Setup

We modify the existing implementations of BIRCH and GMM clustering of ELKI 0.7.5 [22]. By keeping most of the code shared, we try to minimize the effects caused by implementation differences, as recommended for comparing algorithms [18]. All computations are executed on a small cluster with Intel E5-2697v2 CPUs, we do not use multithreading, and we repeated each experiment 10 times with varying random seeds and data input order, and give the average results. All the CF-Trees are built using the variance-increase distance (D4, Eq. 9) in combination with the radius absorption criterion (R, Eq. 10). This combination yields subclusters with low variance as input for the GMM clustering. We do not present results with other distances and absorption criteria here because of redundancy; they were similar. The size of CF-Trees is by default limited to 5000 leaf entries unless specified differently; when this number is exceeded the tree is rebuilt with a bigger threshold as in BIRCH. For the GMM clustering step, all algorithms are initialized by kmeans++ [4]. After 100 iterations or when no further improvement can be made the optimization is stopped.

4.2 Numerical Stability

First, we demonstrate the numerical instability using synthetic data with two Gaussian clusters in $\mathbb{R}^3$ of 150 000 points. Both clusters have standard deviations $[\frac{4}{3}, 1, \frac{3}{4}]$, and the only variable in the test is how far the clusters are shifted away from the mean. For small separation, both clusters overlap but with increasing distance, the clustering gets trivial until numerical stability comes into play.

The impact of the increasing distance between the clusters can be seen in Fig. 1 where all algorithms provide good results until first the Textbook IGMM implementation at $5 \cdot 10^6$ and then BIRCH IGMM at $2 \cdot 10^7$ begin to deteriorate. The degeneration of BIRCH IGMM begins a bit later than Textbook IGMM because of the aggregation in the CF-Tree helping a bit, but it then fails even worse. A deterioration at 10^7 is to be expected from double-precision because of the squared values; with single-precision floating-point, it is to be expected to occur at a separation of 10^3. Both the "Stable" regular GMM and BETULA are not affected and solve this idealized toy problem without difficulties (RQ1).

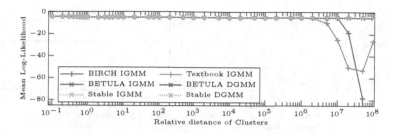

Fig. 1. The log-likelihood goodness of fit of the model with increasing distance between the clusters demonstrates the numerical instability of some algorithms.

4.3 Quality Comparison on Synthetic Data

We now address the question of result quality (RQ2) in a scenario where all algorithms are stable. For the evaluation, two synthetic data sets are used, which are similar to data used for the evaluation of the original BIRCH algorithm [24] but larger and with increased variability. We use the data generator of ELKI [22], which has a convenient size multiplier parameter for this experiment.

The first data set is called "Grid" and consists of a 10 by 10 grid of clusters with a distance of 5 between the means of the clusters on each axis. Each cluster consists of 10 000 points with a variance per attribute randomly drawn from $\mathcal{N}(1, 0.25)$. The second data set, "Random", consists of 100 clusters in a 50 by 50 area with the cluster means distributed by Halton sampling, which produces a pseudo-random uniform distribution. The variance of each cluster is again specified by a normal distribution $\mathcal{N}(1, 0.15)$. This time the size of each cluster varies and is randomly drawn from between 5000 and 15 000 points.

Figure 2 shows the log-likelihood of the models on these data sets. For both, it can be seen that the data set size has next to no influence on the quality of the fit. The models with diagonal variance (Stable DGMM and BETULA DGMM) produce a better fit than the models that are restricted to using the same variance in each attribute. On the "Random" data set, all IGMM approaches perform similar (as expected). On the "Grid" data set, both BETULA IGMM and BIRCH IGMM unexpectedly achieve a higher likelihood than the standard IGMM algorithms. This difference can be explained by the fact that the implementations using cluster features converge faster (because there are fewer objects) than the approaches that use the raw data; the latter have not yet converged within the maximum number of iterations. However, this experiment is designed to test if BETULA performs similar to BIRCH on the same test data that the BIRCH publications used, and to detect programming errors.

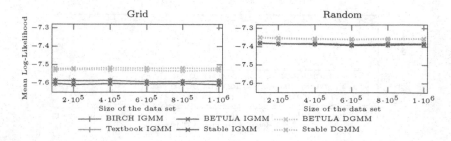

Fig. 2. Log-Likelihood goodness of fit of the model on both synthetic data sets.

4.4 Runtime Evaluation on Synthetic Data

When evaluating the runtime of BETULA with GMM clustering two measurements are of interest: The time to build the CF-Tree only, and the time for the entire clustering procedure. Figure 3 shows the build time BETULA and BIRCH need for various tree sizes. It can be seen that the time for building the tree increases with the size of the data set and also with the size of the tree due to an increasing number of distance calculations for the insertion of new points. The construction time for BETULA is shorter than for BIRCH—despite using a vector to store variances—because of the more efficient distance calculations.

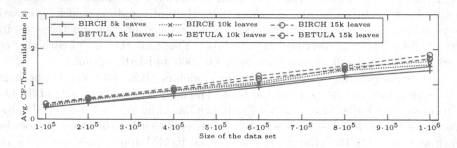

Fig. 3. Build time with a varying number of leaf entries on the random data set.

When looking at the complete runtime of BIRCH (respectively BETULA) including GMM clustering, shown in Fig. 4, we can see that the standard GMM algorithms have a much higher runtime by a factor of 10 to 50 on these data sets, due to the compression achieved by the CF-Tree (which improves with data set size). We use a log-log plot to see the differences between BIRCH and BETULA, which perform very similar (RQ3). BETULA is about 12% faster than BIRCH because the BETULA cluster features can be used directly for clustering, while more additional computations are necessary with BIRCH clustering features to obtain mean and variance on the fly.

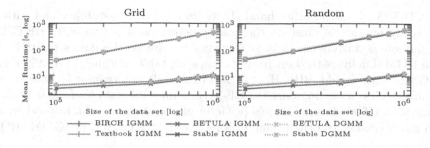

Fig. 4. Runtime of the clustering on both synthetic data sets.

4.5   Clustering Real Data

To test the algorithm on real data, we use the location information of the UK "Road Safety Data" from 1979 to 2004 from data.gov.uk.[1] This data set has about 6.2 million entries and contains data on road accidents from Great Britain. The location information in this data set is given in the OSGR grid reference system which is only used in Great Britain; which we convert to the appropriate UTM coordinate system. For this experiment, we reduced the cluster feature precision from double precision to single precision in both BIRCH and BETULA to demonstrate the numerical instabilities on real data. The regular GMM clustering is performed with double precision to get a more precise reference value.

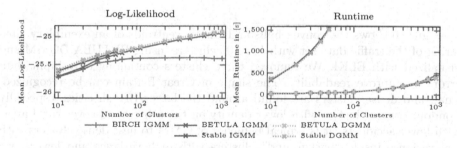

Fig. 5. Log-Likelihood goodness of fit of the models and runtime on the traffic accident data with 15000 leaf entries (Stable GMM only up to 50 clusters because of runtime).

Figure 5 shows that Stable DGMM and Stable IGMM achieve a better fit than the CF-Tree based approximations (which is to be expected, as they use the individual points and double precision). But the runtime of this method is much higher, and hence was only computed up to $k = 50$ clusters. BETULA with DGMM and IGMM obtain only slightly worse results, showing that the BETULA cluster features provide a reasonably close approximation of the data.

---

[1] https://data.gov.uk/dataset/cb7ae6f0-4be6-4935-9277-47e5ce24a11f.

BIRCH IGMM on the other hand shows its numerical instability and with an increasing number of clusters, the quality deteriorates compared to BETULA. For numerous clusters (and a large value makes sense on this data set), BETULA with DGMM delivers the best results at an acceptable runtime: As seen in Fig. 5, all GMM with Stable, BIRCH, and BETULA scale approximately linear in the number of clusters k; but since the CF-Trees reduce the data set from 6.2 million to at most 15000 cluster features (a factor of over 400), we obtain good results at a much smaller run time than with regular Stable DGMM or IGMM (RQ4).

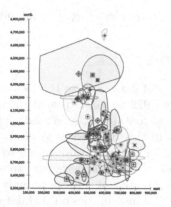

Fig. 6. Convex hulls of clusters with BETULA DGMM on the traffic accident data with 100 clusters (three clusters omitted for a cleaner visualization).

Figure 6 shows the convex hulls and cluster centroids of an exemplary clustering of the traffic data set with $k = 100$ clusters, using BETULA DGMM and visualized with ELKI. We removed three clusters containing only input data errors to improve readability. The shape of Great Britain can be recognized; small and dense clusters are found around the larger British cities, especially London. Larger clusters with lower density on the other hand cover rural areas with fewer accidents (it is typical behavior of GMM to nest dense clusters with low variance inside "background" clusters with high variance and fewer data points).

5   Conclusion

Big data analysis and data stream clustering are hot topics in today's research.

The CF-Tree of BIRCH is a popular technique for this that inspired many subsequent works. Recently, the reliability of machine learning is receiving increased attention; unfortunately, we found that "catastrophic cancellation" is a major problem when calculating variances in BIRCH and derived methods, which can cause the results to deteriorate. In this article, we propose BETULA cluster features, that can serve as a drop-in replacement. These no longer exhibit this numerical problem, while also increasing the performance. We also show how to use BETULA to accelerate Gaussian Mixture Modeling, while using the variance information from the cluster features for improved quality, compared to the standard approach of only using the centroids of each leaf entry.

References

1. Aggarwal, C.C., Han, J., Wang, J., Yu, P.S.: A framework for clustering evolving data streams. In: VLDB, pp. 81–92 (2003). https://doi.org/10.1016/B978-012722442-8/50016-1
2. Aggarwal, C.C., Han, J., Wang, J., Yu, P.S.: A framework for projected clustering of high dimensional data streams. In: VLDB, pp. 852–863 (2004). https://doi.org/10.1016/B978-012088469-8.50075-9
3. Ankerst, M., Breunig, M.M., Kriegel, H., Sander, J.: OPTICS: ordering points to identify the clustering structure. In: SIGMOD, pp. 49–60 (1999). https://doi.org/10.1145/304182.304187
4. Arthur, D., Vassilvitskii, S.: k-means++: the advantages of careful seeding. In: SODA, pp. 1027–1035 (2007)
5. Bonner, R.E.: On some clustering techniques. IBM J. Res. Dev. 8(1), 22–32 (1964)
6. Bradley, P.S., Fayyad, U.M., Reina, C.: Scaling clustering algorithms to large databases. In: KDD, pp. 9–15 (1998)
7. Breunig, M.M., Kriegel, H., Kröger, P., Sander, J.: Data bubbles: quality preserving performance boosting for hierarchical clustering. In: SIGMOD, pp. 79–90 (2001). https://doi.org/10.1145/375663.375672
8. Breunig, M.M., Kriegel, H.-P., Sander, J.: Fast hierarchical clustering based on compressed data and OPTICS. In: Zighed, D.A., Komorowski, J., Żytkow, J. (eds.) PKDD 2000. LNCS (LNAI), vol. 1910, pp. 232–242. Springer, Heidelberg (2000). https://doi.org/10.1007/3-540-45372-5_23
9. Cao, F., Ester, M., Qian, W., Zhou, A.: Density-based clustering over an evolving data stream with noise. In: SDM, pp. 328–339 (2006). https://doi.org/10.1137/1.9781611972764.29
10. Chiu, T., Fang, D., Chen, J., Wang, Y., Jeris, C.: A robust and scalable clustering algorithm for mixed type attributes in large database environment. In: KDD, pp. 263–268 (2001). https://doi.org/10.1145/502512.502549
11. Dempster, A.P., Laird, N.M., Rubin, D.B.: Maximum likelihood from incomplete data via the EM algorithm. J. Royal Stat. Soc. Series B 39(1), 1–38 (1977)
12. Estivill-Castro, V.: Why so many clustering algorithms: a position paper. SIGKDD Explor. 4(1), 65–75 (2002)
13. Fichtenberger, H., Gillé, M., Schmidt, M., Schwiegelshohn, C., Sohler, C.: BICO: BIRCH meets coresets for k-means clustering. In: Bodlaender, H.L., Italiano, G.F. (eds.) ESA 2013. LNCS, vol. 8125, pp. 481–492. Springer, Heidelberg (2013). https://doi.org/10.1007/978-3-642-40450-4_41
14. Fraley, C., Raftery, A.E.: Bayesian regularization for normal mixture estimation and model-based clustering. J. Classif. 24(2), 155–181 (2007). https://doi.org/10.1007/s00357-007-0004-5
15. Ganti, V., Gehrke, J., Ramakrishnan, R.: DEMON: mining and monitoring evolving data. IEEE Trans. Knowl. Data Eng. 13(1), 50–63 (2001). https://doi.org/10.1109/69.908980

16. Han, J., Kamber, M., Pei, J.: Data Mining: Concepts and Techniques, 3rd edn. Morgan Kaufmann, Burlington (2011)
17. Kranen, P., Assent, I., Baldauf, C., Seidl, T.: The ClusTree: indexing micro-clusters for anytime stream mining. Knowl. Inf. Syst. **29**(2), 249–272 (2011). https://doi.org/10.1007/s10115-010-0342-8
18. Kriegel, H.-P., Schubert, E., Zimek, A.: The (black) art of runtime evaluation: are we comparing algorithms or implementations? Knowl. Inf. Syst. **52**(2), 341–378 (2017). https://doi.org/10.1007/s10115-016-1004-2
19. Lang, A., Schubert, E.: BETULA: numerically stable CF-trees for BIRCH clustering. CoRR abs/2006.12881 (2020). https://arxiv.org/abs/2006.12881
20. Ng, R.T., Han, J.: CLARANS: a method for clustering objects for spatial data mining. IEEE Trans. Knowl. Data Eng. **14**(5), 1003–1016 (2002). https://doi.org/10.1109/TKDE.2002.1033770
21. Schubert, E., Gertz, M.: Numerically stable parallel computation of (co-)variance. In: SSDBM, pp. 10:1–10:12 (2018). https://doi.org/10.1145/3221269.3223036
22. Schubert, E., Zimek, A.: ELKI: a large open-source library for data analysis - ELKI release 0.7.5 "Heidelberg". CoRR abs/1902.03616 (2019)
23. Zhang, T.: Data clustering for very large datasets plus applications. Technical report 1355, University of Wisconsin Madison (1996)
24. Zhang, T., Ramakrishnan, R., Livny, M.: BIRCH: an efficient data clustering method for very large databases. In: SIGMOD, pp. 103–114 (1996). https://doi.org/10.1145/233269.233324
25. Zhang, T., Ramakrishnan, R., Livny, M.: BIRCH: a new data clustering algorithm and its applications. Data Min. Knowl. Discov. **1**(2), 141–182 (1997). https://doi.org/10.1023/A:1009783824328
26. Zhang, T., Ramakrishnan, R., Livny, M.: Fast density estimation using CF-kernel for very large databases. In: KDD, pp. 312–316 (1999). https://doi.org/10.1145/312129.312266

Using a Set of Triangle Inequalities
to Accelerate K-means Clustering

Qiao Yu[ID], Kuan-Hsun Chen[✉][ID], and Jian-Jia Chen[ID]

Design Automation for Embedded Systems Group, Department of Computer Science,
TU Dortmund, Dortmund, Germany
kuan-hsun.chen@tu-dortmund.de

Abstract. The k-means clustering is a well-known problem in data mining and machine learning. However, the de facto standard, i.e., Lloyd's k-mean algorithm, suffers from a large amount of time on the distance calculations. Elkan's k-means algorithm as one prominent approach exploits triangle inequality to greatly reduce such distance calculations between points and centers, while achieving the exactly same clustering results with significant speed improvement, especially on high-dimensional datasets. In this paper, we propose a set of triangle inequalities to enhance the filtering step of Elkan's k-means algorithm. With our new filtering bounds, a filtering-based Elkan (FB-Elkan) is proposed, which preserves the same results as Lloyd's k-means algorithm and additionally prunes unnecessary distance calculations. In addition, a memory-optimized Elkan (MO-Elkan) is provided, where the space complexity is greatly reduced by trading-off the maintenance of lower bounds and the run-time efficiency. Throughout evaluations with real-world datasets, FB-Elkan in general accelerates the original Elkan's k-means algorithm for high-dimensional datasets (up to 1.69x), whereas MO-Elkan outperforms the others for low-dimensional datasets (up to 2.48x). Specifically, when the datasets have a large number of points, i.e., $n \geq 5M$, MO-Elkan still can derive the exact clustering results, while the original Elkan's k-means algorithm is not applicable due to memory limitation.

Keywords: K-means · Clustering accelerating · Triangle inequalities

1 Introduction

The k-means clustering is one of the popular problems in data mining and machine learning due to its simplicity and applicability. The de facto k-means algorithm, i.e., Lloyd's k-means algorithm [12], performs two steps repeatedly: 1) the *assignment step* matches each point to its closest center, and 2) the *update step* calibrates the center for each cluster with the assigned points. However, the bottleneck in terms of time complexity, is to identify the closest center for each input data point, which leads to significantly high time complexity, i.e., $O(nkd)$, where n is the number of data points, k is the number of centers and

© Springer Nature Switzerland AG 2020
S. Satoh et al. (Eds.): SISAP 2020, LNCS 12440, pp. 297–311, 2020.
https://doi.org/10.1007/978-3-030-60936-8_23

d is the number of dimensions. In many situations, those numbers are big, e.g., data on health status of patients, on earth observation, on computer vision, etc. Therefore, efficient k-mean clustering algorithms are indeed desired.

In order to accelerate the k-means algorithm, two distinctive categories are widely-studied in the literature. 1) *Approximated solution*: Instead of accelerating the exact k-means algorithm, the proposed techniques in this category perform approximated solutions, e.g., [15,17,18], which indeed accelerate k-means algorithms, but the final clustering results cannot be guaranteed to be the same as Lloyd's k-means algorithm. 2) *Acceleration with exact results*: The proposed techniques in this category accelerate the calculation procedure while preserving the exact results as Lloyd's k-means algorithm. For example, Kanungo et al. [11] and Pelleg et al. [14] propose to accelerate the nearest neighbor search without computing distances to all k centers by using the properties of special data structures. However, the overhead of preprocessing becomes significant when the input datasets are high-dimensional. Alternatively, several acceleration techniques exploit bounds on distance between data points and centers, e.g., [3,5,7,8,10,13,16]. By maintaining lower and upper bounds on the distances to the cluster centers, most of distance calculations can be skipped. In particular, Elkan's k-means algorithm [8] as one prominent approach of them can still dominate the others on high-dimensional datasets [13]. Nevertheless, Elkan's k-means algorithm is apparently infeasible when the number of data point (n) or centers (k) is large due to the size of memory footprint for storing the lower bounds, where the space complexity is $O(nk)$.[1]

With the above pros and cons, we are motivated to revisit Elkan's k-means algorithm and propose a set of new filtering bounds based on triangle inequalities to improve the filtering step.

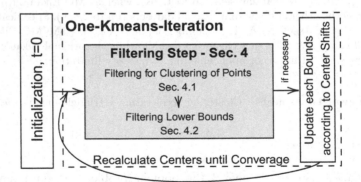

Fig. 1. Overview of the optimized Elkan's k-means algorithm, which illustrates interactions between different components. Our contributions focus on the filtering step, which are highlighted in green. (Color figure online)

---

[1] The $O(nd)$ space complexity of the input points is ignored in our complexity analysis.

Our Contributions: Figure 1 illustrates an overview of our contributions. We aim at the filtering step in Elkan's k-means algorithm highlighted in green, detailed as follows:

- Three filtering bounds are proposed based on triangle inequalities to overcome shortcomings of Elkan's k-means algorithm, by which the most unnecessary distance calculations between points and centers during the iterations of Elkan's k-means algorithm can be pruned (see Sect. 4).
- We present how to optimize the original Elkan's k-means algorithm to alleviate the time and space overheads by applying above filtering bounds. Two optimized algorithms are proposed: runtime optimized Elkan (FB-Elkan) and memory optimized Elkan (MO-Elkan). Specifically, the MO-Elkan has the space complexity $O(n + k^2 + kd)$, whereas Elkan's k-means algorithm requires $O(nk + kd)$, where n is number of the input data points, d is number of dimensions and k is number of clusters (see Sect. 5).
- Throughout evaluation we show that FB-Elkan is faster than the original Elkan's k-means algorithm on high-dimensional datasets in general, whereas MO-Elkan can outperforms the others on low-dimensional datasets considerably. Specifically, MO-Elkan can derive the exact clustering results when the number of data points is large, i.e., $n = 5M$ while the original one and FB-Elkan may not be applicable due to memory limitation. (see Sect. 6).

The rest of this paper is organized as follows: In Sect. 2, we review related work regarding the bound-based accelerated algorithms with the exactly same clustering results as the standard (Lloyd's) k-means algorithm. Section 3 defines the notation used in this paper and presents a short, general overview of Elkan's k-means algorithm as we use it as a backbone. Section 4 presents our new filtering conditions. In Sect. 5 we discuss how to use the proposed bounds to optimize the original Elkan's k-means algorithm. In Sect. 6, extensive evaluation results and discussions on different real-world datasets are presented. Finally, we conclude the paper in Sect. 7.

2   Related Work

In this section, we review related work regarding accelerating Lloyd's k-means algorithm with the triangle inequality so called bound-based acceleration, listed as follows:

- Elkan's k-means algorithm [8] takes advantage of lower bounds and upper bounds to reduce the redundant distance calculations.
- Hamerly in [10] proposes to keep only one lower bound on the distance between each point and its second closest center instead of keeping lower bounds per point. Actually, it is a simplified version of Elkan's k-means algorithm, but it is more efficient for low-dimensional datasets.
- Drake and Hamerly [7] extend the above approach [10] to keep a variable number of lower bounds, which is automatically adjusted on the fly. Drake later on proposes Annulus algorithm in [6] to prune the search space for each point by annular region.

- Yinyang k-means algorithm [5] groups a number of cluster centers, which balances the time of filtering and the time of distance calculations.
- Fast Yinyang k-means algorithm [3] further proposes to approximate Euclidean distances by using block vectors, which can achieve good improvements when the dimension of data is high.
- Newling and Fleuret in [13] simplify Yinyang and Elkan k-means algorithms and provide tighter upper and lower bounds for updating. They also propose an Exponion algorithm, which improves Yinyang and Elkan's k-mean algorithms for low-dimensional datasets.
- Ryšavý and Hamerly in [16] propose a few methods to accelerate all aforementioned algorithms, such as producing tighter lower bounds, finding neighbor centers and accelerating k-means in the first iteration.
- Fission-Fusion k-means algorithm [19] keeps bounds for subgroups of clusters. It performs better for low-dimensional datasets.

Elkan's k-means algorithm is known to suffer from the required space complexity $O(nk)$ to store the lower bounds, which may be infeasible for large k, demonstrated in [5,13]. However, Elkan's k-means algorithm performs the best in terms of run-time among the aforementioned accelerated k-means algorithms, for high-dimensional datasets, as shown in [13], e.g., Gassensor ($d = 128$), KDD-cup98 ($d = 310$), and MNIST784 ($d = 784$).[2] Therefore, we are motivated to continue this same vain to make Elkan's k-means algorithm even faster or with less memory footprint to improve the scalability.

3   K-means Clustering and Elkan's K-means Algorithm

For k-means clustering, we are given a positive integer k and a set $\mathbf{X}$ of n d-dimensional data points. The objective is to partition data points in $\mathbf{X}$ into k clusters while minimizing within-cluster variances, which are defined as Euclidean distance between each data point and the center of the cluster it belongs to. In this paper, we use $t = 0, 1, 2, \ldots$ to identify the discrete iterations, and each of the given data points in $\mathbf{X}$ is classified into one of the k clusters in each iteration t. Specifically, Elkan's k-means algorithm [8] accelerates Lloyd's k-means algorithm using triangle inequality.

We use $C_i(t)$ to denote the set of data points that are classified into the i-th cluster at the end of the t-th iteration. The i-th cluster at the end of the t-th iteration is defined by its cluster center $c_i(t)$. A data point x is classified into the cluster $C_i(t)$ if the Euclidean distance between the data point x and the cluster center is the shortest among all cluster centers. That is, $x \in C_i(t)$ if $\delta(x, c_i(t)) \leq \delta(x, c_j(t))$, ties being broken arbitrarily, where $\delta(x, y)$ is the Euclidean distance between two points x and y. For any t, we have $\cup_{i=1}^{k} C_i(t) = \mathbf{X}$ and $C_i(t) \cap C_j(t) = \emptyset$ when $i \neq j$. In this paper, we assume that calculation of the distance of any two points can be done in $O(d)$ time complexity and $O(1)$ space complexity.

---

[2] In fact, Elkan's k-means algorithm using the ns-bounds derived from the norm of a sum in [13] sometimes outperforms the original Elkan's k-means algorithm.

Initially, when $t = 0$, k seeds are chosen as the initial cluster centers and each of the data points in $\mathbf{X}$ is classified into one of the k clusters. At the beginning of the next iteration, i.e., $t+1$, the i-th cluster center is positioned to $c_i(t+1)$ by calculating the *means* of the data points in $C_i(t)$. The shift of the cluster center is $\delta(c_i(t), c_i(t+1))$. To update the clustering at the end of the t-th iteration, the time complexity is $O(nkd)$ in the above procedure.

Elkan's k-means algorithm can reduce a large number of distance calculations by applying triangle inequality based on the upper bound and lower bounds decided by each point in each cluster. More precisely, in the t-th iteration, for every data point x in $\mathbf{X}$, instead of calculating the distance of x to the k cluster centers, the algorithm maintains two types of bounds:

- An upper bound $ub(x, c_i(t))$ to the cluster center $c_i(t)$ when x is classified into the i-th cluster, i.e., $x \in C_i(t)$.
- $k-1$ lower bounds $lb(x, c_j(t))$ to the other cluster centers $c_j(t)$ for any $x \notin C_j(t)$.

The elegance of Elkan's k-means algorithm is to apply triangle inequality to maintain these bounds without calculating the distances. If x remains in the same cluster, i.e., $x \in C_i(t)$ and $x \in C_i(t+1)$, instead of updating the distance information precisely, we simply apply the triangle inequality by setting

- $ub(x, c_i(t+1))$ to $ub(x, c_i(t)) + \delta(c_i(t), c_i(t+1))$ and
- $lb(x, c_j(t+1))$ to $lb(x, c_j(t)) - \delta(c_j(t), c_j(t+1))$ for any $j \neq i$.

Elkan [8] proves that a data point x in cluster $C_i(t)$ is not going to be assigned to another cluster $C_j(t+1)$ in the following lemma.

Lemma 1 (Elkan [8]). *Suppose that t is a non-negative integer and $x \in C_i(t)$. Then, x is not going to be classified into another cluster $C_j(t+1)$ at the end of the $(t+1)$-th iteration if*

$$ub(x, c_i(t+1)) < \frac{1}{2}\delta(c_i(t+1), c_j(t+1)) \tag{1}$$

or

$$ub(x, c_i(t+1)) \le lb(x, c_j(t)) - \delta(c_j(t), c_j(t+1)) \tag{2}$$

We note that a significant drawback of Elkan's k-means algorithm is that the space complexity is $O(nk)$ due to the storage of the lower bounds, in addition to the $O(nd)$ input data. The algorithm may not be applicable when nk (or even k) is sufficiently large.

4   New Filtering Bounds

Although Elkan's k-means algorithm can greatly avoid unnecessary distance calculations, its has two shortcomings. First, for an iteration, i.e., fixed t, solely applying Eq. (1) to decide the impossibility of relocating a data point to another cluster can be inefficient. In Sect. 4.1, we propose a simple condition, which can be used to filter out centers that no data in $C_i(t)$ will be relocated to, at the end of the $(t+1)$-th iteration. Moreover, the maintained lower bounds $lb(x, c_j(t))$ can be very expensive, i.e., it requires $O(nk)$ space complexity, and even become too inaccurate in some scenarios. In Sect. 4.2, we present two new lower bounds that can be independently applied to improve the space complexity and inaccuracy.

4.1 Filtering for Clusters of Points

The following theorem provides a new filtering condition to ensure that a point that is not assigned to a cluster $C_j(t)$ is not assigned to another cluster $C_j(t+1)$ at the end of the $(t + 1)$-th iteration as well.

Theorem 1. *Suppose that t is a non-negative integer and $x \in C_i(t)$ and some $j \neq i$. Moreover, assume that*

$$ub(x, c_i(t)) < \frac{1}{2}\delta(c_i(t), c_j(t)). \tag{3}$$

The data point x is not going to be classified into another data cluster $C_j(t+1)$ at the end of the $(t + 1)$-th iteration, if

$$\frac{1}{2}\delta(c_i(t), c_j(t)) + \delta(c_i(t), c_i(t+1)) \leq \frac{1}{2}\delta(c_i(t+1), c_j(t+1)). \tag{4}$$

Proof. Recall that the the i-th center is shifted from $c_i(t)$ to $c_i(t+1)$ after one iteration. By triangle inequality, we have

$$
\begin{aligned}
\delta(x, c_i(t+1)) \quad &\leq && \delta(x, c_i(t)) + \delta(c_i(t), c_i(t+1)) \\
&\underset{\text{definition of ub}}{\leq} && ub(x, c_i(t)) + \delta(c_i(t), c_i(t+1)) \\
&\underset{\text{Eq. (3)}}{\leq} && \frac{1}{2}\delta(c_i(t), c_j(t)) + \delta(c_i(t), c_i(t+1)) \\
&\underset{\text{Eq. (4)}}{\leq} && \frac{1}{2}\delta(c_i(t+1), c_j(t+1)).
\end{aligned}
$$

By the above condition, i.e., $\delta(x, c_i(t + 1)) \leq \frac{1}{2}\delta(c_i(t + 1), c_j(t + 1))$, we can apply the key property from Elkan [8] (summarized in Lemma 1), which concludes that the data point x is not going to be classified into cluster $C_j(t+1)$ whenever the conditions in Eq. (3) and Eq. (4) hold.   □

The condition in Eq. (3) is always ensured by applying the original Elkan's k-means algorithm as this property is ensured by Lemma 1. The difference here is to apply a tighter bound if the condition in Eq. (4) holds. This theorem is useful when the distance $\delta(c_i(t + 1), c_j(t + 1))$ is larger than $\delta(c_i(t), c_j(t))$.

Corollary 1. *Suppose that t is a non-negative integer and the upper bound on the Euclidean distance of every data point in cluster $C_i(t)$ is at most $UB_i(t)$, i.e., $UB_i(t) = \max_{x \in C_i(t)} ub(x, c_i(t))$. If $UB_i(t) < \frac{1}{2}\delta(c_i(t), c_j(t))$ and the condition in Eq. (4) holds $\forall x \in C_i(t)$, then none of the data points in cluster $C_i(t)$ is going to be classified into another data cluster $C_j(t + 1)$ at the end of the $(t + 1)$-th iteration.*

Proof. This comes directly from Theorem 1.

4.2   Additional Lower Bounds

In Elkan's k-means algorithm, to ensure the impossibility that a data point x in $C_i(t)$ is going to be classified into a new cluster $C_j(t + 1)$ for some $j \neq i$ in the next iteration is to make sure that the *upper bound* of the distance $\delta(x, C_i(t+1))$ is no more than the *lower bound* of the distance $\delta(x, C_j(t + 1))$. To ensure that, $lb(x, c_j(t)) - \delta(c_j(t), c_j(t + 1))$ is used as a lower bound of $\delta(x, C_j(t + 1))$, as stated in Eq. (2) in Lemma 1.

However, this lower bound becomes very small if the shift of the j-th center is significant. In fact, when $\delta(c_j(t), c_j(t+1))$ is large, it is possible to find a tighter (i.e., larger) lower bound of $\delta(x, C_j(t+1))$, as presented in the following theorem:

Theorem 2. *Suppose that t is a non-negative integer and $x \in C_i(t)$. The data point x is not going to be classified into another data cluster $C_j(t+1)$ at the end of the $(t + 1)$-th iteration for any $j \neq i$, if*

$$ub(x, c_i(t + 1)) \leq \delta(c_i(t), c_j(t + 1)) - ub(x, c_i(t)) \tag{5}$$

Proof. By triangle inequality

$$\delta(c_i(t), c_j(t + 1)) - ub(x, c_i(t)) \leq \delta(c_i(t), c_j(t + 1)) - \delta(x, c_i(t))$$
$$\leq \delta(x, c_j(t + 1))$$

Therefore, if the condition in Eq. (5) holds, we ensure that $\delta(x, c_i(t + 1)) \leq \delta(x, c_j(t + 1))$ and the theorem is proved.   □

Moreover, the lower bound $lb(x, c_j(t))$ may be not available if we do not want to keep tracking the distance between x and the other $k - 1$ cluster centers that x does not belong to. In fact, if $c_i(t)$ and $c_j(t)$ are quite distant, the lower bound in the following lemma can be applied:

Theorem 3. *Suppose that t is a non-negative integer and $x \in C_i(t)$. The data point x is not going to be classified into another data cluster $C_j(t+1)$ at the end of the $(t + 1)$-th iteration for any $j \neq i$, if*

$$ub(x, c_i(t + 1)) \leq \delta(c_i(t), c_j(t)) - ub(x, c_i(t)) - \delta(c_j(t), c_j(t + 1)) \tag{6}$$

Proof. By triangle inequality

$$\delta(c_i(t), c_j(t)) - ub(x, c_i(t)) - \delta(c_j(t), c_j(t+1))$$
$$\leq \delta(x, c_j(t)) - \delta(c_j(t), c_j(t+1))$$
$$\leq \delta(x, c_j(t+1))$$

Therefore, if the condition in Eq. (6) holds, we ensure that $\delta(x, c_i(t+1)) \leq \delta(x, c_j(t+1))$ and the theorem is proved.   □

We note that the two new lower bounds introduced in Theorems 2 and 3 only require the information of $ub(x, c_i(t))$ and distances of the cluster centers. Therefore, they can be used to reduce the space complexity when maintaining the lower bounds $lb(x, c_j(t)), \forall x \in \mathbf{X}$ and $x \neq C_j(t)$ is too expensive, i.e., $O(nk)$, detailed in Sect. 5.

5   Optimized Elkan's K-means

In this section we present how to optimize the original Elkan's k-means algorithm to alleviate the time and space overheads by applying different triangle inequalities presented in Lemma 1, Theorems 1, 2, and 3, and Corollary 1. We note that the triangle inequalities based on $lb(x, c_j(t))$, for all $x \in \mathbf{X}$ and $C_j(t)$ with $x \notin C_j(t)$, are only applicable when these $O(nk)$ lower bounds are maintained, which can be problematic for the memory usage when nk is large. That is, whenever Eq. (2) is applied, the space complexity may become a bottleneck.

Algorithm 1 presents the pseudocode of our optimized algorithms. After the initialization (Line 3), the clustering procedure keeps repeating until the process converges, i.e., all centers stop changing. If Eq. (2) is not used in the algorithm (in Line 27), we can skip the maintenance of the lower bounds in Lines 13, 18, and 35. The pseudocode consists of two procedures, one for the initialization when t is 0 (i.e., Line 8 to Line 13) and one for the $t' \leftarrow (t+1)$-th iteration (i.e., Line 14 to Line 35). We focus our explanation on the latter procedure.

Line 15 updates each of the k centers by calculating the Euclidean *mean* value of the points assigned to the cluster in the previous iteration. Line 16 calculates different distances between different centers in the last iteration t and in this iteration $t' = t + 1$. Line 17 updates the upper bound of the distance from x to its shifted center $c_i(t+1)$ by applying a triangle inequality. The time complexity of the above steps is $O((n + k^2)d)$ and the space complexity is $O(n + k^2 + kd)$.

Moreover, Line 18 updates the lower bounds of the distance from x to other centers with $x \neq C_j(t)$ using a triangle inequality if necessary. Line 18 requires $O(nk)$ space and time complexity.

For the simplicity of presentation, we use an auxiliary set $C_i(t')$ which is initialized as $C_i(t)$ in Line 19 for every $i = 1, \ldots, k$. We then go through each of the i-th center in the loop described between Line 20 and 35. Line 21 defines a set Set$_i$ based on Corollary 1. That is, it is guaranteed that, for any $j \notin$ Set$_i$, there is no possibility that a data point in $C_i(t)$ is classified into $C_j(t+1)$. Line 21

Algorithm 1. Optimized Elkan's K-means

1: **Input**: data set $\mathbf{X}$, integer k
2: **Output** : *cluster centers $c_i...c_k$*

3: INITIALIZATION($\mathbf{X}$,k, $t \leftarrow 0$)
4: **repeat**
5:    ONE-KMEANS-ITERATION($\mathbf{X}$, k, $t+1$)
6:    $t \leftarrow t+1$
7: **until the process converges**

8: **procedure** INITIALIZATION($\mathbf{X}$,k, $t \leftarrow 0$)
9:    Randomly select k initial centers $c_i(0), \dots, c_k(0)$;
10:    **for** every $x \in \mathbf{X}$ **do**
11:        Let $i*$ be the closest center $c_{i*}(0)$ from data point x, ties arbitrarily broken
12:        x is assigned to $C_{i*}(t)$ and $ub(x, c_{i*}(t)) \leftarrow \delta(x, c_{i*}(t))$
13:        If *necessary*, $lb(x, c_j(t)) \leftarrow \delta(x, c_j(t))$ for every $j \neq i*$

14: **procedure** ONE-KMEANS-ITERATION($\mathbf{X}$, k, t')
15:    $c_i(t') \leftarrow mean(C_i(t))$ for every $i = 1, \dots, k$;
16:    Calculate $\delta(c_i(t), c_i(t')), \delta(c_i(t'), c_j(t'))$, and $\delta(c_i(t), c_j(t'))$, for every $i = 1, \dots, k$ and $j = 1, \dots, k$
17:    $ub(x, c_i(t')) \leftarrow ub(x, c_i(t)) + \delta(c_i(t), c_i(t'))$ for every $x \in C_i(t)$ and $i = 1, \dots, k$
18:    If *necessary*, $lb(x, c_j(t')) \leftarrow lb(x, c_j(t)) - \delta(c_j(t), c_j(t'))$ for every point x when x *was* assigned to cluster $C_i(t)$ for every $1 \leq j \leq k$ and $j \neq i$
19:    $C_i(t') \leftarrow C_i(t)$ for every $i = 1, \dots, k$
20:    **for** $i \leftarrow 1, 2, \dots, k$ **do**
21:        $\text{Set}_i \leftarrow \{j | 1 \leq j \leq k, j \neq i$, Corollary 1 does not hold$\}$
22:        **if** $\text{Set}_i \neq \emptyset$ **then**
23:            **for** each data point $x \in C_i(t)$ **do**
24:                $Temp \leftarrow \emptyset$
25:                **for each** center $j \in \text{Set}_i$ **do**
26:                    **if** Eq. (1) holds: **continue**
27:                    **if** Eq. (2) (or Eq. (6)) holds: **continue**
28:                    **if** Eq. (5) holds: **continue**
29:                    **if** $\delta(x, c_i(t')) > \delta(x, c_j(t'))$ **then**
30:                        $Temp \leftarrow Temp \cup \{j\}$
31:                **if** $Temp \neq \emptyset$ **then**
32:                    Let $j*$ be $\text{argmin}_{j \in Temp}\delta(x, c_j(t'))$
33:                    $C_i(t') \leftarrow C_i(t') \setminus \{x\}$ and $C_{j*}(t') \leftarrow C_{j*}(t') \cup \{x\}$
34:                    $ub(x, c_{j*}(t')) \leftarrow \delta(x, c_{j*}(t'))$;
35:                    If *necessary*, $lb(x, c_j(t')) \leftarrow \delta(x, c_j(t'))$ for every $j \neq j*$

---

requires $O(k)$ time/space complexity, provided that $UB_i(t)$ is always maintained. For each $j \in \text{Set}_i$, there are two possibilities in the presented algorithms between Line 26 and Line 30:

- We can apply Eq. (1), Eq. (2), and Eq. (5). For a given x and j, each of them takes $O(1)$ time/space complexity. Line 29 takes $O(d)$ time complexity.

However, this requires the lower bounds maintained in Lines 13, 18, and 35. We denote this option as filtering-based Elkan, **FB-Elkan**.
- We can apply Eq. (1), Eq. (5), and Eq. (6). For a given x and j, this takes $O(1)$ time/space complexity. Line 29 takes $O(d)$ time complexity. This combination does not require the lower bounds maintained in Lines 13, 18, and 35. We denote this option as memory-optimized Elkan, **MO-Elkan**.

In the pseudo-code, for the simplicity of presentation, we use an auxiliary set $Temp$ to store the indexes of the possible new centers for a data point x, which are maintained in Line 24 and Line 30. The data point x is assigned to the closest center in Lines 31 to 35. The time complexity between Line 25 and Line 35 is $O(|Set_i|d) = O(kd)$ and the space complexity is $O(k)$. Please note that $Temp$ is just introduced for better readability in the pseudocode. A simple implementation regarding $Temp$ can directly calculate and store the closest index j^* on the fly using a buffer (instead of calculating the distance again in Line 32).

With the above discussion, we have the following conclusion for *one iteration* when $n \geq k$ and $t \geq 1$:

- **FB-Elkan**: time complexity $O(nkd)$ and space complexity $O(nk + kd)$.
- **MO-Elkan**: time complexity $O(nkd)$ and space complexity $O(n + k^2 + kd)$.

We note that the above time complexity analysis is asymptotic and does not reflect the actual run-time efficiency of these two algorithms and the lower bound in Elkan's k-means algorithm, i.e., Eq. (2), is usually stronger than Eq. (6). Therefore, if the space complexity is affordable, using Eq. (2) is more run-time efficient than using Eq. (6), further explained in Sect. 6.

6   Evaluation and Discussion

In this section, we first present our evaluation setup. Afterwards, we present the evaluation results of normalized speed-up. Specifically, we show the scalability of MO-Elkan on large n datasets, i.e., SUSY and HIGGS. Please note that the ns-bounds provided in [13] can also be included in our algorithms, but we decide not to involve them here due to the page limit.

6.1   Evaluation Setup

We compared two optimized Elkan's k-means algorithms with the original Elkan's k-means algorithm (denoted as Elkan) [8]: FB-Elkan represents the combination of Eq. (1), Eq. (2), and Eq. (5) in Algorithm 1. MO-Elkan represents the combination of Eq. (1), Eq. (5), and Eq. (6) in Algorithm 1. The presented speed-up factors are all normalized according to Elkan. If the normalized value is greater than 1, the considered algorithm is faster than Elkan. Otherwise, it is slower than Elkan.

To evaluate the runtime efficiency, we considered several datasets from the following repositories: the UCI machine learning repository [2], clustering basic datasets [9], and LIBSVM [4]. To show the scalability of MO-Elkan, we specifically consider two additional datasets, i.e., SUSY ($n = 5$M) and HIGGS ($n = 11$M). For each of dataset (excluding SUSY and HIGGS), we tested over 50 times[3] for each $k \in \{10, 50, 100, 500\}$ and calculated the variance to show how spread out the measured results are. Although most datasets are given number of classes, which could be used as a natural choice of k, testing over various k is to demonstrate the computational performance. All tested algorithms were executed under the same initialization via k-means++ [1], and the clustering results of all algorithms were eventually the same as expected. All approaches were implemented in the same programming language, i.e., C++, and executed on the same machine, i.e., Intel Core i7-8550U with 1.8 GHz and 16 GB RAM.

Table 1. Speed-up normalized to Elkan and variances with high-dimensional datasets. For the simplicity of the presentation, the shown variance is set to 0 if the calculated value is less than 10^{-4}.

Dataset	n	d	k	MO-Elkan	Variance	FB-Elkan	Variance
Covtype	150000	54	10	0.33	0.006	1.14	0.0004
			50	0.44	0.49	1.18	0.04
			100	0.56	0.41	1.19	0.23
			500	0.67	4.07	0.70	1.23
KDDcup98	95412	56	10	0.32	0.008	1.40	0.004
			50	0.39	0.12	1.18	0.017
			100	0.49	1.28	1.09	0.22
			500	0.53	6.33	0.83	0.25
KDDcup04	145751	74	10	0.26	0.43	1.05	0.005
			50	0.25	19.81	1.10	0.60
			100	0.24	200.13	1.09	2.35
			500	0.18	339.76	1.17	8.71
Gassenor	14000	128	10	0.36	0	1.25	0
			50	0.37	0.002	1.31	0.0003
			100	0.43	0.002	1.22	0.0009
			500	0.47	0.049	1.06	0.07
Usps	7291	256	10	0.33	0.003	1.11	0.0003
			50	0.27	0.05	1.48	0.002
			100	0.39	0.12	1.69	0.015
			500	0.56	2.28	1.28	0.68
MNIST784	60000	784	10	0.57	0.088	1.19	0.003
			50	0.3	1.36	1.28	0.15
			100	0.43	9.69	1.10	0.13
			500	0.45	15.73	1.38	1.23

---

[3] Due to the amount of required time for each test, we can reach this number for all setups to fairly demonstrate the statistical significance of the differences.

Table 2. Speed-up normalized to Elkan and variances with low-dimensional datasets. For the simplicity of the presentation, the shown variance is set to 0 if the calculated value is less than 10^{-4}.

Dataset	n	d	k	MO-Elkan	Variance	FB-Elkan	Variance
birth	100000	2	10	1.03	0.0001	0.94	0
			50	1.64	0.0006	0.92	0.011
			100	1.90	0.018	0.90	0.0006
			500	1.69	0.038	0.89	0.008
skin_noneskin	245057	3	10	0.95	0	0.90	0
			50	1.42	0.002	0.92	0.002
			100	1.43	0.0036	0.93	0.0036
			500	1.49	0.61	0.95	0.078
3D_spatial_network	434874	4	10	1.36	0	0.91	0
			50	2.30	0	0.86	0
			100	2.48	0.001	0.91	0.004
			500	1.27	0.005	0.93	0.001

6.2 Runtime Efficiency Evaluation

With high-dimensional datasets (see Table 1), FB-Elkan can mostly outperform the others and achieve up to 1.69x. The trends of variances also follow the increase of k for each dataset. However, when the number of clusters k is as large as 500, we observe that the benefit of filtering routines, i.e., avoiding unnecessary distance calculations, is mitigated by the overhead of calculating the filtering bounds. For Covtype dataset, the additional time for calculating Eq. 5 increases from 11% to over 20% when k increases from 50 to 500, whereas the original Elkan's k-means algorithm has no such overhead.

For low-dimensional datasets (see Table 2), we can notice that the variance of the measured results is almost negligible. Moreover, MO-Elkan can reach up to 2.48x, whereas FB-Elkan performs slightly worse than Elkan. In fact, the overhead of checking additional filtering bounds in FB-Elkan is higher than the benefit of filtering unnecessary distance calculations. With a similar reason, MO-Elkan only requires less memory accesses to the filtering bounds. Therefore, it is faster than Elkan for such datasets.

6.3 Scalability Evaluation

In order to demonstrate the improvement of scalability, we specifically evaluated Elkan, FB-Elkan and MO-Elkan with two additional datasets with large n, i.e., SUSY ($n = 5M$) and HIGGS ($n = 11M$). We tested over different numbers of clusters k, where $k = \{5, 10, 50, 100, 500\}$ and report the normalized speed-up factor. In case Elkan halted due to out of memory, we mark the corresponding entry with "v" if FB-Elkan or MO-Elkan can be successfully executed till completion. Otherwise, if FB-Elkan or MO-Elkan also halted, the corresponding entry is marked with "-". As shown in Table 3, Elkan and FB-Elkan essentially out-perform MO-Elkan when their required memory footprints are affordable. When the number of data points n multiplied with k becomes bigger, e.g., SUSY with $k = 500$ or HIGG with $k \geq 100$, the required memory footprints clearly become a critical issue, whereas MO-Elkan can still finish the k-means clustering. We note that the memory footprint of MO-Elkan was mainly dominated by the number of data points, and the increased size respect to k was tolerable, i.e., $\simeq 1.44$ GB for SUSY and $\simeq 2.57$ GB for HIGGS. However Elkan required 4.608 GB for SUSY with $k = 100$ and 6.72 GB for HIGGS with $k = 50$, and FB-Elkan required slightly more than Elkan.

Table 3. Speed-up normalized to Elkan with large n datasets.

Dataset	n	d	k	MO-Elkan	FB-Elkan
SUSY	5M	18	5	0.19	1.17
			10	0.15	1.05
			50	0.20	0.96
			100	0.28	0.95
			500	v	–
HIGGS	11M	28	5	0.14	1.10
			10	0.11	1.08
			50	0.07	1.10
			100	v	–
			500	v	–

7 Conclusion and Outlook

In this paper, we present new filtering bounds to optimize Elkan's k-means algorithm. Specifically, two different combinations of the proposed bounds are proposed to either filter more unnecessary distance calculations (FB-Elkan), or reduce the space complexity (MO-Elkan) to improve the scalability of the original Elkan's k-means algorithm. Throughout extensive evaluations with several real-world datasets, we reach the conclusion that FB-Elkan improves the runtime efficiency of Elkan for high-dimensional datasets and MO-Elkan outperforms the others for low-dimensional datasets while improving the scalability of Elkan, i.e., the memory footprint is mainly dominated by the number of data points.

In the future work, we plan to integrate the proposed filtering bounds into other bounds-based accelerated k-means algorithms. For example, an integration with Fission-Fusion k-means algorithm [19] may additionally refine the bounds not only for each data point but also for each cluster. Integrating our bounds with Yingyang [5] and Fast Yingyang k-means algorithms [3], can be expected to greatly reduce computation time of distance calculations.

Acknowledgement. We thank our colleague Mr. Mikail Yayla for his precious comments at early stages. This paper has been supported by Deutsche Forschungsgemeinschaft (DFG, German Research Foundation), as part of the Collaborative Research Center (SFB 876), "Providing Information by Resource-Constrained Analysis" (project number 124020371), project A1 (http://sfb876.tu-dortmund.de).

References

1. Arthur, D., Vassilvitskii, S.: K-means++: The advantages of careful seeding. In: Proceedings of the Eighteenth Annual ACM-SIAM Symposium on Discrete Algorithms, SODA 2007, pp. 1027–1035. Society for Industrial and Applied Mathematics (2007)
2. Bache, K., Lichman, M.: UCI machine learning repository (2013). http://archive.ics.uci.edu/ml
3. Bottesch, T., Bühler, T., Kächele, M.: Speeding up k-means by approximating euclidean distances via block vectors. In: Proceedings of the 33rd International Conference on International Conference on Machine Learning, ICML 2016, vol. 48, pp. 2578–2586. JMLR.org (2016)
4. Chang, C.C., Lin, C.J.: LIBSVM: a library for support vector machines. ACM Trans. Intell. Syst. Technol. **2**(3), 1–27 (2011)
5. Ding, Y., Zhao, Y., Shen, X., Musuvathi, M., Mytkowicz, T.: Yinyang k-means: a drop-in replacement of the classic k-means with consistent speedup. In: Proceedings of the 32nd International Conference on International Conference on Machine Learning, ICML 2015, vol. 37, pp. 579–587. JMLR.org (2015)
6. Drake, J.: Faster k-means Clustering. Master Thesis in Baylor University (2013)
7. Drake, J., Hamerly, G.: Accelerated k-means with adaptive distance bounds. In: 5th NIPS Workshop on Optimization for Machine Learning (2012)
8. Elkan, C.: Using the triangle inequality to accelerate k-means. In: Proceedings of the Twentieth International Conference on International Conference on Machine Learning, ICML 2003, pp. 147–153. AAAI Press (2003)
9. Fränti, P., Sieranoja, S.: K-means properties on six clustering benchmark datasets (2018). http://cs.uef.fi/sipu/datasets/
10. Hamerly, G.: Making k-means even faster. In: SDM, pp. 130–140 (2010)
11. Kanungo, T., Mount, D.M., Netanyahu, N.S., Piatko, C.D., Silverman, R., Wu, A.Y.: An efficient k-means clustering algorithm: analysis and implementation. IEEE Trans. Pattern Anal. Mach. Intell. **24**, 881–892 (2002)
12. Lloyd, S.: Least squares quantization in PCM. IEEE Trans. Inf. Theor. 28(2), 129–137 (2006). https://doi.org/10.1109/TIT.1982.1056489
13. Newling, J., Fleuret, F.: Fast k-means with accurate bounds. In: Balcan, M.F., Weinberger, K.Q. (eds.) Proceedings of the 33rd International Conference on Machine Learning. Proceedings of Machine Learning Research, vol. 48, 20–22 Jun 2016, New York, USA, pp. 936–944
14. Pelleg, D., Moore, A.: Accelerating exact k-means algorithms with geometric reasoning. In: Proceedings of the Fifth ACM SIGKDD International Conference on Knowledge Discovery and Data Mining, KDD 1999, pp. 277–281. Association for Computing Machinery, New York (1999). https://doi.org/10.1145/312129.312248
15. Philbin, J., Chum, O., Isard, M., Sivic, J., Zisserman, A.: Object retrieval with large vocabularies and fast spatial matching. In: 2007 IEEE Conference on Computer Vision and Pattern Recognition, pp. 1–8 (2007)

16. Ryšavý, P., Hamerly, G.: Geometric methods to accelerate k-means algorithms. In: Proceedings of the 2016 SIAM International Conference on Data Mining, pp. 324–332 (2016)
17. Sculley, D.: Web-scale k-means clustering. In: Proceedings of the 19th International Conference on World Wide Web, pp. 1177–1178. Association for Computing Machinery, New York (2010)
18. Wang, J., Wang, J., Ke, Q., Zeng, G., Li, S.: Fast approximate k-means via cluster closures. In: IEEE Conference on Computer Vision and Pattern Recognition, pp. 3037–3044 (2012)
19. Yu, Q., Dai, B.-R.: Accelerating K-Means by grouping points automatically. In: Bellatreche, L., Chakravarthy, S. (eds.) DaWaK 2017. LNCS, vol. 10440, pp. 199–213. Springer, Cham (2017). https://doi.org/10.1007/978-3-319-64283-3_15

Angle-Based Clustering

Anna Beer[✉], Dominik Seeholzer, Nadine-Sarah Schüler, and Thomas Seidl

Ludwig-Maximilians-Universität München, Munich, Germany
{beer,schueler,seidl}@dbs.ifi.lmu.de,
d.seeholzer@campus.lmu.de

Abstract. The amount of data increases steadily, and yet most clustering algorithms perform complex computations for every single data point. Furthermore, Euclidean distance which is used for most of the clustering algorithms is often not the best choice for datasets with arbitrarily shaped clusters or such with high dimensionality. Based on ABOD, we introduce ABC, the first angle-based clustering method. The algorithm first identifies a small part of the data as border points of clusters based on the angle between their neighbors. Those few border points can, with some adjustments, be clustered with well-known clustering algorithms like hierarchical clustering with single linkage or DBSCAN. Residual points can quickly and easily be assigned to the cluster of their nearest border point, so the overall runtime is heavily reduced while the results improve or remain similar.

1 Introduction

If there are clusters in a dataset, most of the points lie rather in the middle of a cluster than at its border, and if the clusters of the border points are known, the assignment of inner points is easy and fast using a simple 1NN classification. To identify border points we suggest an angle based approach inspired by *Angle-Based Outlier Detection* (ABOD) [4], which is robust even for higher dimensionalities.

Our new clustering method ABC (Angle-Based Clustering), consists of three steps: First, by assessing the angles between difference vectors of points to their kNN, we can reliably identify points located at the boundaries of clusters. Secondly, we apply existing clustering techniques on those border points only, which allows us to reduce the number of points to be clustered severely. Finally, inner points are assigned to the same cluster as their nearest border point. As clustering has a higher complexity than the angle-based border point extraction as well as inner point assignment, the total runtime is dramatically reduced by clustering only a small fraction of all data points.

Our main contributions are as follows:

- Based on angles between a point and its kNN we detect the border points bounding clusters
- We apply adapted versions of DBSCAN and Hierarchical Single-Linkage Clustering on the border points

S. Satoh et al. (Eds.): SISAP 2020, LNCS 12440, pp. 312–320, 2020.
https://doi.org/10.1007/978-3-030-60936-8_24

– In experiments we show not only the speedup of algorithms using only the border points, but also the improvement of results regarding quality

2 Related Work

ABOD [4] was the first algorithm to use angles for outlier detection by regarding the variance of angles between the difference vectors of a point to all pairs of other points. Several works extended it regarding, e.g., acceleration [8], streams [13], and or stability [5]. ABSAD [14] uses angles between points and axis-parallel lines for an angle-based subspace anomaly detection method.

We could only find one work which uses angles in the field of clustering: SCUBI [11] combines classical clustering with detecting boundary information using angles to create a highly scalable clustering scheme. In contrast to our approach, they use the angles only for an approximation to an intrinsically density-based boundary extraction. Furthermore, we consider the previously calculated angles also for the clustering step by improving the distance function.

There are diverse approaches to identify border points: density based [11,12], hull based [7], and graph based [6] . Nevertheless, they lead to problems for higher dimensionalities, either regarding meaningfulness, or complexity.

3 Mathematical Background

Angles Between Data Points. Angles in a finite-dimensional real Euclidean vector space $\mathbb{V}^{\mathbb{R}}(\simeq \mathbb{R}^d, d \in \mathbb{N}, d \geq 2)$ are defined between any pair of vectors $A, B \in \mathbb{V}^{\mathbb{R}}$ with:

$$cos\Theta(A, B) = \frac{(A, B)_R}{|A| \, |B|}, \tag{1}$$

where $(A, B)_R = \sum_{k=1}^{d} A_k B_k$ is the scalar product between the two vectors and $|A| = \sqrt{(A, A)_R}$ [9]. For the resulting (real) angle $\Theta(A, B)$ the following holds true: $0 \leq \Theta \leq \pi$.

Directional Angle and Enclosing Angle. Figure 1 (left) shows the minimal angle for a point X between two difference vectors to its neighboring points which "encloses" all other neighboring points (green shape). We call it the *enclosing angle* Θ_{enc} of a point. One way to calculate the *enclosing angle* in two dimensions requires to calculate the directional angle between two vectors. In a 2d vector space with vectors $\overrightarrow{XY} = (u_1, u_2)$, $\overrightarrow{XZ} = (v_1, v_2) \in \mathbb{V}_2$, the counter-clockwise directional angle from $\overrightarrow{XY}$ to $\overrightarrow{XZ}$ is $\Theta_{YZ}(X) = atan2(u_2, u_1) - atan2(v_2, v_1)$. If the resulting Θ_{dir} is negative, we add 2π to receive only positive values between 0 and 2π. Figure 1 shows an example directional angle Θ_{dir}. Note, that if the directional angle is less than π, it will be equal to the cosine angle.

To obtain the enclosing angle of a point X, we calculate the directional angle between difference vectors to all pairs of neighbors and differentiate two cases:

First, if $\exists Y \in kNN(X) : \forall Z \in kNN(X) : \Theta_{YZ} \geq \pi$ as illustrated in Fig. 1 (middle), the enclosing angle can be calculated as $2\pi - min(\{\Theta_{YZ}|Y, Z \in kNN(X)\})$. Otherwise, the enclosing angle can be calculated as $2\pi - max(\{min(\{\Theta_{YZ}|Z \in kNN(X)\})|Y \in kNN(X)\})$, as shown in Fig. 1 (right). We can use the concept behind *enclosing angle* to characterize the relative position of neighboring points. Points in the center of a cluster tend to have much larger enclosing angles.

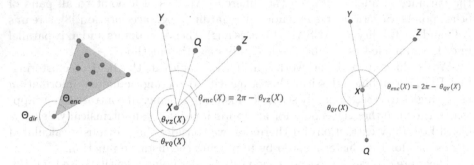

Fig. 1. Left: Enclosing Angle θ_{enc} and counter-clockwise Directional Angle θ_{dir}. Middle and Right: Example calculation of the enclosing angle θ_{enc}. (Color figure online)

4   ABC: Angle-Based Clustering Approach

ABC consists of three steps: First we calculate an angle-based border degree, see Sect. 4.1. The top β points with the highest border degree are the border points. Secondly, we cluster the border points using either an adapted DBSCAN or Hierarchical-Single Linkage Clustering, see Sect. 4.2. Finally, inner non-border points are assigned to cluster of their nearest border point. With a k-d tree this can be done in $O(n \log n)$.

4.1   Border Point Detection Based on Enclosing Angles

Because the nearest neighbors are all located in a similar direction for border points, their enclosing angle (see Sect. 3) tends to be much smaller compared to inner points. As we work with higher dimensionalities we use the following approximation: The *enclosing angle based border degree* is calculated as the maximum of all angles between the vector formed by query point to the kNN-mean and the vector from query point to one of the neighbors. Figure 2 (left) shows a simplified 2d example. The approximated enclosing angle θ_{enc} for border points tends to be much smaller than for inner points. The green shape encompasses the enclosed points.

The complete enclosing angle based border point extraction process proceeds as follows: For each point the kNN, the average distance to them, and the enclosing angles are calculated. For the *direction* of a border point, we use the vector from the query point to the kNN-mean. Border points are then sorted by border degree and the $\beta \cdot n$ points with the highest border degree are returned as the *Boundary*. Figure 2 (middle and right) shows an example on a two dimensional dataset, where darker points imply a higher border degree.

Parameter Analysis. Small values for k can lead to inner points being falsely identified as border points, high values can lead to inter-cluster border points not being recognized as such, i.e., we only find the global boundary of all clusters. For datasets with many close clusters a small k should be preferred, while far separated clusters yield better results with a larger k.

The parameter β determines the separation threshold between border and inner points. Too high values yield more border points leading to a longer execution time of the subsequent clustering step. Too small values will fail to correctly identify enough cluster boundaries. In general, we have found values for β between 5–20% to yield optimal results.

4.2   ABC-DBSCAN/ABC-Hierarchical-SL

To cluster the boundary points we can use an adaption of DBSCAN [1] in which we regard also the *direction* of each border point to its neighbors. As border points that lie close to each other but have opposing directions are unlikely to belong to the same cluster, we use the following new the distance function instead of the Euclidean:

Definition 1. *Direction-Angle modified Distance Function*
*Given two border points $A, B \in \mathcal{D}$ and their respective direction vectors a, b as well as the Euclidean distance $d(A,B)_{eucl}$ between the points and the angle $\Theta(A,B)$ between their direction vectors. Then, given a direction-angle modifier σ_{mod}, the **direction-angle modified distance** $d(A,B)_{mod}$ is calculated as:*

$$d(A,B)_{mod} = d(A,B)_{eucl} * (1 + (\frac{\sigma_{mod} - 1}{\pi}) * cos\Theta(A,B)) \tag{2}$$

A larger angle between the direction vectors a and b results in a larger modified distance, where σ_{mod} controls the maximum. A higher σ_{mod} leads to more influence of direction-angle similarity compared to the Euclidean distance. When $\sigma_{mod} = 1$, then $d(A,B)_{mod} = d(A,B)_{eucl}$. A value of $\sigma_{mod} < 1$ increases the distance between points with different angle. Note, that this distance function does not represent a metric, since the triangle inequality does not always hold.

Another well suited approach to cluster border points is hierarchical agglomerative clustering using single linkage (Hierarchical-SL) [3]. Again with a complexity of $O(n^2)$, potential time savings using Angle-Based border point clustering are high. Also here we use the modified distance as described in Definition 1.

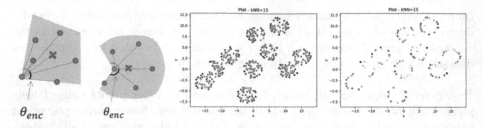

Fig. 2. Left: Approximated Enclosing Angles for border point and inner point. The red cross marks the mean of the blue kNN of the regarded gray point. Right: Border degree and selected border points ($k = 15$, $\beta = 0.2$). (Color figure online)

Complexity Analysis. Calculating the border degree requires an kNN query with complexity $O(n \log n)$ using a k-d tree [10]. The border degree calculation itself has complexity $O(n * k)$, as an angle between each nearest neighbor of each point and the mean of all its kNN is calculated. The sorting and selection of border points is $O(n \log n)$. In total, we get $O(n \log n + nk)$. As k is typically very small ($k \leq \log n$) the overall complexity is then $O(n \log n)$.

5   Experiments and Results

The following Sect. 5.1 covers results of experiments analyzing the runtime of algorithms. The quality on different kinds of datasets, both synthetic and real, are compared in Sect. 5.2 based on the Adjusted Rand Index (ARI).

5.1   Runtime

As ABC only requires to cluster a small fraction of all data points it is highly scalable and well suited for big datasets. Figure 3 (left) summarizes the experiments on how long each of the main three steps (border degree calculation, border point clustering and inner point assignment) take for an increasing number of points. As clustering is the most time consuming task with growing number of observations, reducing the amount of points having to be clustered significantly saves time.

As seen in Fig. 3 (right), ABC-DBSCAN outperforms the naive implementation of DBSCAN with time complexity $O(n^2)$. Even with the use of optimized index structures, the complexity of DBSCAN cannot be reduced below $O(n^{4/3})$ for higher dimensional data [2]. Thus, for large enough datasets, the ABC version with $O(n \log n)$ outperforms even optimized variants of DBSCAN.

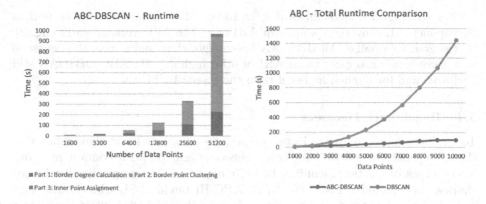

Fig. 3. Left: ABC-DBSCAN components runtime with $d = 5$, $\beta = 0.2$, $k = 10$. Right: Total runtime of DBSCAN and ABC-DBSCAN.

5.2 Quality

Datasets. First, we compare the quality of results on synthetic Gaussian data while modifying either cluster count, dimension count or standard deviation (the last one was left out due to the lack of space, even though ABC constantly outperformed the competitors slightly). The default dataset consists of $n = 1000$ data points, $c = 5$ clusters, $d = 5$ dimensions and a standard deviation $\sigma = 0.1$. Then, we test the algorithms on synthetic complex shaped data sets with and without noise. Finally, we investigate how they perform on real data sets.

Algorithms. We compare **ABC-DBSCAN** to the classic DBSCAN. Additionally, we compare it to **ABC-SCUBI-DBSCAN**, for which we adapt the idea of [11] and exclude a point from the DBSCAN ε-range if its angle is greater than $\pi/2$ (instead of our combined distance measure), but still use our border-degree measurement. Then, we compare the **ABC-Hierarchical-SL** approach to the classic Hierarchical-SL algorithm.

For ABC-DBSCAN and DBSCAN the same range of parameters is tested and the best result is kept. ABC-Hierarchical-SL and Hierarchical SL get the correct amount of clusters given as the maximum cluster parameter. For the border point calculation, we used parameters $\beta = 0.3$ and $k = 15$. For the direction-angle modifier for ABC-DBSCAN and ABC-Hierarchical, we tested values $\sigma_{mod} \in \{0.1, 0.2, 0.3, 0.5, 1, 2, 5\}$ for different weightings of the angle compared to distance and kept the best result.

5.3 Synthetic Gaussian Distributed Data

Based on the dataset described above we varied the number of clusters c from 1 to 500, as shown in Fig. 4 (top). ABC-Hierarchical-SL outperforms the classical Hierarchical-SL, especially for higher c, where the latter only performs poorly. For the DBSCAN versions, the overall performance decreases with increasing c, but the ABC versions yield constantly better results than the original DBSCAN.

For varying dimensionalities $d \in [2, 1000]$. ABC-Hierarchical-SL as well as Hierarchical-SL converge towards an ARI of 1. The ABC version works slightly better even for small d. All DBSCAN based algorithms suffer from the *"curse of dimensionality"*, dropping to an ARI of 0 for high $d \geq 70$. ABC-DBSCAN still performs well for a much higher d than the classic DBSCAN.

5.4   Benchmark Datasets

To evaluate more complex cluster shapes, we also tested our algorithms with the *Complex9* dataset and its noisy version *Cluto-t7*. Both contain nine different types of clusters including blobs, moons and anisotropically distributed shapes. As depicted in Fig. 4 (bottom), ABC-Hierarchical-SL achieves near perfect results and outperforms the original, since the single link effect connecting two different Hierarchical-SL clusters is prevented by using our adapted distance measure. ABC-DBSCAN and ABC-SCUBI-DBSCAN are slightly outperformed by the original DBSCAN. In such cases, ABC could still be chosen with a trade-off between a huge improvement of the runtime and a rather small decrease of the quality. Results for the noisy dataset *Cluto-t7* show similar behavior, except for a significant improvement from ABC-Hierarchical-SL over the original.

Finally, we applied all algorithms on the real datasets *Iris*, *Seed*, and *Ecoli* from the UCI Machine Learning Repository (http://archive.ics.uci.edu/ml). In summary, the ABC versions performed at least comparatively well, in many cases even better than the original, as shown exemplarily in Fig. 4.

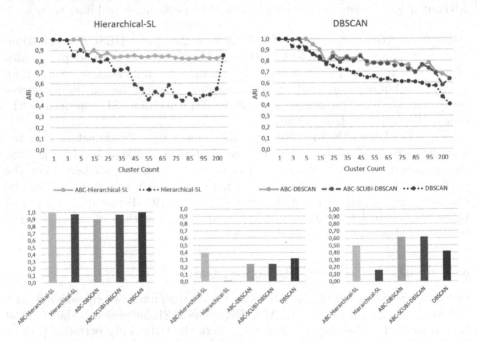

Fig. 4. Top: ARI of synthetic Gaussian distributed data for increasing number of clusters Bottom: ARI of Complex9 (left), Noisy Cluto-t7 (middle) and Ecoli (right)

6 Conclusion

We developed ABC, an angle-based clustering method, which is based on common clustering algorithms like DBSCAN and hierarchical Single-Link clustering, but many times faster as only the few cluster border points, have to be clustered by the respective algorithm. The points lying in the middle of a cluster can easily be assigned to the cluster of their nearest border point. We developed a method to detect those border points based on the angle enclosing their nearest neighbors, which is significantly smaller for points bordering a cluster than for those lying in the inner part. Experiments show that the results are similar or slightly better than those of the original algorithms on synthetic as well as on real world data.

Acknowledgments. This work has been funded by the German Federal Ministry of Education and Research (BMBF) under Grant No. 01IS18036A. The authors of this work take full responsibilities for its content.

References

1. Ester, M., Kriegel, H.P., Sander, J., Xu, X., et al.: A density-based algorithm for discovering clusters in large spatial databases with noise. In: KDD, vol. 96, pp. 226–231 (1996)
2. Gan, J., Tao, Y.: DBSCAN revisited: mis-claim, un-fixability, and approximation. In: Proceedings of the 2015 ACM SIGMOD International Conference on Management of Data, pp. 519–530. ACM (2015)
3. Johnson, S.C.: Hierarchical clustering schemes. Psychometrika **32**, 241–254 (1967). https://doi.org/10.1007/BF02289588
4. Kriegel, H.P., Zimek, A., et al.: Angle-based outlier detection in high-dimensional data. In: Proceedings of the 14th ACM SIGKDD International Conference on Knowledge Discovery and Data Mining, pp. 444–452. ACM (2008)
5. Li, X., Lv, J.C., Cheng, D.: Angle-based outlier detection algorithm with more stable relationships. In: Handa, H., Ishibuchi, H., Ong, Y.-S., Tan, K.C. (eds.) Proceedings of the 18th Asia Pacific Symposium on Intelligent and Evolutionary Systems, Volume 1. PALO, vol. 1, pp. 433–446. Springer, Cham (2015). https://doi.org/10.1007/978-3-319-13359-1_34
6. Liu, D., Nosovskiy, G.V., Sourina, O.: Effective clustering and boundary detection algorithm based on Delaunay triangulation. Pattern Recogn. Lett. **29**(9), 1261–1273 (2008)
7. Moreira, A., Santos, M.Y.: Concave hull: a k-nearest neighbours approach for the computation of the region occupied by a set of points (2007)
8. Pham, N., Pagh, R.: A near-linear time approximation algorithm for angle-based outlier detection in high-dimensional data. In: Proceedings of the 18th ACM SIGKDD International Conference on Knowledge Discovery and Data Mining, pp. 877–885. ACM (2012)
9. Scharnhorst, K.: Angles in complex vector spaces. Acta Applicandae Mathematica **69**(1), 95–103 (2001). https://doi.org/10.1023/A:1012692601098
10. Sproull, R.F.: Refinements to nearest-neighbor searching ink-dimensional trees. Algorithmica **6**(1-6), 579–589 (1991). https://doi.org/10.1007/BF01759061

11. Tong, Q., Li, X., Yuan, B.: A highly scalable clustering scheme using boundary information. Pattern Recogn. Lett. **89**, 1–7 (2017)
12. Xia, C., Hsu, W., Lee, M.L., Ooi, B.C.: Border: efficient computation of boundary points. IEEE Trans. Knowl. Data Eng. **18**(3), 289–303 (2006)
13. Ye, H., Kitagawa, H., Xiao, J.: Continuous angle-based outlier detection on high-dimensional data streams. In: Proceedings of the 19th International Database Engineering & Applications Symposium, pp. 162–167. ACM (2015)
14. Zhang, L., Lin, J., Karim, R.: An angle-based subspace anomaly detection approach to high-dimensional data: with an application to industrial fault detection. Reliab. Eng. Syst. Safety **142**, 482–497 (2015)

Artificial Intelligence and Similarity

Improving Locality Sensitive Hashing by Efficiently Finding Projected Nearest Neighbors

Omid Jafari[✉][iD], Parth Nagarkar[iD], and Jonathan Montaño[iD]

New Mexico State University, Las Cruces, USA
{ojafari,nagarkar,jmon}@nmsu.edu

Abstract. Similarity search in high-dimensional spaces is an important task for many multimedia applications. Due to the notorious *curse of dimensionality*, approximate nearest neighbor techniques are preferred over exact searching techniques since they can return *good enough* results at a much better speed. *Locality Sensitive Hashing* (LSH) is a very popular random hashing technique for finding approximate nearest neighbors. Existing state-of-the-art Locality Sensitive Hashing techniques that focus on improving performance of the overall process, mainly focus on minimizing the total number of IOs while sacrificing the overall processing time. The main time-consuming process in LSH techniques is the process of finding neighboring points in projected spaces. We present a novel index structure called radius-optimized Locality Sensitive Hashing (*roLSH*). With the help of sampling techniques and Neural Networks, we present two techniques to find neighboring points in projected spaces efficiently, without sacrificing the accuracy of the results. Our extensive experimental analysis on real datasets shows the performance benefit of *roLSH* over existing state-of-the-art LSH techniques.

Keywords: Approximate nearest neighbor search · High-dimensional spaces · Locality Sensitive Hashing · Neural Networks

1 Introduction

Finding nearest neighbors is an important problem in many domains such as information retrieval, computer vision, machine learning, multimedia retrieval, etc. For low-dimensions (<10), popular tree-based index structures, such as KD-tree, Quad-tree, etc. are effective, but for higher number of dimensions, these index structures suffer from the well-known problem, *curse of dimensionality* (where the performance of these index structures is often out-performed even by linear scans) [3]. One solution to this problem is to search for approximate results instead of exact results. In many applications where strictly correct results are not necessary, approximate results can produce *good enough* results while achieving much better running times. The goal of the *c*-approximate version of the Nearest Neighbor problem (ANN) is to find nearest neighbors for a given

© Springer Nature Switzerland AG 2020
S. Satoh et al. (Eds.): SISAP 2020, LNCS 12440, pp. 323–337, 2020.
https://doi.org/10.1007/978-3-030-60936-8_25

query point that are within $c * R$ distance (where $c > 1$ is the approximation ratio and R is the search radius).

1.1 Locality Sensitive Hashing

Locality Sensitive Hashing (LSH) [8] is a very popular technique for solving the Approximate Nearest Neighbor problem in high-dimensional spaces. LSH uses *random* projections to map high-dimensional points to lower dimensional representations. The intuition behind LSH is that nearby points in high-dimensional spaces will map to same (or nearby) hash buckets in the projected lower dimensional space with a high probability (and vice-versa). Since the original LSH index structure was proposed for Hamming distance, LSH families have been proposed for other popular distances such as the Euclidean distance [6]. The main benefits of LSH are three-fold: 1) LSH provides theoretical guarantees on the accuracy of the results, 2) LSH can answer ANN queries in sub-linear time with respect to the dataset size, and 3) LSH can be easily implemented as external memory-based index structures, thus making them more scalable [13]. While the original LSH design suffered from large index sizes [16], recent works [4,7,9] have either improved theoretical bounds or introduced techniques such as *Collision Counting* (Sect. 3) to reduce the number of required hash functions. Due to the popularity of LSH in diverse applications [18,22], several research works have been proposed to improve the search efficiency and/or accuracy of LSH techniques [4,7,9,10,13,14,16,23].

1.2 Motivation of Our Work: Improving the Efficiency of Existing State-of-the-Art LSH Techniques

One of the important benefits of LSH is their ease of implementation as external storage based algorithms. State-of-the-art external memory-based algorithms (namely C2LSH [7], QALSH [9], and I-LSH [13]) use a bucket-expansion strategy to find points from neighboring buckets. C2LSH and QALSH use a bucket exponential expansion strategy, whereas I-LSH uses an incremental expansion strategy. While I-LSH is the state-of-the-art algorithm that minimizes disk I/Os, it achieves this optimization at the expense of a costly overall processing time as shown in Sect. 6.[1] Additionally, random I/Os (disk seeks) are known to be bottleneck in query processing [11] and much more expensive than sequential I/Os [12]. I-LSH reduces overall I/Os by mainly reducing sequential I/Os. In this paper, our goal is to design an LSH external memory technique, *roLSH*, that can reduce overall IOs, *mainly random I/Os*, (by finding neighboring points efficiently) which improves the overall query processing time.

---

[1] There is no existing work that compares the overall performance of C2LSH, QALSH, and I-LSH. We present a detailed performance analysis between these works as a technical report (https://arxiv.org/abs/2006.11285).

1.3 Contributions of This Paper

In this paper, we propose a novel approach, called radius-optimized Locality Sensitive Hashing (*roLSH*) for efficiently finding top-k approximate nearest neighbors in high-dimensional spaces. Our main contributions are as follows:

- We present a sampling-based technique, *roLSH-samp*, that reduces the overall random disk I/Os which improves the query processing time while satisfying the theoretical guarantees of LSH. We provide the theoretical analysis for the correctness of *roLSH-samp*.
- We further improve the efficiency by proposing a Neural Network-based technique, *roLSH-NN*, for an improved prediction of projected radiuses (and thus further reduction in random disk I/Os), and hence further improving the performance without affecting the query accuracy. To the best of our knowledge, we are the first work to improve LSH parameters by using Neural Networks. Lastly, we experimentally evaluate both techniques of *roLSH* on real high-dimensional datasets and show that *roLSH* can outperform the state-of-the-art solutions in terms of performance while providing similar query accuracy.

2 Related Work

Locality Sensitive Hashing is a popular technique for solving the Approximate Nearest Neighbor (ANN) problem in high-dimensional spaces. It was first introduced in [8] for the Hamming distance and later extended to the Euclidean distance (E2LSH) [6]. These structures suffered from large index sizes due to the need to have large number of hash functions in multiple hash tables [7]. Additionally, a *magic radius* need to be inputted to find the neighboring projected points, and in order to find the desired number of results, this *magic radius* was arbitrarily chosen to be very high. Multi-Probe LSH [16] presented a technique to probe neighboring buckets if enough number of results were not found. C2LSH [7] introduced a *Collision Counting* approach that reduced the need to have multiple hash tables, and hence reduced the overall index size. SK-LSH [14] introduced a linear ordering on the disk pages with the help of Z-order curve in order to reduce the overall I/Os. The drawback of SK-LSH was that it was created on the original LSH design, and hence also suffered from the *magic radius* problem. QALSH [9] introduced query-aware hash functions and further reduced the number of hash functions necessary to achieve theoretical guarantees. The work closest to our proposed idea is I-LSH [13], which introduces an incremental strategy for finding nearest neighbors in the projected space, as explained in the next Sect. 3.1. Recently, PM-LSH [23] developed a novel tunable confidence interval while using a PM-tree to solve c-ANN queries.

3 Background and Key Concepts

In this section, we describe the key concepts behind LSH. We mainly use the notations and formulations described in the seminal paper on Euclidean LSH families [6] and C2LSH [7].

Hash Functions: A hash function family H is (R, cR, p_1, p_2)-sensitive if it satisfies the following conditions for any two points x and y in a d-dimensional dataset $D \subset \mathbb{R}^d$: if $||x-y|| \leq R$, then $Pr[h(x) = h(y)] \geq p_1$, and if $||x-y|| > cR$, then $Pr[h(x) = h(y)] \leq p_2$.

Here, p_1 and p_2 are probabilities and c is an approximation ratio. In order for LSH to work, $c > 1$ and $p_1 > p_2$. The above definition states that the two points x and y are hashed to the same bucket with a very high probability $\geq p_1$ if they are close to each other (i.e. the distance between the two points is less than or equal to R), and if they are not close to each other (i.e. the distance between the two points is greater than cR), then they will be hashed to the same bucket with a low probability $\leq p_2$. In the original LSH scheme for Euclidean distance, each hash function is defined as $h_{a,b}(x) = \lfloor \frac{a.x+b}{w} \rfloor$, where a is a d-dimensional random vector with entries chosen independently from the standard normal distribution $N(0, 1)$ and b is a real number chosen uniformly from $[0, w)$, such that w is the width of the hash bucket [6]. This leads to the following collision probability function [6], which states that if $||x - y|| = r$, then the probability that x and y map to the same hash bucket for a given hash function $h_{a,b}(x)$ is: $P(r) = \int_0^w \frac{1}{r} \frac{2}{\sqrt{2\pi}} e^{\frac{-t^2}{2r^2}} (1 - \frac{t}{w}) dt$. Here, the collision probability $P(r)$ is decreasing on r for a given w. For a t, which is the largest absolute value of a coordinate of point in D, and for every b uniformly drawn from the interval $[0, c^{\lceil \log_c td \rceil} w^2]$ and $R = c^n$ for some $n \leq \lceil \log_c td \rceil$ we have that $h^R(x) = \lfloor \frac{h_{a,b}(x)}{R} \rfloor$ is (R, cR, p_1, p_2)-sensitive, where $p_1 = p(1)$ and $p_2 = p(c)$ [7].

Collision Counting: In [7], authors theoretically show that two close points x and y collide in at least l hash layers (out of m hash layers) with a probability $1 - \delta$. Further, only those points that collide at least l times with the query point, where l is the collision count threshold, are chosen as candidates. We refer the reader to [7] for further details. Since C2LSH creates only one hash function per hash layer, the number of hash functions are equal to the number of hash layers.

3.1 Existing Techniques for Finding Neighboring Projected Points

C2LSH [7] also introduced the concept of *Virtual Rehashing* that finds neighboring points that collide in neighboring hash buckets. The naive solution to finding neighboring points is to use a large projected radius such that enough neighboring points are found to return top-k results. The projected radius is entirely dependent on the data distribution, and as we show in Fig. 2, these projected radiuses can vary significantly. Hence, using an arbitrarily large radius results in wasted I/Os and unnecessary processing. Instead, *Virtual Rehashing* starts with a very small radius (R = 1), and then exponentially increases the radius in the following sequence: $R = 1, c, c^2, c^3$.... If at *level-R*, enough candidates are not found, the radius is increased until enough query results are found. C2LSH [7] and QALSH [9] follow this exponential expansion strategy. I-LSH [13] introduces an incremental strategy where, instead of expanding the search radius exponentially, they find the nearest point to the query in each projection. C2LSH and

QALSH store points in hash buckets which are stored in disk pages. I-LSH stores each data point separately and hence instead of reading disk pages that store a group of data points, only reads a point (which effectively is the same as reading a disk page of 4 bytes). While they save on disk I/O operations by this method, this is a very costly operation (as shown in Sect. 6) since this process has to be done thousands of times for larger radiuses.

4 Problem Specification

The approximate version of the nearest neighbor problem, also called $c\text{-}\delta$ -*approximate Nearest Neighbor search*, aims to return points that are within $c*R$ distance from the query point with probability at least $1 - \delta$, where $c > 1$ is a user-defined approximation ratio, R is the distance of the query point from its nearest neighbor, and δ is a user-defined error probability.

In this paper, our goal is to return c-δ-ANNs for a given query q while reducing the overall processing time and satisfying the theoretical guarantees. In Sect. 5, we present the processing cost breakdown of the LSH process based on which we design our proposed index structure, radius-optimized Locality Sensitive Hashing (*roLSH*).

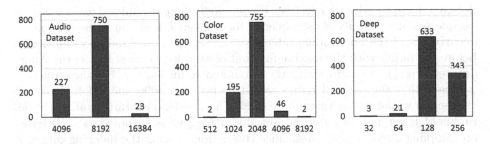

Fig. 1. Frequency (Y-axis) of Final Radius Values (X-axis) for finding Top-100 Points for 1000 Point Queries on Different Datasets using C2LSH

5 roLSH

In this section, we present the design of *roLSH*, which consists of two strategies for efficiently finding neighboring points in the hash functions. We introduce and describe these two strategies in this section: a sampling-based strategy, called *roLSH-samp*, and a Neural Network-based strategy, called *roLSH-NN*.

5.1 Sampling-Based Improved Virtual Rehashing Strategy

In Sect. 2, we explained the original Virtual Rehashing strategy (denoted as *oVR* strategy) as proposed in C2LSH [7]. The initial radius is set to 1, and if sufficient results are not found, then the radius is increased in an exponential sequence: $R = 1, c, c^2, c^3 \ldots$ until sufficient number of results are found. The main drawback of this approach is when the values of R become larger (i.e. when the difference between two consecutive radius values is large - e.g., 4096 and 8192). In such situations, it happens frequently that very few (or no) nearest neighbor points are found at radius value 4096 but all (and lot more) are found at radius 8192. Thus, for example, if the actual radius of the kth-nearest point was near 5000, then index files corresponding to radius 5000–8192 will be read unnecessarily from the disk, leading to expensive wasted IO operations. Instead, we propose a sampling-based improved Virtual Rehashing strategy (denoted as *roLSH-samp*) based on the following observation:

Observation 1. *For high-dimensional datasets, the required radius values for a k value are similar to each other for different query points for a given dataset.*

This observation was also noted by a very recent paper [23] where the authors show that the homogeneity of the distance distributions of data points in different high-dimensional datasets is very high. Figure 1 shows our observation on popular real high-dimensionsal datasets with varying cardinalities and dimensionalities (Audio [1], Color [5], Deep [2]). For 1000 randomly chosen query points, we report the final radius values (using the Virtual Rehashing technique from C2LSH [7]) for top-100 points. By leveraging the above observation, we design an improved and effective *Virtual Rehashing* technique: we execute a sample set of randomly chosen queries for a given k and count the number of occurrences of the final radius value. We choose our initial radius value that is before the radius with the maximum count of sampled queries. E.g. in the Audio dataset (Fig. 1), the radius with the maximum count is 8192. For these queries, it means that the optimal radius would be between 4096 and 8192. Hence we choose our initial radius value to be 4096. Thus, instead of starting at the initial radius of 1, we find an improved initial starting radius (denoted as $i2R$) based on sampling queries. Note that, since this is done during the indexing phase, it has no overhead during query execution. Once the initial starting radius ($i2R$) is found, we leverage the same exponential sequence strategy as C2LSH (using x as the expansion step counter), such that:

$$R = \begin{cases} i2R + 2^x & 0 \leq x \leq \log_2 i2R \\ 2^x & x > \log_2 i2R \end{cases}$$

Thus, for 1000 random queries on the Audio dataset, using the oVR technique, the average final radius is 7450 for $c = 2$ and $k = 100$. On the other hand, using our improved strategy, the average final radius is 6083, which leads to significant savings in the IO.

Note that, one disadvantage of this approach is that there potentially can be queries that finish with a radius value much lower than the chosen initial radius. E.g., in the Color dataset (Fig. 1), our strategy will choose $i2R = 1024$. As you can see, there were 2 out of 1000 queries whose final radius value to find top-100 points was 512. In this case, the improved Virtual Rehashing strategy (denoted as iVR strategy) will do wasted work by starting (and ending) at 1024.

Lemma 1. *For those queries whose required radius in oVR is at least $(2 \times i2R)$, iVR strategy will generate less IOs than the oVR strategy.*

Proof. Set $R = i2R$. By construction of the sequence of radii in oVR, it is enough to assume that the required radius is $2R$, that is, the actual radius r of the kth-nearest point satisfies $R < r \leq 2R$. In the oVR, the sequence of radii needed to find the kth-nearest point has $\log_2 R + 2$ elements, that is, $1, 2, 4, \ldots, 2R$. On the other hand, for the same query q, iVR analyzes at most $\log_2 R + 1$ radii, that is, $R + 1, R + 2, \ldots, 2R$. This finishes the proof.

While I-LSH [13] still generates less disk I/Os than $roLSH$-$samp$, $roLSH$-$samp$ is significantly faster than I-LSH (due to less overall processing time) and also generates less disk seeks than I-LSH for bigger datasets (Sect. 6).

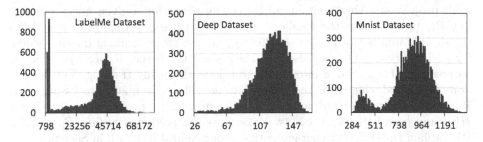

Fig. 2. Frequency (Y Axis) of radiuses (X Axis) for 10,000 Top-100 Queries

5.2 Drawbacks of $roLSH$-$samp$

The main benefit of $roLSH$-$samp$ is that it is effective in reducing the disk I/Os, especially when the radiuses are large (e.g., the Audio dataset in Fig. 1). There is a minor overhead of utilizing the sampling-based method during the indexing phase. Additionally, we found that we also get good sampling representatives even with a small sampling size (e.g., 100). There are two main drawbacks of $roLSH$-$samp$: 1) $roLSH$-$samp$ works best when Observation 1 holds true. We found out that Observation 1 holds true for many datasets, but not all. For example, as seen in Fig. 2, the radiuses for top-100 queries on the LabelMe dataset are quite different leading to inefficient performance of $roLSH$-$samp$ (as shown in Sect. 6), 2) It is not easy to do sampling for different k values since the radius changes for different k values. It is not trivial to build a single model and extend it to multiple k values to find the radius for a particular k value. Instead a model needs to be built for *each* k value.

Table 1. Performance comparison of learning techniques

	MLP	Linear Reg.	RANSAC	Decision tree	Gradient boosting
MSE	0.0265	0.3543	0.3542	0.7057	0.2117
R2	0.9687	0.5826	0.5827	0.1698	0.7504

5.3   Neural Network-Based Prediction of Projected Radiuses

To remedy these two drawbacks, we present a Neural Network-based strategy, *roLSH-NN*, that can better predict starting radiuses based on the query location (in each hash function) for any given k value. The main intuition behind *roLSH-NN* is that nearby points in the original space will have similar projected radiuses to find the desired number (k) of nearest neighbors. Hence, our goal is to predict the projected radiuses given the hash locations of a query for a given k.

Formally, let $h_i(q)$ denote the bucket location of q in the ith hash projection. Thus, $H(q) = h_1(q), ..., h_m(q)$ denotes a vector of size m (since there are m hash projections) that contains m bucket locations for a given query point q. Let $R_{act}(q, k)$ denote the smallest radius in the projected space that satisfies the desired number of results (k). Let Q_{tr} be the set of training queries, where for each query $q \in Q_{tr}$, we also find the ground truth (i.e. $R_{act}(q, k)$), which is entered as a target value into the neural network during training. This step is done in the indexing step, and hence does not affect the query processing time. We train a Neural Network with Q_{tr} queries such that for each query q, we input $H(q)$ and k and the Neural Network outputs the predicted radius, $R_{pred}(q, k)$. We explain the different characteristics of our Neural Network in Sect. 6.

Justification for Choosing Neural Networks: Since the problem of predicting radiuses given the hash function is a regression problem, we tried several machine learning techniques. Table 1 shows that Neural Networks (denoted by *MLP* since we use a Multilayer Perceptron Neural Network) have the best MSE and R2 for a sample dataset (Deep) for $Q_{tr} = 10,000$ among different machine learning techniques (using 10-fold cross validation). Hence, we choose Neural Networks over other techniques in the design of *roLSH-NN*.

Underestimation of Radius: When the radius is underestimated (i.e. $R_{pred}(q, k) < R_{act}(q, k)$), the desired number of results are not found and hence we have to enlarge the radius in all projections. One strategy is to follow the same expansion pattern of *roLSH-samp* presented in Sect. 5.1, where the predicted radius is set as $i2R$. We call this strategy *roLSH-NN-iVR*. The drawback of this strategy is that it can lead to excessive (and expensive) disk seeks if the predicted radius is much lower than the actual radius. Since we observe that $R_{pred}(q, k)$ is close to $R_{act}(q, k)$, we also adopt another strategy where we increase the predicted radius, $R_{pred}(q, k)$, linearly by R_{inc} such that $R_{inc} = R_{pred}(q, k) \times \lambda$. This strategy is referred to as *roLSH-NN-λ* in the rest of this paper.

Overestimation of Radius: While overestimation of the projected radius by the Neural Network leads to wasted disk I/Os during query processing, we experimentally show in Sect. 6 that these wasted disk I/Os are still less than the exponential strategy of C2LSH/QALSH and the improvement in the query processing time (as compared with I-LSH) offsets the disk I/Os significantly.

Extension to any k: In order to train the Neural Network to work for any number of desired results (k), we need to include k as an input feature in the training set. In order to simplify the training procedure, we only consider few values of k in the training set Q_{tr}. In Sect. 6, we experimentally show that as more diverse k values are included in the training set, the MSE decreases. Also, we explain the training setup and the different k values in our training set.

6   Experimental Evaluation

In this section, we evaluate the effectiveness of our proposed index structure, *roLSH*, on three real diverse high-dimensional datasets. All experiments were run on the nodes of the Bigdat cluster[2] with the following specifications: two Intel Xeon E5-2695, 256 GB RAM, and CentOS 6.5 operating system. We implement our work on top of C2LSH [7] since we found it to be the fastest external memory-based LSH algorithm (while achieving high accuracy for high-dimensional datasets). Note that, our method is orthogonal to the LSH algorithm and can be used in any state-of-the-art LSH algorithms. All codes were written in C++11 and compiled with gcc v4.7.2 with the -O3 optimization flag. We compare our three strategies, *roLSH-samp*, *roLSH-NN-iVR*, and *roLSH-NN-λ* with the state-of-the-art LSH algorithms C2LSH [7] and I-LSH [13].[3,4]

Table 2. Comparison of (a) Index Construction Time (in sec) and (b) roLSH Index Time Breakdown (in sec) on Different Datasets

Index Time	LabelMe	Deep	Mnist
roLSH-samp	83.5	93.5	1480.8
roLSH-NN	88.6	98.4	1488
C2LSH	80.6	69	1430.2
I-LSH	20.8	25.7	1359.9

Index Time Breakdown	LabelMe	Deep	Mnist
Base Index Time	80.6	69	1430.2
Sampling	2.9	24.5	50.6
NN Training	8	29.4	57.8

6.1   Datasets

We use the following three popular real datasets to evaluate the proposed method:

---

[2] Supported by NSF Award #1337884.

[3] PM-LSH code is not yet released; hence, we do not compare with it.

[4] The source code for *roLSH* is available at: https://3m.nmsu.edu/rolsh.

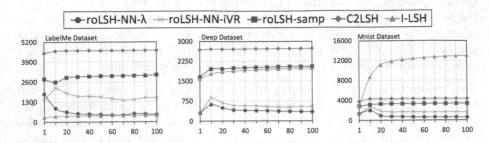

Fig. 3. Number of Disk Seeks (Y axis) for different k (X Axis) on 3 datasets

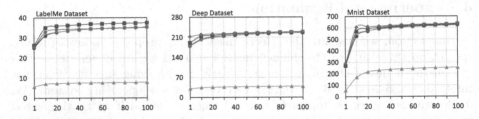

Fig. 4. Amount of Data Read (in MB) (Y axis) for k (X Axis) on 3 datasets

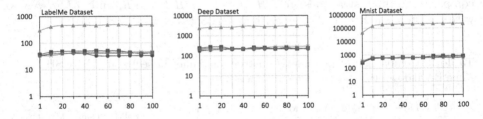

Fig. 5. Algorithm Time (in ms, *log scale*) (Y axis) for k (X Axis) on 3 datasets

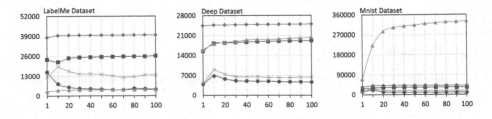

Fig. 6. Query Processing Time (in ms) (Y axis) for k (X Axis) on 3 datasets

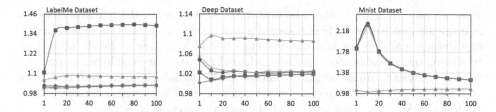

Fig. 7. Accuracy Ratio (Y axis) for different k (X Axis) on 3 datasets

- **LabelMe** [19] consists of $181,093$ 512-dimensional points which were generated by running the GIST feature extraction algorithm on annotated images.
- **Deep** consists of $1,000,000$ 96-dimensional points that were randomly chosen from the Deep1B dataset introduced in [2].
- **Mnist** [15] This dataset contains $8,100,000$ 784-dimensional points that represent images of the digits 0 to 9 which are grayscale and of size 28×28.

6.2   Evaluation Criteria and Parameters

The goal of *roLSH* is to improve the performance efficiency without sacrificing the accuracy of existing LSH techniques. The performance and accuracy of the technique used in this paper are evaluated using the following metrics:

- **Query Processing Time** (QPT): We break down the Query Processing Time into the Index I/O cost, the Algorithm time ($AlgTime$), and the negligible false positive removal cost (denoted by $FPRemTime$, which consists of the cost of reading the data point candidates and computing their exact Euclidean distance for removing false positives). Following [13], we further break down Index I/O cost into the number of disk seeks (i.e. random I/O reads, $noDiskSeeks$) and the amount of data (i.e. index files, $dataRead$) read in MB. We observed that the index I/O times were not consistent (i.e. running the same query multiple times, which needed the same index I/Os, would return drastically different results, mainly because of disk cache and instruction cache issues). Therefore, following [20], for a Seagate 1TB HDD with 7200 RPM, we assume a random seek to cost 8.5 ms on average, and the average time to read data to be 0.156 MB/ms. Thus, we have $QPT = noDiskSeeks * 8.5 + dataRead * 0.156 + AlgTime + FPRemTime$.
- **Accuracy:** We follow the accuracy ratio definition followed by many previous works [7,9,14]: $\frac{1}{k} \sum_{i=1}^{k} \frac{||o_i - q||}{||o_i^* - q||}$. Here, o_i is the ith point returned by the technique and o_i^* is the true ith nearest point from q (ground truth). Ratio of 1 means the returned results have the same distance from the query as the ground truth. The closer the ratio is to 1, the higher is the accuracy.

For the state-of-art methods, we used the same parameters suggested in their papers ($w = 2.719$ for QALSH and $w = 2.184$ for C2LSH). Also, as *roLSH* is built on top of C2LSH, it uses the same parameters as C2LSH. We set the allowed error probability, δ, to be 0.1. The Multilayer Perceptron (MLP) Neural Network is implemented using the Scikit-learn Python package [17]. In this paper, we use the default parameters and options (i.e. 100 hidden layers, $ReLU$ activation function, and the $Adam$ optimization algorithm). We leave the hyper-parameter tuning analysis to future work. We choose 10,000 training queries randomly from the dataset. 50 different queries were randomly chosen from the dataset for the evaluation. We report an average of the results on these 50 queries.

6.3   Effect of Different Parameters on Performance of $roLSH\text{-}NN\text{-}\lambda$

In this section, we present the performance of $roLSH\text{-}NN\text{-}\lambda$ under different parameters for the Deep dataset.

Effect of Training Size: We consider three different training sizes (5K, 10K, 50K). In our experiments, the MSE reduces (by 18.4% between 5K and 50K training size) as the training size increases and since the MSE decreases (i.e. the predicted radius is close to the actual radius), the overall Query Processing Time (QPT) also decreases (by 33% between 5K and 50K training size). In the following experiments, we choose 10K as the default training size. Due to space limitations, we do not present the results in Sect. 6.3 in detail.

Effect of Number of Different k in Training: We analyze the performance of $roLSH\text{-}NN\text{-}\lambda$ for different values of k that are present in the training data while keeping the total training size and λ constant. We chose $\{1, 50, 100\}$, $\{1, 25, 50, 75, 100\}$, and $\{1, 10, 25, 50, 75, 90, 100\}$ as three different settings. The MSE reduces as more diverse k are included: by 33% between the first two settings, but only by 14% between the last two settings since the neural networks are capable of adequately predicting the radiuses for different k even for the second setting (which is our default in the following experiments).

Effect of Different Radius Increment (λ): We experiment using λ values of 5%, 10%, and 20%. As λ increases, the number of disk seeks decrease (by 32%) since a higher λ eventually results in a larger radius and in turn makes the algorithm stop sooner without processing all projections, but the algorithm time and the amount of I/O increases (by 4% and 1% respectively) since more hash buckets are processed. We choose 10% as our default in further experiments.

6.4   Discussion of the Results

Table 2 (a) shows the time taken to finish the index construction. The reported times show that the sampling and training overhead for $roLSH\text{-}samp$ and $roLSH\text{-}NN$ are only 3.4% and 3.9% for the largest dataset (Mnist). Table 2 (b) shows the break-down of index construction time for $roLSH\text{-}samp$ and $roLSH\text{-}NN$ (i.e. overhead of the techniques used in each method) for all datasets. The index size of our techniques are similar to C2LSH since we use C2LSH as our underlying LSH implementation. The index size overhead of $roLSH\text{-}samp$ is 0.1 MB for all datasets, and overhead of *roLSH-NN* is 0.4 MB for LabelMe and 0.5 MB for Deep and Mnist datasets.

Number of Disk Seeks: Figure 3 shows the number of disk seeks (random I/Os) required by these different techniques. It is very interesting to note that while I-LSH performs the best (*roLSH-NN-λ* is a close second) for LabelMe, their performance degrades as the dataset size increases. I-LSH produces significant more disk seeks as the dataset size increases. We believe this is mainly due to the fact that more points need to be accessed incrementally to find the candidates. *roLSH-NN-λ* significantly performs the best for Deep and Mnist datasets because it can accurately predict the radius for different k. Every time *roLSH-NN-λ* underestimates the radius (Sect. 5.3), it has to increment the radius by $λ$ resulting in a disk seek in each projection. Also, as expected, *roLSH-NN-iVR* produces more disk seeks due to radius underestimation (Sect. 5.3).

Amount of Data Read: Figure 4 shows the total amount of data (index files) read. Since I-LSH incrementally increases the search to the nearest point in the projected space (instead of an empirically chosen number, such as $λ$), it results in the least amount of data read for all datasets. These savings in the I/O are offset due to the expensive search for the nearest point as shown in Fig. 5. Especially for lower k, *roLSH-NN-iVR* and *roLSH-NN-λ* read less data than C2LSH, but as k increases the overall data read is similar for both techniques. It is interesting to note that *roLSH-samp* reads significantly more data for LabelMe dataset. This is due to choosing of a bad starting radius due to the unique distribution of the LabelMe radiuses (Fig. 2 (a)). Moreover, *roLSH-NN-iVR* and *roLSH-NN-λ* read similar amount of data since their starting radius is the same.

Algorithm Time: Figure 5 shows the time needed by the algorithms to find the candidates (excluding the time taken to read the index files). Note the log scale of this figure because the algorithm time for I-LSH was orders of magnitude more than the other techniques. This is because I-LSH expands the radius incrementally in each projection which creates a significant overhead. Figure 5 shows that the overhead of our methods is negligible when compared with C2LSH.

Query Processing Time: Figure 6 shows the overall time required to solve a given k-NN query. I-LSH works well for smaller datasets (LabelMe) but is significantly slower as the dataset size increases (due to high overhead in incrementally finding the next neighbor in each projection). *roLSH-samp* is always faster than C2LSH because of the savings in disk seeks. *roLSH-NN-iVR* and *roLSH-NN-λ* are always much faster than *roLSH-samp* and C2LSH because of their ability to accurately predict radiuses, resulting in significantly less disk seeks and lesser (or similar in some cases) data read than C2LSH. *roLSH-NN-λ* has better performance compared to *roLSH-NN-iVR*, mainly because of having lesser disk seeks as discussed before. This figure shows the performance benefit of *roLSH-NN-λ* over its competitors for different datasets, and confirms that the design of *roLSH-NN-λ* leads to improvement in overall efficiency.

Accuracy: Figure 7 shows the accuracy of all techniques. *roLSH-samp* gives the worst accuracy for LabelMe dataset. We found that this is due to the fact that LabelMe dataset has queries with very different large radiuses. *roLSH-samp* is unable to work well for datasets that have differing radiuses because if the starting radius is chosen wrong, then *roLSH-samp* can significantly overestimate the radius for larger radiuses leading to lower accuracy. *roLSH-NN-λ* always returns similar accuracy to that of C2LSH. I-LSH returns a better accuracy for Mnist dataset due to their usage of *query-aware* hash functions, but the performance is significantly slower as shown in Fig. 6.

7 Conclusion

Locality Sensitive Hashing is a popular technique for efficiently solving Approximate Nearest Neighbor queries in high-dimensional spaces. State-of-the-art LSH techniques improve the overall disk I/Os at the expense of algorithm time. In this paper, we present a unique index structure called radius-optimized Locality Sensitive Hashing (*roLSH*). The goal of *roLSH* is to improve the efficiency of LSH techniques by improving the random disk seeks without any significant overhead in algorithm time. We propose two novel strategies, *roLSH-samp* and *roLSH-NN* that are based on sampling and Neural Networks respectively. Experimental results on real datasets show the benefit of *roLSH* in improving overall performance over existing state-of-the-art techniques, C2LSH and I-LSH.

References

1. Audio dataset. http://www.cs.princeton.edu/cass/audio.tar.gz
2. Babenko, A., Lempitsky, V.: Efficient indexing of billion-scale datasets of deep descriptors. In: CVPR (2016)
3. Chávez, E., Navarro, G., Baeza-Yates, R., Marroquín, J.L.: Searching in metric spaces In: CSUR (2001)
4. Christiani, T.: Fast locality-sensitive hashing frameworks for approximate near neighbor search. In: SISAP (2019)
5. Color dataset. http://kdd.ics.uci.edu/databases/CorelFeatures
6. Datar, M., Immorlica, N., Indyk, P., Mirrokni, V.S.: Locality-sensitive hashing scheme based on p-stable distributions. In: SOCG 2004 (2004)
7. Gan, J., Feng, J., Fang, Q., Ng, W.: Locality-sensitive hashing scheme based on dynamic collision counting. In: SIGMOD (2012)
8. Gionis, A., Indyk, P., Motwani, R.: Similarity search in high dimensions via hashing. In: VLDB (1999)
9. Huang, Q., Feng, J., Zhang, Y., Fang, Q., Ng, W.: Query-aware locality-sensitive hashing for approximate nearest neighbor search. VLDB **9**, 1–12 (2015)
10. Jafari, O., Ossorgin, J., Nagarkar, P.: qwLSH: cache-conscious indexing for processing similarity search query workloads in high-dimensional spaces. In: ICMR (2019)
11. Kim, A., Xu, L., Siddiqui, T., Huang, S., Madden, S., Parameswaran, A.: Optimally leveraging density and locality for exploratory browsing and sampling. In: HILDA (2018)
12. Leis, V., et al.: Query optimization through the looking glass, and what we found running the join order benchmark. VLDB **27**, 643–668 (2018)

13. Liu, W., Wang, H., Zhang, Y., Wang, W., Qin, L.: I-LSH: I/O efficient c-approximate nearest neighbor search in high-dimensional space. In: ICDE (2019)

14. Liu, Y., Cui, J., Huang, Z., Li, H., Shen, H.T.: SK-LSH: an efficient index structure for approximate nearest neighbor search. VLDB **7**, 745–756 (2014)

15. Loosli, G., Canu, S., Bottou, L.: Training invariant support vector machines using selective sampling. Large Scale Kernel Mach. (2007)

16. Lv, Q., Josephson, W., Wang, Z., Charikar, M., Li, K.: Multi-probe LSH: efficient indexing for high-dimensional similarity search. In: VLDB (2007)

17. Pedregosa, F., et al.: Scikit-learn: machine learning in Python. JMLR **12**, 2825–2830 (2011)

18. Rong, K., et al.: Locality-sensitive hashing for earthquake detection: a case study of scaling data-driven science. In: VLDB (2018)

19. Russell, B.C., Torralba, A., Murphy, K.P., Freeman, W.T.: LabelME: a database and web-based tool for image annotation. IJCV **77**, 157–173 (2008)

20. Seagate ST2000DM001 Manual. https://www.seagate.com/files/staticfiles/docs/pdf/datasheet/disc/barracuda-ds1737-1-1111us.pdf

21. Sift dataset. http://corpus-texmex.irisa.fr

22. Yang, Z., Ooi, W.T., Sun, Q.: Hierarchical, non-uniform locality sensitive hashing and its application to video identification. In: ICME (2004)

23. Zheng, B., Zhao, X., Weng, L., Hung, N.Q.V., Liu, H., Jensen, C.S.: PM-LSH: a fast and accurate LSH framework for high-dimensional approximate NN search. VLDB **13**, 643–655 (2020)

SIR: Similar Image Retrieval for Product Search in E-Commerce

Theban Stanley, Nihar Vanjara, Yanxin Pan, Ekaterina Pirogova,
Swagata Chakraborty, and Abon Chaudhuri[✉]

Walmart Labs, Sunnyvale, CA, USA
{theban.stanley,yanxin.pan,achaudhuri}@walmartlabs.com

Abstract. We present a similar image retrieval (SIR) platform that is used to quickly discover visually similar products in a catalog of millions. Given the size, diversity, and dynamism of our catalog, product search poses many challenges. It can be addressed by building supervised models to tagging product images with labels representing themes and later retrieving them by labels. This approach suffices for common and perennial themes like "white shirt" or "lifestyle image of TV". It does not work for new themes such as "e-cigarettes", hard-to-define ones such as "image with a promotional badge", or the ones with short relevance span such as "Halloween costumes". SIR is ideal for such cases because it allows us to search by an example, not a pre-defined theme. We describe the steps - embedding computation, encoding, and indexing - that power the approximate nearest neighbor search back-end. We also highlight two applications of SIR. The first one is related to the detection of products with various types of potentially objectionable themes. This application is run with a sense of urgency, hence the typical time frame to train and bootstrap a model is not permitted. Also, these themes are often short-lived based on current trends, hence spending resources to build a lasting model is not justified. The second application is a variant item detection system where SIR helps discover visual variants that are hard to find through text search. We analyze the performance of SIR in the context of these applications.

Keywords: Content-based image retrieval · Information retrieval · Visual search · e-Commerce · Deep learning

1 Introduction

Product data in catalogs owned by online retailers consist of text (title-description etc.), key-value pairs (attributes), and images. A number of internal systems and customer-facing applications leverage images to search and discover product(s) of interest. In some cases, the inherent nature of the application warrants a search through images. Also, the results of an image search usually complement that of a text search in most use cases.

In this paper, we present a visual similarity-based product search and retrieval system built and deployed to address a number of business use cases at

© Springer Nature Switzerland AG 2020
S. Satoh et al. (Eds.): SISAP 2020, LNCS 12440, pp. 338–351, 2020.
https://doi.org/10.1007/978-3-030-60936-8_26

Walmart. Image search has come a long way with the recent advances in deep learning. However, building such a system that has the ability to scan through millions of images in a few seconds is a challenging task. The deep learning based fingerprints (or embeddings) created from images contain rich and complex information, but creating them is a compute-intensive task until GPUs are available in excess. Creating a search index on top of such large floating point arrays is not straightforward either. We present in this paper our process of encoding the embeddings so that they lend themselves well to popular search indexes like Elasticsearch and can be used to retrieve approximate nearest neighbors.

Our system is currently used in two business-critical applications. In this application-focused paper, we highlight how image similarity search plays a central role such applications.

- **Offensive or non-compliant product search:** The quality and compliance of our catalog are maintained through scheduled and on-demand searches for potentially offensive products. This discovery process demands a quick turnaround which makes the path of building supervised classifiers unattractive. Rather, a search tool that would accept one or a few known examples as a query and return more products with similar images is needed.
- **Variant grouping:** In this classic e-commerce problem, items varying by color, size etc. are grouped together and presented to the customer at once on a single page. To create such groups from the catalog, we often start with a seed item and try to limit the search space to a pool of similar items. This pool of similar candidates can be created by text or image search or both. Our experiments suggest that image search often retrieves candidates that complement the ones retrieved by text search.

Our system SIR has the potential to be used in other applications as well. With visual exploration emerging as an upcoming trend in retail, our image search index based on catalog product images can eventually become the back-end of a customer-facing visual search system. Also, SIR is used by data scientists to augment their training datasets with similar images. They often deal with machine learning problems where the data distribution across classes is highly skewed. This tool helps find training examples for poorly represented classes.

We optimize SIR for two objectives: search accuracy and query performance. We achieve high search accuracy by finding the most optimal deep learning based embedding after examining a few candidates. We achieve near real-time performance by encoding and indexing those embeddings in a scalable manner. The following sections of the paper delve into the technical details of the system and showcase its performance with appealing case studies of real applications.

2   Related Work

In recent years, content-based image retrieval from large data sets has bifurcated into two distinct approaches. Systems like FAISS [12] and NMSLib [13] treat embeddings as first class citizens. At the time of this writing, these systems are typically scaled vertically by taking advantage of GPU based parallelisms. On the other hand, the older, mature search systems like Elasticsearch and Solr come with built-in support for scaling text-based searches to millions of documents. The above two approaches have been empirically compared by Mu et al. [15]. As combining image and text searches (multi-modality) was an integral part of our overall solution, we decided on leveraging the second approach. Also, given the size of our catalog, a distributed system with in-built sharding was preferred.

Many state-of-the-art image retrieval systems rely on very high dimensional features, known as *embeddings* extracted either from a pre-trained network or by fine-tuning a deep neural network [2]. Our system has experimented with a number of popular models such as VGG16 [17], Resnet50-v2 [10], Inception-v2 [18] and EfficientNet [19]. Deep learning based hashing, binarizing or a combination of them [3,4,23] are applied to the embeddings to reduce their storage cost and to improve the retrieval performance of indexes built on them.

From core functionality perspective, our system is a close neighbor to the visual search systems developed by various e-commerce companies [11,20,22]. However, a very important difference between those and SIR is that we apply our system to internal stakeholders; hence, the user interaction flow and other design choices are optimized for them. The actions taken with our system's results are very different from that of customer-facing visual search platforms.

3 Technical Details

The core of the system (Fig. 1) revolves around fingerprinting every image to capture salient features and persisting it in a search index that would allow for efficient search and retrieval of nearest neighbors. We have the ability to use shallow fingerprinting techniques like phash [21] or deep learning based embeddings [17]. Most of our use cases require the ability to be invariant to slight changes in the image including positional and rotational variations. Also, the deeper and semantic aspects to embedding based fingerprinting was preferred.

3.1 Embedding Generation

In this step, we convert each image into a *fingerprint* or *signature* or *unique descriptor*. Under the hood, the fingerprints are essentially embeddings computed from a suitable deep neural network. We have experimented with a large number of techniques for embedding generation and settled down on VGG16 as our primary network. The embeddings are taken off the final fully connected layer of VGG16.

3.2 Index Creation

A typical embedding is a high dimensional vector consisting of floating point numbers. At search time, both the high dimensionality and the need to numerous floating point comparisons are big hindrances to a scalable, near real-time

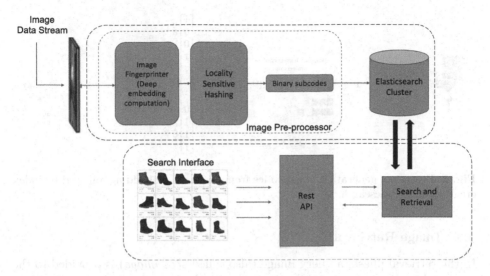

Fig. 1. System architecture of similar image retrieval tool (SIR). The top block outlines the back-end process of computing, encoding, and indexing embeddings. The bottom block shows the search interface.

implementation. In order to solve this, we employ a variant of locality sensitive hashing (LSH) [9] that binarizes the embedding vector by partitioning the embedding space using random projections. By constraining the dimensionality of this new binarized space, we can also mimic the effects of dimensionality reduction on the scalability of the search system. In its binarized form, the embeddings are further split into smaller subcodes [16] and ingested into the Elasticsearch index (Fig. 2). The subcoding enables us to take advantage of the pigeonhole principle and enforce early abandonment of search candidates which in turn helps us achieve sub-linear search times [16]. At retrieval time, we also take advantage of Elasticsearch's ability to compute efficient hamming distance calculations in the form of bit operations.

Our product catalog is an ever-changing system. The business applications focus mostly on the images that were added to the catalog in last few weeks. Hence, we have designed the index creation as a rolling process so that the new and recently updated images are always indexed. The current deployed system listens to a Kafka [1] topic that streams new and updated images. On receiving an image, we compute its embedding, transform it into the suitable format and store in Elasticsearch [7]. A rolling index (last 3 months) of newly created images is maintained for subsequent search and retrieval.

The rolling nature of the application makes hash-based indexing a preferable choice over techniques that learn representations collectively from a static dataset such as principal component analysis (PCA). As the catalog changes, the optimal principal components change as well, requiring frequent re-computation of them.

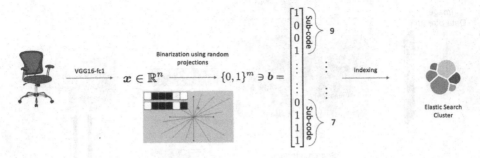

Fig. 2. Process of generation of subcodes from an image embedding followed by index creation on Elasticsearch.

3.3 Image Retrieval

In the retrieval phase, a query image (also called *seed image*) is provided to the system through the front end. In the back end, the query image is converted into an embedding and its nearest neighbors are retrieved from the indexed store. The retrieved images are presented in a grid in order of similarity with the query. Each result image is shown with a checkbox, allowing the user to select only the relevant ones from the grid.

4   Applications

SIR is designed as a generic image-based similarity platform. The analysis of the core algorithm can be found in Mu et al. [16]. In this paper, we focus on two implementations of SIR deployed to address two business application. Its performance is a function of a number of factors including the data on which the index is created. Hence, we present the system's performance in the context of specific applications.

4.1   Non-compliant Product Detection

In a large company like Walmart, it is a common practice to identify offensive themes in products and mark them on a regular basis. Given the size and the diversity of our catalog, this daunting task is akin to finding a needle in a haystack. Product search deals with two distinct types of themes. The first one is characterized by well-defined requirements, with a decent availability of training data. Also, these themes are usually relevant throughout the year. Hence, we address this type of themes by building supervised models [5,8]. The second type of themes is characterized by ill-defined requirements. They are usually volatile and relevant for a short period of time (e.g., unauthorized sale of products at a specific time of the year). Given the sense of urgency which they come with, training and bootstrapping a new model is often too slow. Also, only one or two examples are usually available, hence finding enough data for training a model is nearly impossible. SIR is an ideal solution for addressing these ephemeral themes. We have built and deployed a platform with SIR at the core to address such issues. The platform consists of following two modes of operation:

- **Streaming:** The new products that get added to our catalog need to be constantly monitored for various issues. We provide a version of SIR that leverages the image similarity technique described in Sect. 3 to quickly identify such issues in new products. We accomplish this by listening to triggers that are generated as new products get ingested into our catalog. We fingerprint each new product and store them in an Elasticsearch index. The index is engineered to have a rolling window (currently set at past 3 months) of new products.
- **Full Catalog Scan:** In addition of checking new products, business often needs to scan large parts of the catalog to find products similar to an example at hand. For this purpose, we provide a portal where an analyst can define rules based on image and text. An example rule would be an image of an e-cigarette and a filter that says "product title contains e-cigarette". We use these rules to fetch parts of the catalog and then scan them in more detail. The fingerprinting technique discussed in Sect. 3 is used in two distinct ways in this portal. **Simulation:** Given a rule, we first scan a rolling index of sampled products to provide real-time feedback on the effectiveness of a rule as it get defined by the analyst. This is accomplished by maintaining an index of a good representation of the catalog. The simulation results help the analyst fine tune her rule and also the similarity thresholds that would lead to expected precision and recall. **Sweep:** Once the simulation is done, a full-fledged scan of the catalog is triggered, preferably on a GPU cluster, where we stream and compare every product to the set of finalized rules defined by the analyst. Empirically, the combination of image fingerprinting and text based filtering has proven to be very effective in identifying and flagging offensive products.

Analysis of Search Quality. We present the precision-recall characteristics of the image similarity technique in the context of our trust and safety application.

Given an application, search quality of the image similarity technique is dependent on two factors: the model used to generate embeddings and the level of binarization. We have repeated the following experiment for five different types of embeddings. For each embedding type, we populated the Elasticsearch with the embeddings from 1.5 million images of top-selling products. In order to test, we also ingested 10000 offensive images that were related to 3600 query images. For each query image, we compared the retrieved results against the ground truth to compute three metrics: Mean R-Precision, Mean Average Precision@K and Recall.

- R-Precision [6] is useful when the number of relevant images varies from query to query. For example, if R images are relevant to a query image, R-precision (r/R) is computed based only on the top R images returned by the system. Mean R-Precision is the average of R-Precisions of all the queries.

- Precision@K [14] is computed based on the first K returned results. Average Precision@K is the average of AP for 1 to K. Average precision takes into account the position of the relevant documents, making it very useful for measuring the quality of search systems. Mean average precision@K is the average of AP@K over all the queries.
- We also compute approximate recall@1000 with the assumption that all the relevant documents are either returned within top 1000 or not.

The results of this experiment is shown in Table 1. As the table indicates, we achieved best search quality with VGG16 embeddings. We then experimented with two binarized variations of VGG16 to understand the impact of binarization on search quality. The results, presented in Table 2, indicate that both MAP and R-Precision is impacted by only about 2% with the subcoded embeddings.

Table 1. Comparison of SIR search quality for different embedding types

Embedding type	MAP@1	MAP@5	MAP@10	Mean R-Precision	Approx. recall
VGG16	0.993	0.79	0.779	0.827	0.989
Inception-v2	0.993	0.688	0.663	0.711	0.801
ResNet-50v2	0.993	0.774	0.761	0.81	0.986
EfficientNet-b4	0.995	0.592	0.576	0.62	0.631
Custom	0.992	0.772	0.76	0.806	0.993

Table 2. Comparison of SIR search quality for different levels of subcodings

Embedding type	MAP@1	MAP@5	MAP@10	Mean R-Precision	Approx. recall
VGG16	0.993	0.79	0.779	0.827	0.989
VGG16 with 512 subcodes	0.993	0.784	0.774	0.824	0.984
VGG16 with 256 subcodes	0.993	0.772	0.758	0.806	0.979

Analysis of Query Response Time: We also study the trade-off between search quality and query performance in the context of the same application. As Table 3 indicates, The mean query time of VGG16 is higher than Inception-v3 and Resnet50-v2. However, the mean query time reduces by 20% as we switch to 512 subcodes of VGG16 from original VGG16 embeddings. It reduces by a massive 70% as we move to 256 subcodes of VGG16. Table 2 has shown that this performance gain has been achieved with less than 2% reduction in precision and recall.

Table 3. Comparison of SIR Elastic Search Query times for different embeddings.

Embedding type	Min time (ms)	Max time (ms)	Mean time (ms)	Total time (hrs)
ResNet-50v2	4764	5608	5201.644	5.538
Custom	4448	5920	5458.125	5.500
Inception-v3	4227	5552	4622.075	4.525
VGG16	4892	6117	5600.235	5.560
VGG16 with 512 subcodes	4112	4827	4481.130	4.449
VGG16 with 256 subcodes	1568	1862	1631.270	1.112

Performance Improvement with Text Filters: In real applications, both image and text (mainly product title) contain information valuable for search. We have experimented with composite indexes. In this version, we create an image-based index as described earlier as well as a traditional Elasticsearch index of textual keywords. At the query retrieval phase, the image search space is first narrowed down by applying keyword-based text filters. Figure 3 shows the trends of query performance with and without the text filter. As the data size on which the indexes are built increases, the benefit of text filters on top of images becomes apparent. This early but promising result has opened up possibilities of turning this application into a multi-modal one.

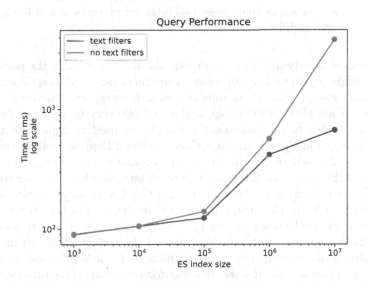

Fig. 3. Query performance on different index sizes with and without text based filtering

4.2   Variant Detection

When shopping online, customers expect all item variants, for example, the same T-shirt in different sizes, on one item page so that they can easily make a well-informed shopping decision. Incorrect variant assortment in the item pages could result in poor customer experience and affect GMV due to an increased bounce rate as customers leave the site without any action. Traditionally, internal experts have been manually creating, consolidating, and updating the variant groups. This task is error-prone and time-consuming due to the volume of our catalog. To increase the variant grouping accuracy and efficiency, we developed a machine learning system to automatically generate variant suggestions so that experts only need to review a set of suggested variants instead of exploring the entire catalog.

This variant grouping system consists of two stages: high-recall stage and high-precision stage. In the first stage, given a query product, a set of candidate variants is generated to narrow down the variant search space from the entire catalog to a few hundreds or thousands of products. In the second stage, high-precision classifiers are used to identify variants from the candidate set previously generated. In the first stage, a text similarity search was originally in place to retrieve candidates with similar product name and descriptions as the reference item. We deployed an implementation of the similar image retrieval (SIR) system to retrieve candidates that are visually similar to the reference item. Our hypothesis was that these two retrieval systems would fetch complementary variant candidates.

Performance Analysis: To test our hypothesis, we measured the performance of the candidate generation system on a production-level dataset consisting of about 5,000 groups from thousands of product categories. For each reference item, we fetched about 1000 image and about 500 text based candidates (this discrepancy is due to the limitation of the library used to implement the text-retrieval system) independently and then combined them as well. It turned out that the recall based on the image-based candidates were already 13% higher than that of the text-alone retrieval. The recall increased by 24% after combining text and image-based candidate. Even with the discrepancy mentioned above, the numbers indicate that the image-based retrievals add significant value to the system. Figure 4 shows an example where the variant is retrieved in the image SIR but is missed in text-alone retrieval. Though the text information for both products are semantically similar, the actual words, phrases, and writing style are so different that it is challenging for text-alone retrieval system. This challenge is prevalent in marketplace settings where multiple sellers for a single product are active. For such cases, product images are less subjective and harder to modify, hence image-base retrievals are critical.

We take a *suggestion of suggestions* strategy where a few image candidates are retrieved for each text candidate. This strategy has two parameters: number of text-based candidates (shown as NMSLIB neighbors in the plots), namely N and the number of image-based nearest neighbors for each text-based candidate,

Attribute	Reference	Product
title	Progressive Furniture Melrose Panel Bed	Progressive Melrose Queen Panel Bed in Driftwood
image		
product_id	6DD0O2O1U1IE	6X02L27UZ4V8
brand	Progressive Furniture	Progressive Furniture
manufacturer	Progressive Furniture	Progressive Furniture Inc
manufacturer_part_number	P604-94/95/78	P604-34/35/78
model	P604-94	P604-34/35/78
size	King	King
actual_color	Multicolor	Queen: 85 in. L x 64 in. W x 53 in. H
product_short_description	Surround yourself in the natural beauty of pine. The rustic **Progressive Furniture Melrose Panel Bed** will give your master bedroom a log cabin feel and with a serene setting comes a more pleasant night's sleep. This panel bed includes headboard, footboard, and rails crafted from solid pine that flaunts its natural woodgrain, color, notches, and "flaws" that give it unique character. Multiple size options are available to match your mattress and box springs.	The natural beauty of pine is obvious when allowed to stand on its own. Melrose was designed to let the wood tell its own story. Architectural lines and the natural color of pine provide a unique look. Flaws in the wood are valued and considered to be of positive character. The flaws create character and allows the wood to look natural. The bed features hand applied distressing on rough hewn and planed pine solids to achieve a natural, vintage look.
	Dimensions • Queen Headboard: 64W x 2D x 53H in. • Queen Footboard: 64W x 2D x 20H in. • King Headboard: 81W x 2D x 52H in. • King Footboard: 81W x 2D x 20H in. • Rails: 01W x 1D x 6H in.	**Features:** • Finish: Driftwood • Materials and Construction: Salvaged Pine solids • Style: Transitional • Box Spring Required **Specifications:** • Overall Product Dimensions: 53" H x 64" W x 85" D • Overall Product Weight: 89.6 lbs

Fig. 4. The query product (on the right) and the variant candidate (on the left) have similar product image, but different product text information, especially the product descriptions. This variant candidate is only retrieved by SIR.

namely K. If we keep increasing N, a single retrieval system tends to saturate as the blue line shown in Fig. 5. The recall increases little even though the number of searched items is doubled. Our experiment shows that a "suggestion of suggestions" approach can efficiently break through this saturation. Specifically, each retrieved text-based candidate becomes the reference item for SIR. The top K neighbors retrieved in SIR are added into the candidate set. Figure 5 shows how recall is increased for k = 1, 2, 3, 4. Figure 6 shows that this "suggestion of suggestions" strategy significantly increase the recall that surpass the saturated point with insignificant increment in the number of fetched candidates. The y-axis of this plot shows the number of candidates averaged over all the queries used for testing.

4.3   Visual Examples from Applications

Both the above mentioned applications regularly use SIR to discover products similar to a query example that is of interest to one of the users. The retrieval is based on an index of products that entered the catalog in the last one month or so. Figure 7 showcases a few such examples and corresponding search results.

The top one with a table lamp, Fig. 7a, shows how SIR retrieves similar products with subtle variations in shape, size, and color. Such variations either make them variants of the queried product or help in discovering products with a certain shape or style catering to the customer's choice.

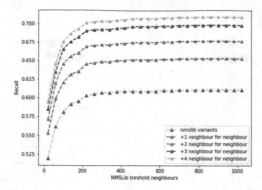

Fig. 5. Recall of any retrieval system saturates as we keep increasing number of candidates. Combining image-based retrieval with text elevates recall significantly over a text-alone system.

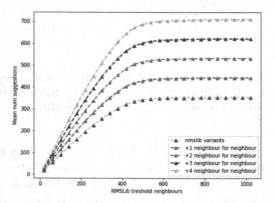

Fig. 6. Image and text together increases recall of text retrieval system with relatively small increase in number of candidates fetched.

Figure 7b showcases a search for sports t-shirts where variations in the text on the t-shirt are captured. This is a relatively difficult example since our underlying embeddings were not trained to detect letters and numbers. However, SIR still identifies the graphic as a key feature and is able to fetch t-shirts with a similar graphic.

The third example in Fig. 7c is an attempt to find cosmetic products, most likely coming from different brands. All the results returned are near duplicates to the query, but not identical.

The last example in Fig. 7d retrieves tablet computers based on an example. This example actually highlights a limitation of our system. Since the underlying embeddings do not recognize the content being shown inside the screen of the tablet, it retrieved images primarily based on the overall shape of the object. If the user were looking for other tablets showing similar content on the screen, this search result would not satisfy her.

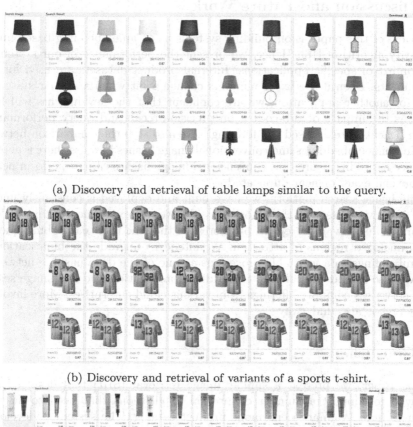

(a) Discovery and retrieval of table lamps similar to the query.

(b) Discovery and retrieval of variants of a sports t-shirt.

(c) Retrieval of cosmetic products that are likely to be variants from different brands.

(d) Retrieval of tablets similar to an example

Fig. 7. A number of examples demonstrating product search capability of SIR

5   Discussion and Future Work

The focus of this paper is on building systems using an embedding based image similarity algorithm [16]. Hence, we present the performance of our system in the context of specific applications. Internal product datasets are used for the experiments. Even if the exact numbers change a bit when a similar system is built for another application, we are confident that the key insights will hold (such as the impact of sub-coding on precision-recall and query performance, or the benefit of text-based pre-filtering). We also skip the comparison between deep learning embeddings and conventional image hashes because there is enough evidence in the literature that conventional image hashes cannot perform nearly well beyond exact or near duplicated.

The image search platform we have built and deployed is constantly undergoing improvements. On the algorithm side, we are experimenting on making the embeddings aware of regions of interest so that the users can submit queries with annotated regions of attention. We intend to upgrade the binarization of the embeddings to a learnable process using one of the deep hashing networks. Can we also intend to scale the embedding computation and the image search using serverless compute offerings from various cloud enterprises. More involved text search is also underway.

6   Conclusion

In this paper, we present a similar image retrieval (SIR) tool designed and deployed to support a number of internal applications that need to discover products from Walmart's enormous product catalog. The system is developed by skillfully combining knowledge of deep learning, data management, and user experience. The core idea behind SIR can be used to build similar visual search tools for many other domains.

References

1. Apache Software Foundation: Apache Kafka (2011). https://kafka.apache.org/
2. Babenko, A., Slesarev, A., Chigorin, A., Lempitsky, V.: Neural codes for image retrieval (2014)
3. Cao, Y., Long, M., Liu, B., Wang, J.: Deep cauchy hashing for hamming space retrieval. In: 2018 IEEE/CVF Conference on Computer Vision and Pattern Recognition, pp. 1229–1237 (2018)
4. Cao, Y., Long, M., Wang, J., Liu, S.: Deep visual-semantic quantization for efficient image retrieval. In: 2017 IEEE Conference on Computer Vision and Pattern Recognition (CVPR), pp. 916–925 (2017)
5. Chaudhuri, A., et al.: A smart system for selection of optimal product images in e-commerce. CoRR abs/1811.07996 (2018). http://arxiv.org/abs/1811.07996
6. Craswell, N.: R-Precision, p. 2453. Springer Boston (2009)
7. Elastic.co: Elasticsearch (2010). https://www.elastic.co/elasticsearch/

8. Gandhi, S., et al.: Scalable detection of offensive and non-compliant content/logo in product images. In: IEEE Winter Conference on Applications of Computer Vision, WACV 2020, Snowmass Village, CO, USA, 1–5 March 2020, pp. 2236–2245. IEEE (2020)

9. Gionis, A., Indyk, P., Motwani, R.: Similarity search in high dimensions via hashing, pp. 518–529 (1999)

10. He, K., Zhang, X., Ren, S., Sun, J.: Deep residual learning for image recognition. CoRR abs/1512.03385 (2015). http://arxiv.org/abs/1512.03385

11. Jing, Y., et al.: Visual search at pinterest. CoRR abs/1505.07647 (2015). http://arxiv.org/abs/1505.07647

12. Johnson, J., Douze, M., Jégou, H.: Billion-scale similarity search with GPUs. arXiv preprint arXiv:1702.08734 (2017)

13. Malkov, Y.A., Yashunin, D.A.: Efficient and robust approximate nearest neighbor search using hierarchical navigable small world graphs (2016)

14. Manning, C.D., Raghavan, P., Schütze, H.: Introduction to Information Retrieval. Cambridge University Press, Cambridge (2008)

15. Mu, C., Yang, B., Yan, Z.: An empirical comparison of FAISS and FENSHSES for nearest neighbor search in hamming space. CoRR abs/1906.10095 (2019). http://arxiv.org/abs/1906.10095

16. Mu, C., Zhao, J., Yang, G., Yang, B., Yan, Z.: Fast and exact nearest neighbor search in hamming space on full-text search engines (2019)

17. Simonyan, K., Zisserman, A.: Very deep convolutional networks for large-scale image recognition. CoRR abs/1409.1556 (2014)

18. Szegedy, C., Vanhoucke, V., Ioffe, S., Shlens, J., Wojna, Z.: Rethinking the inception architecture for computer vision. CoRR abs/1512.00567 (2015). http://arxiv.org/abs/1512.00567

19. Tan, M., Le, Q.V.: EfficientNet: rethinking model scaling for convolutional neural networks. CoRR abs/1905.11946 (2019). http://arxiv.org/abs/1905.11946

20. Yang, F., et al.: Visual search at eBay. CoRR abs/1706.03154 (2017). http://arxiv.org/abs/1706.03154

21. Zauner, C.: Implementation and benchmarking of perceptual image hash functions (2010)

22. Zhang, Y., et al.: Visual search at alibaba. In: Proceedings of the 24th ACM SIGKDD International Conference on Knowledge Discovery & Data Mining, KDD 2018, pp. 993–1001 (2018)

23. Zhu, H., Long, M., Wang, J., Cao, Y.: Deep hashing network for efficient similarity retrieval. In: Proceedings of the Thirtieth AAAI Conference on Artificial Intelligence, AAAI 2016, pp. 2415–2421. AAAI Press (2016)

Cross-Resolution Deep Features Based Image Search

Fabio Valerio Massoli[✉][ID], Fabrizio Falchi[ID], Claudio Gennaro[ID],
and Giuseppe Amato[ID]

ISTI-CNR, via G. Moruzzi 1, 56124 Pisa, Italy
{fabio.massoli,fabrizio.falchi,claudio.gennaro,
giuseppe.amato}@isti.cnr.it

Abstract. Deep Learning models proved to be able to generate highly discriminative image descriptors, named deep features, suitable for similarity search tasks such as Person Re-Identification and Image Retrieval. Typically, these models are trained by employing high-resolution datasets, therefore reducing the reliability of the produced representations when low-resolution images are involved. The similarity search task becomes even more challenging in the cross-resolution scenarios, i.e., when a low-resolution query image has to be matched against a database containing descriptors generated from images at different, and usually high, resolutions. To solve this issue, we proposed a deep learning-based approach by which we empowered a ResNet-like architecture to generate resolution-robust deep features. Once trained, our models were able to generate image descriptors less brittle to resolution variations, thus being useful to fulfill a similarity search task in cross-resolution scenarios. To asses their performance, we used synthetic as well as natural low-resolution images. An immediate advantage of our approach is that there is no need for Super-Resolution techniques, thus avoiding the need to synthesize queries at higher resolutions.

Keywords: Cross resolution · Similarity search · Deep convolutional neural networks · Image retrieval

1   Introduction

Content-Based Image Retrieval (CBIR) is one of the most active research fields in the multimedia community [9,20]. Key aspects that greatly affect a CBIR system performance are the quality of the used image descriptors and its ability to scale. Before the advent of Deep Learning (DL) techniques, the Scale-Invariant Feature Transform (SIFT) [16] based methods were among the most frequently used to generate image descriptors. It has only been after the breakthrough in 2012 [14] that the scientific community has turned its attention towards DL techniques as a possible approach to the Image Retrieval (IR) problem [3,19].

S. Satoh et al. (Eds.): SISAP 2020, LNCS 12440, pp. 352–360, 2020.
https://doi.org/10.1007/978-3-030-60936-8_27

Since then, DL algorithms were employed in a variety of other fields such as object recognition [10], speech recognition [7], natural language processing [8], etc. Among the various architectural designs, Convolutional Neural Networks (CNNs) experienced the greatest success in the field of computer vision-related tasks. These models are largely used to fulfill CBIR tasks, too, thanks to their ability to create image representations, called deep features, that can be employed as global descriptors for similarity searches [2,9]. Despite their success, a well-known problem of CNN models is that the discriminative ability of the extracted features degrades when a model is fed with low-resolution images [17,18,22]. A reasonable explanation for this issue is that the datasets typically used to train the CNNs to contain images predominantly at high resolutions.

To overcome this issue, in this paper, we presented the approach we employed to solve the cross-resolution IR task. Specifically, we experimented with the effectiveness of our method in the scenario of Face Image Retrieval (FIR), which is of particular concern, for example, for surveillance systems [6,22] that rely on probe images extracted from cameras with limited resolution. To conduct our experiments, we leveraged a ResNet-50 architecture [11], equipped with Squeeze-and-Excitation blocks [12], pre-trained on the VGGFace2 dataset [5]. Starting from the state-of-the-art model [5], we fine-tuned it to make its deep features less brittle to resolution variations of the input images.

The remaining part of the paper is organized as follows. In Sect. 2, we briefly reviewed some related works. In Sect. 3, we presented the experimental procedure alongside the results. In Sect. 4, we concluded the paper with a summary of the main results and future perspectives.

2 Related Works

Before the advent of the Machine Learning (ML), CBIR was based on the extraction and use of low-level feature descriptors, such as color and edge features [13] or local features [4,16]. With the advent of DL models, researchers started to use their inner activations, called deep features, as descriptors for the input images. The similarity search was then directly carried out among features employing specific metrics.

Lin et al. [15] proposed a two steps-based framework in which they did a first coarse-level search followed by a fine-level one. For each query image, the authors considered two features vectors: a binary and global ones. The former was employed to perform a quick search in the database, while the latter was used to perform the fine-level search. In Ahmad et al. [1], the authors proposed a bilinear model in which image features were accumulated at various locations and scales using the convolutional activations extracted from different inner layers of a CNN. With such an approach, the authors were able to extract image descriptors with high discriminative power. Tzelepi et al. [21] proposed a method to retrain a model to empower it to generate descriptors better suited · for CBIR applications. Specifically, they proposed three basic model retraining approaches: Fully Unsupervised Retraining; Retraining with Relevance Information; Relevance Feedback based Retraining.

3 Experiments

Our starting point was the state-of-the-art SeNet-50 architecture [5] trained on the VGGFace2 [5] dataset. Subsequently, we fine-tuned it employing our training procedure, and finally, we tested the performance of the model on the FIR task using the deep features extracted from its penultimate layer.

3.1 Dataset

The VGGFace2 [5] dataset consists of ~3.31 million images shared among ~9K identities. It is divided into two splits, one for training and one for test purposes only. The latter contains ~170K images divided into 500 identities, while all the other images belong to the remaining ~8K classes available for training. The entire dataset is characterized by a very low label noise and by a high intra-class variance, especially among head poses. These characteristics make it a suitable choice for training DL models on face-related tasks. Despite these qualities, the dataset mainly comprises high-resolution images. Indeed, training set images have an average resolution of ~137 × 180 pixels with less than 1% at a resolution below 32 pixels. As it is shown later, this makes the internal representations of a neural network trained on this dataset brittle to resolution variations in the input data.

3.2 Training Details

To fine-tune the model, we made a first trial in which we kept the entire net frozen except for the last fully connected layer, and we fed it with low-resolution images. To obtain the desired inputs, we leveraged the bilinear interpolation algorithm, implemented in the PIL python library, to down-sample the images at a specific (low) resolution. However, we obtained better results by fine-tuning the entire neural network. The intuition was that there were patterns in the new low-resolution images that the model needed to adjust for. We initially set the value of the learning rate at $5 \cdot 10^{-4}$ and dropped its value by a factor of 5 every time the loss reached a plateau. We used a batch size of 256, a weight decay of 10^{-5}, and a momentum of 0.9.

By only using low-resolution input images, we noticed that the models lost the ability to recognize images at high resolution. For this reason, we introduced a new hyperparameter to control the probability with which to down-sample an image. In this way, the CNNs were trained on both low and high-resolution images at the same time. More specifically, the algorithm we employed to resize images was based on two random extractions: the first one used to decide whether or not to give the image at full resolution, and the second one utilized to set the final resolution at which the image would then be resized in the [8] pixels range.

After this first preprocessing phase, the input images were resized so that the size of the shortest side was 256 pixels, later a random crop was applied to select a 224 × 224 pixels region which matches the input of the network. We split the training dataset into training and validation sets. Specifically, during the training phase, we employed two versions of the latter to monitor the performance of the model on both low and high-resolution domains.

3.3   Similarity Search

After the training phase, we assessed the models' performance on the FIR task. We trained several models with different values of the hyperparameter that controlled the probability with which an image was down-sampled. Specifically, we considered probabilities of 0.1, 0.3, 0.5, 0.7 and 1.0. For each scenario, we took the best models and compared their performance on the FIR task with the baseline model.

3.4   Experimental Results

In Fig. 1, we reported a comparison between the queries results obtained by the original pre-trained model and by our fine-tuned ones. Specifically, the results correspond to the first five images returned when the query was downsampled at resolutions of 16 pixels (shortest side). Clearly, the deep representation produced by our model was much more discriminative compared to the one generated by the original model.

Fig. 1. Comparison of the top query results returned by the original and a fine-tuned model considering a query resolution of 16 pixels.

In Fig. 2 and Fig. 3, we reported the precision and the recall scores, for each fine-tuned model and for the original pre-trained one. Each plot in the figures corresponds to a specific query resolution. As it is clear from Fig. 2 and Fig. 3, the original pre-trained model experienced a noticeable degradation of its performance when tested against a cross-resolution scenario. With our approach, we have been able to improve upon its performance up to about one order of magnitude, considering queries with resolutions down to 8 pixels, with a negligible loss at higher resolutions.

Finally, in Table 1, we showed the results from the mean Average Precision (mAP) measurements, as a function of the query resolution, for each of the fine-tuned models. Moreover, the first column of the table reported the mAP for the original pre-trained model as a term of comparison. According to Table 1, it is remarkable to notice that our models had higher performance concerning the original pre-trained model in the range between 8 and 24 pixels.

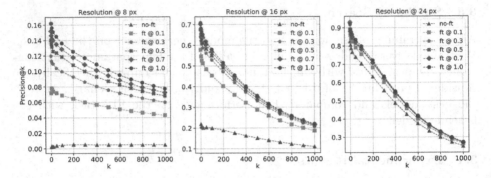

Fig. 2. Precision@k. The baseline model has been reported as "no ft" while the "ft" models are the fine-tuned ones. The value after the "@" symbol represents the probability we used during training to decide whether reducing or not the input resolution.

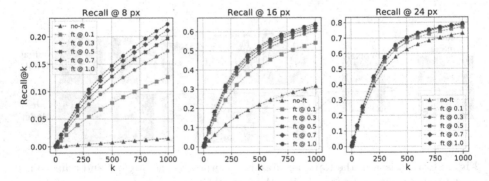

Fig. 3. Recall@k. The baseline model has been reported as "no ft" while the "ft" models are the fine-tuned ones. The value after the "@" symbol represents the probability we used during training to decide whether reducing or not the input resolution.

So far, we have considered down-sampled images that have been generated by interpolation methods. Other than that, it is interesting to test our models' performance on datasets composed of native low-resolution images. As a preliminary study, we considered the training set available from the TinyFace dataset[1].

---

[1] https://qmul-tinyface.github.io/.

Table 1. mAP results, as function of the query resolution, for each model. The baseline model has been reported as "no ft" while the "ft" models are the fine-tuned ones. The value after the "@" symbol represents the probability we used during training to decide whether reducing or not the input resolution.

		no ft	ft @ 0.1	ft @ 0.3	ft @ 0.5	ft @ 0.7	ft @ 1.0
Query resolution (pixels)	8	0.01	0.05	0.08	0.09	0.09	**0.10**
	16	0.15	0.34	0.40	0.43	0.42	**0.44**
	24	0.56	0.62	0.64	0.65	0.65	**0.66**
	32	0.74	0.75	0.73	0.74	0.75	**0.76**
	64	**0.85**	0.84	0.84	0.84	0.84	0.84
	128	**0.86**	0.85	0.84	0.85	0.84	0.85
	256	**0.86**	0.85	0.84	0.85	0.84	0.85

It contains about 8K images with an average height of 20 pixels, distributed among ∼2K different identities. Following the same procedure adopted in the previous measurements, we randomly selected one image for each class as a query, thus extracting the deep features from them and all the other images in the set, to construct the descriptors database. Differently from what was done previously, we did not apply any down-sampling algorithm in this case, since the images were already at low resolution. The precision and recall results were reported in Fig. 4, while in Table 2 we reported the mAP values for the fine-tuned models as well as for the original pre-trained model.

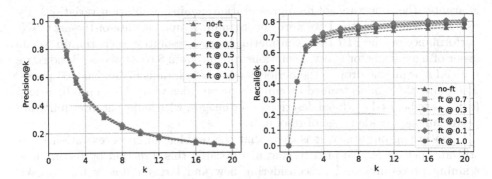

Fig. 4. Precision@k (left) and Recall@k (right) for each fine tuned model. The baseline model has been reported as "no ft" while the "ft" models are the fine-tuned ones. The value after the "@" symbol represents the probability we used during training to decide whether reducing or not the input resolution.

From the results showed in Table 2 it is noticeable that, even though we trained our models on synthetic low-resolution images, their performance was consistent when tested against native low-resolution images and still higher than the original pre-trained model, thus confirming the effectiveness of our training procedure.

Table 2. mAP results obtained for each model.

no ft	ft @ 0.1	ft @ 0.3	ft @ 0.5	ft @ 0.7	ft @ 1.0
0.68	0.73	0.72	0.74	0.74	**0.75**

4   Conclusion and Future Perspectives

In this paper, we proposed a strategy to train a DL model to generate image descriptors robust against a cross-resolution scenario that can be used to fulfill IR tasks. Specifically, to assess the effectiveness of our method, we considered the task of the FIR, being of particular interest for applications such as surveillance systems. Indeed, in such cases, a probe image that is typically acquired from a security camera has to be matched against a database of known identities, typically characterized by high-resolution image descriptors. Since the security cameras do not usually shoot at very high resolution, and considering that they are often far from the scene, the extracted image can be at resolutions as low as 16 pixels, or even below.

We showed that training a model on a vast dataset, even though it has low noise level and high intra-class variance such as VGGFace1 [5], does not guarantee the robustness of its representations against resolution variations. By using our training method, we were able to improve upon a state-of-the-art CNNs performance on the FIR task, considering a cross-resolution scenario, up to one order of magnitude for query resolutions ranging from 8 to 24 pixels. Besides, we noticed a negligible drop in the performance at resolutions higher than 64 pixels. Therefore, the models trained by embodying our idea were able to produce deep features to be used as global descriptors for images, with sufficient discrimination power among a wide range of resolutions.

Concerning our study, it is clear that the problem of cross-resolution IM is still an open issue. We plan to continue towards this direction by testing new training procedures and by considering new and larger datasets that consist of native low-resolution images to train and test the CNNs in more realistic situations.

Acknowledgments. This research has been funded by the Istituto di Scienza e Tecnologie dell'Informazione "A. Faedo" (ISTI) of the National Research Council (CNR). We gratefully acknowledge the support of NVIDIA Corporation with the donation of the Titan V GPU used for this research.

References

1. Alzu'bi, A., Amira, A., Ramzan, N.: Content-based image retrieval with compact deep convolutional features. Neurocomputing **249**, 95–105 (2017)
2. Amato, G., Falchi, F., Gennaro, C., Vadicamo, L.: Deep permutations: deep convolutional neural networks and permutation-based indexing. In: Amsaleg, L., Houle, M.E., Schubert, E. (eds.) SISAP 2016. LNCS, vol. 9939, pp. 93–106. Springer, Cham (2016). https://doi.org/10.1007/978-3-319-46759-7_7
3. Babenko, A., Slesarev, A., Chigorin, A., Lempitsky, V.: Neural codes for image retrieval. In: Fleet, D., Pajdla, T., Schiele, B., Tuytelaars, T. (eds.) ECCV 2014. LNCS, vol. 8689, pp. 584–599. Springer, Cham (2014). https://doi.org/10.1007/978-3-319-10590-1_38
4. Bay, H., Tuytelaars, T., Van Gool, L.: SURF: speeded up robust features. In: Leonardis, A., Bischof, H., Pinz, A. (eds.) ECCV 2006. LNCS, vol. 3951, pp. 404–417. Springer, Heidelberg (2006). https://doi.org/10.1007/11744023_32
5. Cao, Q., Shen, L., Xie, W., Parkhi, O.M., Zisserman, A.: VGGFace2: a dataset for recognising faces across pose and age. In: 2018 13th IEEE International Conference on Automatic Face & Gesture Recognition (FG 2018), pp. 67–74. IEEE (2018)
6. Cheng, Z., Zhu, X., Gong, S.: Surveillance face recognition challenge. arXiv preprint arXiv:1804.09691 (2018)
7. Chiu, C.C., et al.: State-of-the-art speech recognition with sequence-to-sequence models. In: 2018 IEEE International Conference on Acoustics, Speech and Signal Processing (ICASSP), pp. 4774–4778. IEEE (2018)
8. Deng, L., Liu, Y. (eds.): Deep Learning in Natural Language Processing. Springer, Singapore (2018). https://doi.org/10.1007/978-981-10-5209-5
9. Donahue, J., et al.: DeCAF: a deep convolutional activation feature for generic visual recognition. In: International Conference on Machine Learning, pp. 647–655 (2014)
10. Han, J., Zhang, D., Cheng, G., Liu, N., Xu, D.: Advanced deep-learning techniques for salient and category-specific object detection: a survey. IEEE Signal Process. Mag. **35**(1), 84–100 (2018)
11. He, K., Zhang, X., Ren, S., Sun, J.: Deep residual learning for image recognition. In: Proceedings of the IEEE Conference on Computer Vision and Pattern Recognition, pp. 770–778 (2016)
12. Hu, J., Shen, L., Sun, G.: Squeeze-and-excitation networks. In: Proceedings of the IEEE Conference on Computer Vision and Pattern Recognition, pp. 7132–7141 (2018)
13. Jain, A.K., Vailaya, A.: Image retrieval using color and shape. Pattern Recogn. **29**(8), 1233–1244 (1996)
14. Krizhevsky, A., Sutskever, I., Hinton, G.E.: ImageNet classification with deep convolutional neural networks. In: Advances in Neural Information Processing Systems, pp. 1097–1105 (2012)
15. Lin, K., Yang, H.F., Hsiao, J.H., Chen, C.S.: Deep learning of binary hash codes for fast image retrieval. In: Proceedings of the IEEE Conference on Computer Vision and Pattern Recognition Workshops, pp. 27–35 (2015)
16. Lowe, D.G.: Distinctive image features from scale-invariant keypoints. Int. J. Comput. Vision **60**(2), 91–110 (2004)
17. Massoli, F.V., Amato, G., Falchi, F.: Cross-resolution learning for face recognition. Image Vision Comput. 103927 (2020)

18. Massoli, F.V., Amato, G., Falchi, F., Gennaro, C., Vairo, C.: Improving multi-scale face recognition using VGGFace2. In: Cristani, M., Prati, A., Lanz, O., Messelodi, S., Sebe, N. (eds.) ICIAP 2019. LNCS, vol. 11808, pp. 21–29. Springer, Cham (2019). https://doi.org/10.1007/978-3-030-30754-7_3
19. Tolias, G., Sicre, R., Jégou, H.: Particular object retrieval with integral max-pooling of CNN activations. arXiv preprint arXiv:1511.05879 (2015)
20. Torresani, L., Szummer, M., Fitzgibbon, A.: Efficient object category recognition using classemes. In: Daniilidis, K., Maragos, P., Paragios, N. (eds.) ECCV 2010. LNCS, vol. 6311, pp. 776–789. Springer, Heidelberg (2010). https://doi.org/10.1007/978-3-642-15549-9_56
21. Tzelepi, M., Tefas, A.: Deep convolutional learning for content based image retrieval. Neurocomputing **275**, 2467–2478 (2018)
22. Zou, W.W., Yuen, P.C.: Very low resolution face recognition problem. IEEE Trans. Image Process. **21**(1), 327–340 (2011)

Learning Distance Estimators
from Pivoted Embeddings
of Metric Objects

Fabio Carrara[✉], Claudio Gennaro, Fabrizio Falchi, and Giuseppe Amato

ISTI-CNR, via G. Moruzzi 1, 56124 Pisa, Italy
{fabio.carrara,claudio.gennaro,fabrizio.falchi,
giuseppe.amato}@isti.cnr.it

Abstract. Efficient indexing and retrieval in generic metric spaces often translate into the search for approximate methods that can retrieve relevant samples to a query performing the least amount of distance computations. To this end, when indexing and fulfilling queries, distances are computed and stored only against a small set of reference points (also referred to as pivots) and then adopted in geometrical rules to estimate real distances and include or exclude elements from the result set. In this paper, we propose to learn a regression model that estimates the distance between a pair of metric objects starting from their distances to a set of reference objects. We explore architectural hyper-parameters and compare with the state-of-the-art geometrical method based on the n-simplex projection. Preliminary results show that our model provides a comparable or slightly degraded performance while being more efficient and applicable to generic metric spaces.

Keywords: Distance estimation · Metric spaces · Regression · Deep neural networks · Pivoted embeddings

1 Introduction

Thanks to the impetus given by recent developments in deep learning, machine learning has gained unprecedented popularity and spread into unimaginable number domains of computing. Perhaps one of the most unexpected application domains is that of index structures. In an exploratory research paper, Kraska et al. [8] showed how machine learning models, including deep learning ones, can fully or partially replace existing index structures, such as B-Tree or Bloom filters. This work paved the way for a whole new area of research in index structure, termed *learned index*. The core idea of learned indexes is to obtain a more compact index representation or performance gains by learning from data distribution.

In particular, Kraska et al. show that an index can be seen as a model f that predicts the position y of a record x, i.e. $y = f(x)$. Seen with the eyes of machine learning, this problem is known as a regression problem. Therefore, in

© Springer Nature Switzerland AG 2020
S. Satoh et al. (Eds.): SISAP 2020, LNCS 12440, pp. 361–368, 2020.
https://doi.org/10.1007/978-3-030-60936-8_28

this context a learned index is simply an ML model (such as linear regression or a neural network) that replaces a B-Tree and predicts the position of query key. In this work, we would like to generalize the concept of learned index by providing a broader definition, in which the regression allows us to predict the representation of metric objects in vector form from the knowledge of distances from some anchors. This type of transformation of data is useful in the problems of searching in large-scale metric spaces, where a compromise between accuracy and speed of response to queries is often required. In large-scale scenarios, the amount of distance computations between objects needed for an exact search tends to saturate the available computational budget for obtaining reasonable response times, considering also that in metric spaces, distance functions are often expensive to compute.

The idea of reconstructing the distance between any pair of objects in a metric space by exploiting distances with a group of reference objects was probably first addressed in [9]. The authors proposed an *embedding* into another metric space where it is possible to deduce upper and lower bounds on the actual distance of any pair of objects. Connor et al. [4–6] observed that for a large class of metric spaces, distances to a set of n pivots can be used to project the data objects into a n-dimensional Euclidean space such that in the projected space the Euclidean distance between any two points is an upper or lower bound of the actual distance. They called this approach n-*Simplex projection*, and they proved that it can be used in all the metric spaces meeting the n-*point property* [2]. As also pointed out in [3], many common metric spaces meet the desired property, like Cartesian spaces of any dimension with the Euclidean, cosine or quadratic form distances, probability spaces with the Jenson-Shannon or the Triangular distance, and more generally any Hilbert-embeddable space [2,10]. This approach has recently been used in an inverted index for approximate research on the n-nearest neighbors to obtain an estimate of the real distances of the objects present in the results of a query [12].

We intend to develop a similar approach to the n-simplex projection, however instead of using a handcrafted deterministic algorithm, in this paper we test a machine learning approach based on deep neural networks to probe their capabilities in this context. The rest of the paper is structured as follows. Section 2 describes the general idea of method developed and the model used. Section 3 presents the dataset and experimental evaluation. Section 4 concludes.

2   Method

2.1   Model Definition

Let $\mathcal{X}$ a metric space with distance function $d : \mathcal{X} \times \mathcal{X} \to \mathbb{R}^+$ and $\mathcal{P} = \{ p_i \in \mathcal{X} : i = 1 \ldots N \}$ a set of reference points (or pivots) in $\mathcal{X}$. We define the *pivoted embedding* $e(x, \mathcal{P})$ of an object $x \in \mathcal{X}$ w.r.t $\mathcal{P}$ as an N-dimensional real-valued vector where the i-th component is the distance between x and the i-th pivot, i.e.

$$e(x, \mathcal{P}) = [\, d(x, p_1), \ldots, d(x, p_i), \ldots, d(x, p_N) \,] \in \mathbb{R}^N. \tag{1}$$

We are interested in estimating the distance $d(x, y)$ between a pair of objects $x, y \in \mathcal{X}$ given their pivoted embeddings $\mathbf{e}_x = e(x, \mathcal{P}), \mathbf{e}_y = e(y, \mathcal{P})$ with respect to a common set of reference objects $\mathcal{P}$. We formulate this task as a regression problem: we define a parametric model f that outputs the estimated distance given the pivoted embeddings and optimize its parameters on a training set via gradient descent. In addition to $\mathbf{e}_x$ and $\mathbf{e}_y$, we include the distances between pivots $\{d(p_i, p_j) : i = 1 \ldots N, j = 1 \ldots N, i < j\}$ in the inputs of our model as a real-valued vector $\mathbf{p} \in \mathbb{R}^{\frac{N(N-1)}{2}}$ is commonly computed once offline and available. Formally,

$$\tilde{d}(x, y) = f(\mathbf{e}_x, \mathbf{e}_y, \mathbf{p}; \theta), \tag{2}$$

where $\tilde{d}(x, y)$ indicates the estimate for $d(x, y)$ of the model f having parameters θ. Following common practice in metric learning, we define

$$f(\mathbf{e}_x, \mathbf{e}_y, \mathbf{p}; \theta) = |\Phi(\mathbf{e}_x, \mathbf{p}; \theta) - \Phi(\mathbf{e}_y, \mathbf{p}; \theta)|_2, \tag{3}$$

where $\Phi(\mathbf{e}, \mathbf{p}; \theta)$ is a neural network that takes as input a pivoted embedding $\mathbf{e}$ and the distances between pivots $\mathbf{p}$ and outputs a real-valued vector representation.

As the architecture of $\Phi(\mathbf{e}, \mathbf{p}; \theta)$, we choose a two-branch fully-connected residual network: $\mathbf{e}$ and $\mathbf{p}$ are independently processed by two MLPs with residual connections whose outputs are then merged by concatenation and followed by one or more additional fully-connected layer. Each branch comprises multiple residual blocks [7] having the structure reported in Fig. 1a. We explore and evaluate architectural hyperparameters such as depth and branch merging point in Sect. 3.

2.2   Model Training

We train our model with mini-batch gradient descent. Given a training set $\mathcal{X}_{tr} \subset \mathcal{X}$, to form a training batch we randomly draw N objects as pivots $\mathcal{P}$ and B pairs of objects $(x_i, y_i), i = 1 \ldots B, x_i, y_i \in \mathcal{X}_{tr}$, and we adopt the original metric distance d to obtain the inputs (the pivoted embedding of the objects $\mathbf{e}_{x_i}, \mathbf{e}_{y_i}$ and distances between pivots $\mathbf{p}$) and the target (exact distances between objects $d(x_i, y_i)$) of our model. We optimize the loss function

$$\mathcal{L} = \frac{1}{B} \sum_{i=1}^{B} \text{SmoothL1}\left(f(\mathbf{e}_{x_i}, \mathbf{e}_{y_i}, \mathbf{p}), d(x_i, y_i)\right), \tag{4}$$

with

$$\text{SmoothL1}(a, b) = \begin{cases} \frac{1}{2}(a - b)^2, & \text{if } |a - b| < 1 \\ |a - b| - \frac{1}{2}, & \text{otherwise}. \end{cases} \tag{5}$$

We choose the SmoothL1 function to avoid huge gradients in the early phase of training that often lead to numerical instabilities. At each batch, we sample new pivots and couples and repeat the procedure. We periodically evaluate our model on a test set $\mathcal{X}_{ts}$ by measuring the estimation error, and we adopt early stopping when the test error stops decreasing.

3 Experiments

3.1 Dataset

Throughout all experiments, we adopt a subset of the YFCC100M-HNfc6 deep features dataset [1]—a dataset of 100M 4096-dimensional features extracted from YFCC100M [11] images using the HybridCNN [13] deep convolutional pretrained network and selecting the output of the *fc6* layer. In the original space, features are compared with the L_2 distance, i.e. $d = L_2$. We select the first 1M features and divide them in training, validation and test sets with a 750K/150K/100K split. As a metric of performance, we report the mean absolute error and the mean absolute percentage error (MAPE) computed on the test set together with their standard deviations.

3.2 Choice of the Φ Network

We perform experiments to evaluate two main architectural hyperparameters of Φ—the *depth* of the network and the *fusion strategy* of the two branches. For the former, we test a number of intermediate layers in $\{1, 2, 4\}$, while for the latter, we test concatenation of the two branches at the input level (*early fusion*), at half depth (*mid fusion*), and right before the final layer (*late fusion*). Figure 1 depicts the tested architectures for each parameter combination. We decided not to apply any bottleneck layer to reduce the dimensionality of the input, thus every layer keeps the dimensionality of its input except for the last projection. We are aware this leads to prohibitive memory requirements as N increases, as the dimensionality of $\mathbf{p}$ is $O(N^2)$, but in this preliminary phase, this reduces the architectural search space and enables us to evaluate the model without introducing performance caps. We train all models with SGD with momentum 0.9, learning rate of 0.05 (divided by 10 when the validation loss plateaus), batch size 100 for 10K iterations, validating every 100 iterations. We adopt early stopping by monitoring the MAPE on the validation set and selecting the model reaching the minimum[1]. On our single-GPU configuration, we were able to test all variations with N up to 128, as larger values are prohibitively expensive in terms of GPU memory required for training (+10GB); we left the exploration of larger values with reduced models to future work. Results in terms of MAPE and MAE are reported in Table 1. We notice that on average, an early fusion strategy is able to reach slightly better results with more performance gains with deeper networks. On the other hand, deeper networks with other fusion strategy suffer from numerical instabilities leading to divergence; we left to future work the tuning of optimization hyperparameters that may alleviate this phenomenon. Among shallower (and thus more efficient) ones, the model with late fusion and depth $= 1$ shows good overall performance on all tested N values with respect to other variants.

---

[1] The code for reproducing our results is available at https://github.com/fabiocarrara/pivoted-estimation.

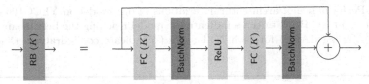

(a) Composition of a Residual Block. Output dimensionality is reported in brackets for fully connected layers.

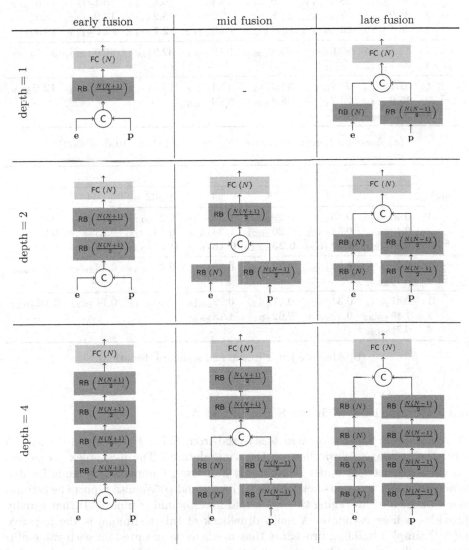

(b) Architectural variations tested. The output dimensionality (K in Fig. (1a)) of each layer is reported in brackets. C indicates concatenation.

Fig. 1. Explored architectures for $\Phi(\mathbf{e}, \mathbf{p})$

Table 1. Performance of architectural variations of our model on YFCC100M-HNfc6 features test subset. **Bold entries** indicate the model reaching the best mean absolute percentage error (MAPE) for each N. Dashes (-) indicate configurations that do not converge.

		N						
arch		2	4	8	16	32	64	128
early	1	$36.7_{\pm24.3}$	$28.1_{\pm20.1}$	$23.0_{\pm16.4}$	$18.4_{\pm13.1}$	$16.4_{\pm12.1}$	$16.1_{\pm12.1}$	$14.8_{\pm11.3}$
	2	$36.9_{\pm24.7}$	$27.2_{\pm20.5}$	$22.8_{\pm17.1}$	$19.2_{\pm14.3}$	$18.2_{\pm14.2}$	$17.9_{\pm16.7}$	$18.0_{\pm14.3}$
	4	$36.3_{\pm25.0}$	$\mathbf{25.0_{\pm20.0}}$	$\mathbf{18.7_{\pm15.2}}$	$\mathbf{14.0_{\pm11.7}}$	$\mathbf{12.7_{\pm11.5}}$	$\mathbf{12.9_{\pm11.8}}$	$15.3_{\pm15.7}$
mid	2	$37.4_{\pm26.2}$	$39.6_{\pm28.2}$	$33.8_{\pm26.1}$	$25.9_{\pm18.8}$	$32.9_{\pm30.1}$	$30.0_{\pm29.2}$	$31.1_{\pm28.1}$
	4	$39.1_{\pm26.4}$	-	-	-	-	-	-
late	1	$37.1_{\pm24.9}$	$27.0_{\pm19.9}$	$21.3_{\pm15.8}$	$17.4_{\pm12.4}$	$15.6_{\pm11.4}$	$13.6_{\pm10.3}$	$\mathbf{12.9_{\pm9.9}}$
	2	$\mathbf{35.6_{\pm24.7}}$	$25.7_{\pm21.9}$	$48.8_{\pm46.1}$	$88.9_{\pm70.9}$	-	-	-
	4	$56.3_{\pm36.3}$	-	-	-	-	-	-

(a) Absolute Percentage Error (%, mean and standard deviation)

		N						
arch		2	4	8	16	32	64	128
early	1	$0.46_{\pm0.30}$	$0.35_{\pm0.25}$	$0.29_{\pm0.20}$	$0.23_{\pm0.16}$	$0.21_{\pm0.15}$	$0.20_{\pm0.15}$	$0.19_{\pm0.14}$
	2	$0.46_{\pm0.31}$	$0.34_{\pm0.25}$	$0.29_{\pm0.21}$	$0.24_{\pm0.18}$	$0.23_{\pm0.18}$	$0.23_{\pm0.21}$	$0.23_{\pm0.18}$
	4	$0.46_{\pm0.31}$	$\mathbf{0.32_{\pm0.25}}$	$\mathbf{0.23_{\pm0.19}}$	$\mathbf{0.18_{\pm0.15}}$	$\mathbf{0.16_{\pm0.14}}$	$\mathbf{0.16_{\pm0.15}}$	$0.19_{\pm0.19}$
mid	2	$0.47_{\pm0.33}$	$0.50_{\pm0.36}$	$0.43_{\pm0.33}$	$0.33_{\pm0.24}$	$0.42_{\pm0.39}$	$0.38_{\pm0.37}$	$0.39_{\pm0.35}$
	4	$0.49_{\pm0.33}$	-	-	-	-	-	-
late	1	$0.47_{\pm0.31}$	$0.34_{\pm0.25}$	$0.27_{\pm0.20}$	$0.22_{\pm0.15}$	$0.20_{\pm0.14}$	$0.17_{\pm0.13}$	$\mathbf{0.16_{\pm0.12}}$
	2	$\mathbf{0.45_{\pm0.31}}$	$0.32_{\pm0.27}$	$0.62_{\pm0.57}$	$1.13_{\pm0.89}$	-	-	-
	4	$0.71_{\pm0.46}$	-	-	-	-	-	-

(b) Absolute Error (mean and standard deviation)

3.3 Comparison with the State of the Art

We compare in Fig. 2 our models selected from Table 1 (the best for each N) with the *n-Simplex projection* on the same dataset. The n-Simplex projection provides geometrical upper (simplex-U) and lower (simplex-L) bounds for distance estimates in super-metric spaces given **e** and **p**. We also report the estimation obtained by averaging the upper and lower bound (simplex-M) that usually provides a finer estimate. A main drawback of this technique is the iterative $O(N^3)$ simplex building procedure that needs to be executed for each value of **p** we are willing to use. Our approach provides a comparable or slightly degraded performance, but once trained, it can cope with different values of **p** without the need of expensive procedures. Moreover, our approach can be applied to generic metric spaces.

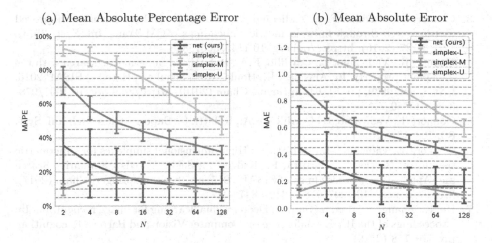

Fig. 2. Comparison with the state-of-the-art n-Simplex estimator: simplex suffixes -U, -L, and -M indicate estimation using respectively the upper bound, lower bound, and their mean.

4   Conclusion

We explored the use of neural regressors for estimating distances from pivoted embedding in generic metric spaces. Preliminary experiments on deep-learned image descriptors suggest that the proposed approach can be used in approximated regimes providing a performance comparable to exact geometrical bounds while being more efficient. Moreover, our formulation is not limited to supermetric spaces and can be applied seamlessly to different set of reference points— properties that pave the way to advanced indexing structures including dynamically chosen reference points sets. Future work comprises the development of more compact architectures for higher dimensionalities and extended experimentation on additional metric and retrieval datasets.

Acknowledgment. This work was partially supported by "Automatic Data and documents Analysis to enhance human-based processes" (ADA, CUP CIPE D55F1 7000290009) and the AI4EU project (funded by the EC, H2020 - Contract n. 825619). We also gratefully acknowledge the support of NVIDIA Corporation with the donation of the Tesla K40 GPU used for this research.

References

1. Amato, G., Falchi, F., Gennaro, C., Rabitti, F.: YFCC100M hybridnet fc6 deep features for content-based image retrieval. In: Proceedings of the 2016 ACM Workshop on Multimedia COMMONS, pp. 11–18 (2016)
2. Blumenthal, L.M.: Theory and Applications of Distance Geometry. Clarendon Press, Oxford (1953)

3. Connor, R., Cardillo, F.A., Vadicamo, L., Rabitti, F.: Hilbert exclusion: improved metric search through finite isometric embeddings. ACM Trans. Inf. Syst. **35**(3), 17:1–17:27 (2016). https://doi.org/10.1145/3001583

4. Connor, R., Vadicamo, L., Cardillo, F.A., Rabitti, F.: Supermetric search with the four-point property. In: Amsaleg, L., Houle, M.E., Schubert, E. (eds.) SISAP 2016. LNCS, vol. 9939, pp. 51–64. Springer, Cham (2016). https://doi.org/10.1007/978-3-319-46759-7_4

5. Connor, R., Vadicamo, L., Cardillo, F.A., Rabitti, F.: Supermetric search. Inf. Syst. **80**, 108–123 (2019)

6. Connor, R., Vadicamo, L., Rabitti, F.: High-dimensional simplexes for supermetric search. In: Beecks, C., Borutta, F., Kröger, P., Seidl, T. (eds.) Similarity Search and Applications, Proceedings of SISAP 2017, pp. 96–109. Springer, Cham (2017). https://doi.org/10.1007/978-3-319-68474-1_7

7. He, K., Zhang, X., Ren, S., Sun, J.: Deep residual learning for image recognition. In: Proceedings of the IEEE Conference on Computer Vision and Pattern Recognition, pp. 770–778 (2016)

8. Kraska, T., Beutel, A., Chi, E.H., Dean, J., Polyzotis, N.: The Case for Learned Index Structures. Association for Computing Machinery, New York (2018)

9. Micó, M.L., Oncina, J., Vidal, E.: A new version of the nearest-neighbour approximating and eliminating search algorithm (AESA) with linear preprocessing time and memory requirements. Pattern Recogn. Lett. **15**(1), 9–17 (1994). https://doi.org/10.1016/0167-8655(94)90095-7

10. Schoenberg, I.J.: Metric spaces and completely monotone functions. Ann. Math. **39**(4), 811–841 (1938)

11. Thomee, B., et al.: YFCC100M: the new data in multimedia research. Commun. ACM **59**(2), 64–73 (2016)

12. Vadicamo, L., Gennaro, C., Falchi, F., Chávez, E., Connor, R., Amato, G.: Re-ranking via local embeddings: a use case with permutation-based indexing and the nSimplex projection. Inf. Syst. 101506 (2020)

13. Zhou, B., Lapedriza, A., Xiao, J., Torralba, A., Oliva, A.: Learning deep features for scene recognition using places database. In: Advances in Neural Information Processing Systems, pp. 487–495 (2014)

Demo and Position Papers

Visualizer of Dataset Similarity Using Knowledge Graph

Petr Škoda[ID], Jakub Matějík, and Tomáš Skopal[✉][ID]

SIRET Research Group, Faculty of Mathematics and Physics,
Charles University, Prague, Czech Republic
{skoda,skopal}@ksi.mff.cuni.cz, jakub.matejik@centrum.cz

Abstract. Many institutions choose to make their datasets available as Open Data. Open Data datasets are described by publisher-provided metadata and are registered in catalogs such as the European Data Portal. In spite of that, findability still remain a major issue. One of the main reasons is that metadata is captured in different contexts and with different background knowledge, so that keyword-based search provided by the catalogs is insufficient. A solution is to use an enriched querying that employs a dataset similarity model built on a shared context represented by a knowledge graph. However, the "black-box" dataset similarity may not fit well the user needs. If an explainable similarity model is used, then the issue can be tackled by providing users with a visualisation of the dataset similarity. This paper introduces a web-based tool for dataset similarity visualisation called ODIN (Open Dataset INspector). ODIN visualises knowledge graph-based dataset similarity, offering thus an explanation to the user. To understand the similarity, users can discover additional datasets that match their needs or reformulate the query to better reflect the knowledge graph. Last but not least, the user can analyze and/or design the similarity model itself.

1 Introduction and Motivation

The number of datasets available in so-called Open Data Catalogs[1] increases dramatically. Every dataset in the open data catalog has associated metadata provided by the dataset publisher. However, many of the open data catalogs offer only basic search tools using the metadata like facets or fulltext search. This may be inefficient as the user context, in which the user interest is specified (query), often does not fit the context used by publisher to create the metadata. The simplest context mismatch can be caused by using different languages, synonyms or using different level of abstraction. To tackle this issue, dataset metadata and user query can be mapped to a shared knowledge graph while a similarity of the two mappings can be evaluated for querying purposes [2].

However, the general paradigm of content-based similarity search assumes a black-box model, in sense that it hides the features from the user and only

---

[1] e.g., European Data Portal, https://www.europeandataportal.eu.

© Springer Nature Switzerland AG 2020
S. Satoh et al. (Eds.): SISAP 2020, LNCS 12440, pp. 371–378, 2020.
https://doi.org/10.1007/978-3-030-60936-8_29

provides overall similarity scores [5]. An imperfect model of similarity may introduce false dismisals and/or false positives in the search process, and thus it may discourage the user to keep using the similarity-based search method. This is a classic issue of trust, explainability and understanding that is gaining attention over the last few years, especially in the field of artificial intelligence [1]. At the same time, there are fields like bioinformatics, where we can find examples of visually explainable similarity. These similarity models allow user not only to trust and understand them, but also to be an active participant in the process of modeling. From certain perspective it is not the model that on its own creates the final outcome. Instead, the final outcome is created by model-user interaction, where the model provides visually explainable results and the users use them to gain the final outcome, i.e., understanding.

1.1 Visually Explainable Similarity Models

In general, to provide visual explanation of a particular similarity model, the black-box similarity model abstraction needs to be replaced by a model structured for a user. Within such a similarity model the (features of) data entities are mutually mapped using a data structure that is suitable for visualizations.

Sequence Alignment. In many fields (e.g., bioinformatics) the data entities are modeled by sequences (strings, time series) while the basic task is to match the sequences by means of alignment. Elements of the sequences are "aligned" (monotonously mapped) in order to achieve the best overall match of the sequences. A similarity value could be derived from a particular alignment as a measure of how good the alignment is, however, the primary outcome is the alignment structure itself. The alignment is achieved by dynamic programming where the alignment structure is represented by a matrix. Various algorithms for alignment were developed [3], for example the dynamic time warping (DTW), longest common subsequence, etc. The resulting alignment could be visualized directly as a path in the matrix (see Fig. 1a left) or could be shown as links between the sequences' elements (see Fig. 1a right). Hence, the sequence alignment represents an explainable similarity model that is easy to visualize.

The concept of alignment might be applied for datasets, where the alignment would be performed through the shared concepts in a knowledge graph. However, unlike sequences, Open Data datasets do not have linear representations and thus monotonous sequence alignment cannot be simply reused.

Alluvial Diagram. An alternative could be to consider Open Data datasets as changes to the knowledge graph. For this scenario we can define subset of the knowledge graph induced by the dataset mapping. We can cluster the mapped entities using a hierarchical distance, utilizing the fact that close concepts should be close to each other. We can then employ alluvial diagram (see Fig. 1b) to visualise similarity of multiple datasets as a change of the induced knowledge graph. The advantage of this methods is that the alluvial diagram can easily

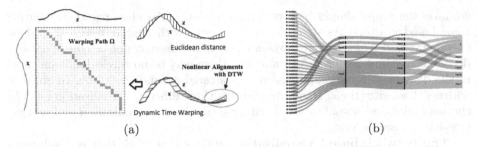

Fig. 1. Examples of visually explainable similarity models

capture similarity of multiple datasets. The main limitations to this method are: the visualisation dependency on the clustering, bad visualisation when only few entities are shared.

1.2 Paper Contribution

In this paper we present Open Dataset Inspector (ODIN); an open-source[2] tool for visualisation of the explainable dataset similarity. ODIN builds upon our previously proposed navigational similarity using knowledge graph [2] and allows user to visually explore the similarity. As the visually explainable dataset similarity is a new topic, there are no standard visualisation techniques. That is why ODIN provides three visualisation modes. Two of them are easy to understand but suffer from certain drawbacks. To address those drawbacks we designed a new dataset similarity visualisation based on vertical tree-like graph projection.

2 Model of Explainable Dataset Similarity

In our previous work [2], we have introduced a framework for modeling explainable dataset similarity with the help of a knowledge graph. The framework introduced mapping of datasets' features onto nodes of a knowledge graph (such as Wikidata). A navigational similarity was then defined on top of the mapped features (the sets of nodes, respectively) and provided explainable similarity in the form of paths in the graph "om one dataset to the other". In this paper, we continue by introducing a visualization model for the framework within the ODIN tool.

2.1 Visualisation

All visualisations take as an input the datasets' mappings into a knowledge graph nodes and a structure of the graph. Nodes in the graph are connected with edges that represent specialization, generalization or is-instance relation. In the ideal

---

[2] https://github.com/mff-uk/open-dataset-inspector.

scenario the graph should be a tree with related nodes close together, creating topic-based clusters. As a result, two datasets with similar topics should be mapped to nodes in the same parts of the graph. This may not be true, if dataset do not share similar topics. Alternatively, there may be no single dominant topic in the dataset, causing mappings to be scattered around the graph. In order to address those situations, ODIN implements three different visualisation models: the network-based visualisation, horizontal tree-based visualisation and vertical tree-based visualisation.

The network-based visualisation utilizes a network that is a subgraph of the entire knowledge graph the nodes of which are involved in the dataset mapping and their common ancestors. The network utilizes force field layout. Nodes in the network that are mapped for a given dataset are colored by the color of the dataset. If two datasets are similar they should map to nodes that are close together in the network. The network visualisation should be easy to understand and provide good visual hint of the dataset similarity. On the contrary, it does not allow user to easily see to what the datasets are mapped. That is why it may be harder for a user to actually understand the content of the dataset. Nevertheless, the main issue is that some datasets map to nodes scattered in the knowledge graph. As a result, the visualisation contains a lot of visual noise making it almost useless for the user.

The horizontal tree-based visualisations is similar to the network visualisation, but utilizes different visual layout. One of the main issues of the network visualisations is that force field layout focuses on spacing out the visualisation while preserving proximity to nodes that are connected with edges. That is a problem if the mapped nodes are scattered around the graph and there are two many connected nodes. In such situation, the network visualisation fails to capture the tree-like nature of the graph. The main objective of the horizontal tree visualizations is to promote this information. It utilizes tree layout with the root being the nodes' shared ancestor and leaves being nodes the datasets are mapped to. The main problem of this visualisation is similar to the network visualisation but in a different way. If the hierarchy is too wide, the tree becomes too big to easily visually comprehend. However, if we focus only on a subset of mapped entities, it can provide a great way to visualise the relations as it provides clear hint of the hierarchy.

The vertical tree-based visualisation is designed to address the issue with visual noise and reduce the amount of information shown to the user. The main idea is to utilize node nesting for visualization of hierarchy, i.e., visualisation of a given node is enclosed in its parent node. Another way of how to describe the visualisation is that we have a tree hierarchy as in the horizontal tree visualisation, but instead of "looking at the tree from a side" we "look on the tree from above". At the beginning the user sees only root of the hierarchy and the first level of nodes. User can then navigate by zooming down/up the nodes. We employ arrows to show mapping of dataset features into the nodes/subtrees. For given pair of nodes this allows user to see common ancestors (topics) on given level or zoom down to gain additional details. The main drawback of this visualisation is its novelty, where user needs to learn how to navigate and perceive the hierarchy.

3 Open Dataset Inspector (ODIN)

ODIN allows user to visualise the explainable dataset similarity with the above described visualisation methods using Wikidata[3] as the knowledge graph. ODIN consists of three components, the web-based application (vis-component), the component for computing dataset similarity (sim-component), and the component responsible for the data preparation (prep-component). The prep-component takes a dataset and Wikidata graph and produces mapping of features into the Wikidata graph nodes. The sim-component takes mappings of two datasets and computes similarity using paths between the mapped nodes. The vis-component implements three visualisation models described in Subsect. 2.1: network-based, horizontal tree-based and vertical tree-based.

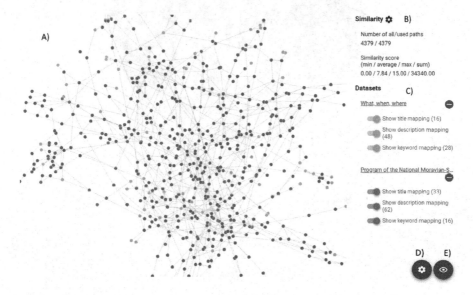

Fig. 2. Visualisation of $D_{Bohumin}$, $D_{Theater}$ datasets. A) network visualisation, B) similarity details, C) loaded datasets, D) show options dialog, E) visualisation menu.

We illustrate the use of ODIN on examples from our previous work [2]. The first dataset $D_{Bohumin}$[4] is called "What, when, where" and it covers cultural, sports and free-time events in the Bohumín city in Moravian-Silesian Region. The second dataset $D_{Theater}$[5] is called "Program of the National Moravian-Silesian Theater" and contains the program of the National Moravian-Silesian Theater. Both datasets are registered in Czech National Open Data portal with

---

[3] https://wikidata.org.

[4] https://data.gov.cz/zdroj/datove-sady/Bohumin/3384768.

[5] https://data.gov.cz/zdroj/datove-sady/https---opendata.ostrava.cz-api-3-action-package_show-id-program-narodniho-divadla-moravskoslezskeho.

all metadata in Czech. All data processing was executed in Czech, here we translate some of the metadata to English for international readers of the paper. For a more detailed presentation of the ODIN functionality we refer readers to the screencast that is reachable from the ODIN web app (see Footnote 6).

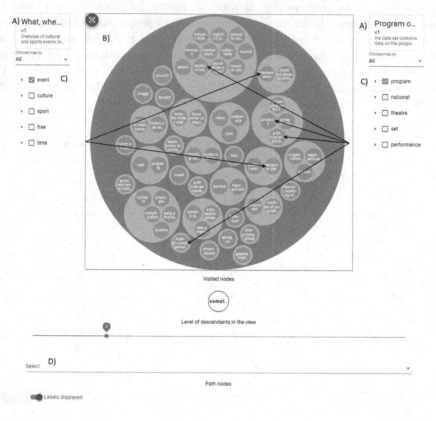

Fig. 3. Visualisation of $D_{Bohumin}$, $D_{Theater}$ datasets in vertical tree visualization. A) loaded datasets, B) vertical tree visualisation, C) list of features, D) path selector.

First we employ the prep-component to prepare the datasets. The prep-component adds to main extensions to the early prof of concept from our previous work [2]. We utilize UdPipe [4] to get canonical tokenization of all features (keywords from the title, description and keywords of a dataset's metadata) as a bag of words in order to increase effectiveness of the mapping procedure. In addition we also perform node reduction, where every mapped node with only *instanceof* relation is replaced with the value of *instanceof*; we call those node *Reduced*, while other *Directly mapped*. The second modification addresses the issue where features map to street names or nodes connected to *Wikimedia disambiguation page*.

Once both datasets are processed they can be loaded into the ODIN visualisation component. $D_{Bohumin}$, $D_{Theater}$ datasets can be added one at the time or they can be pre-loaded using URL query; to make it easy we prepared a link to load both datasets automatically[6].

Fig. 4. ODIN configuration dialogs.

As we can see from Fig. 2 the $D_{Bohumin}$ maps to 92 nodes and $D_{Theater}$ maps to 111 nodes. Nodes are colored red for $D_{Theater}$, cyan for $D_{Bohumin}$ and yellow for shared. We can toggle over the nodes to get their descriptions, and additional mapping information. Many of the nodes are mapped by short features, we can filter them using the options dialog (Fig. 4a) with the following filter function:

```
(mapping) => mapping.metadata.group.map(item => item.length)
    .reduce((left, right) => Math.max(left, right)) > 2
```

This reduces the number of mapped nodes to 44 for $D_{Bohumin}$ and 98 for dataset $D_{Theater}$. This is about 50% decrease in number of nodes for $D_{Bohumin}$, still the number is quite large.

Many of the nodes are of type *Reduced*, i.e., none of the features is directly mapped to them. We can filter them out using the *Keep only directly mapped* options in the options dialog (Fig. 4a). With this additional filter the number of nodes drops to 10 and 41. As it is not clear how close they are in the graph, we can switch to horizontal (Fig. 5) or vertical (Fig. 3) tree visualisation to get more details. As the horizontal tree visualisation is suitable only for small hierarchies (with small nodes' fan-out), we choose the vertical tree visualisation. Here we can select the mapped features and manually explore the navigational dataset similarity by browsing the hierarchy along the paths involved. We could also reduce the set of paths used to compute the similarity to get only nodes that are close together using similarity options menu (Fig. 4b). We choose to select only the shortest path for each mapped node with maximum path size of 2. There is actually only one such path between nodes Q1656682 (event)

---

[6] http://skoda.projekty.ms.mff.cuni.cz/dataset-similarity/example-000.

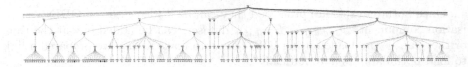

Fig. 5. Example of horizontal tree visualisation. Example of horizontal tree visualisation.

and Q35140 (performance), that we can select from the path selector to see its detail. By doing so we get the same similarity explanation: one dataset contains information about performances, i.e., cultural events, while the other contains information about events related to the culture.

4   Conclusions

In this demo paper we introduced a visualization model for explainable dataset similarity using a knowledge graph. We implemented this model into a tool called Open Dataset INspector (ODIN) that is available to the readers as a web application. We described the user interface of ODIN on a real-world example.

Acknowledgments. This research has been supported by Czech Science Foundation (GAČR) project Nr. 19-01641S.

References

1. Samek, W., Montavon, G., Vedaldi, A., Hansen, L.K., Müller, K.-R. (eds.): Explainable AI: Interpreting, Explaining and Visualizing Deep Learning. LNCS (LNAI), vol. 11700. Springer, Cham (2019). https://doi.org/10.1007/978-3-030-28954-6
2. Škoda, P., Klímek, J., Nečaský, M., Skopal, T.: Explainable similarity of datasets using knowledge graph. In: Amato, G., Gennaro, C., Oria, V., Radovanović, M. (eds.) SISAP 2019. LNCS, vol. 11807, pp. 103–110. Springer, Cham (2019). https://doi.org/10.1007/978-3-030-32047-8_10
3. Skopal, T., Bustos, B.: On nonmetric similarity search problems in complexdomains. ACM Comput. Surv. **43**(4) (2011). https://doi.org/10.1145/1978802.1978813
4. Straka, M., Straková, J.: Tokenizing, POS tagging, lemmatizing and parsing UD 2.0 with udpipe. In: CoNLL 2017 Shared Task: Multilingual Parsing from Raw Text to Universal Dependencies, pp. 88–99. Association for Computational Linguistics (2017)
5. Zezula, P., Amato, G., Dohnal, V., Batko, M.: Similarity Search: The Metric Space Approach, Advances in Database Systems, vol. 32. Springer, Boston (2006). https://doi.org/10.1007/0-387-29151-2

Vitrivr-Explore: Guided Multimedia Collection Exploration for Ad-hoc Video Search

Silvan Heller[1]([⊠])(iD), Mahnaz Parian[1,2](iD), Maurizio Pasquinelli[1](iD),
and Heiko Schuldt[1](iD)

[1] University of Basel, Basel, Switzerland
{silvan.heller,mahnaz.parian-scherb,maurizio.pasquinelli,
heiko.schuldt}@unibas.ch
[2] University of Mons, Mons, Belgium

Abstract. vitrivr is an open-source system for indexing and retrieving multimedia data based on its content and it has been a fixture at the Video Browser Showdown (VBS) in the past years. While vitrivr has proven to be competitive in content-based retrieval due to the many different query modes it supports, its functionality is rather limited when it comes to exploring a collection or searching result sets based on content. In this paper, we present vitrivr-explore, an extension to the vitrivr stack that allows to explore multimedia collections using relevance feedback. For this, our implementation integrates into the existing features of vitrivr and exploits self-organizing maps. Users initialize the exploration by either starting with a query or just picking examples from a collection while browsing. Exploration can be based on a mixture of semantic and visual features. We describe our architecture and implementation and present first results of the effectiveness of vitrivr-explore in a VBS-like evaluation. These results show that vitrivr-explore is competitive for Ad-hoc Video Search (AVS) tasks, even without user initialization.

Keywords: Self-organizing maps · Relevance feedback · Ad-hoc video search

1 Introduction

With the tremendous increase of video recording devices and the resulting abundance of digital video, finding a particular video sequence in ever-growing collections is a major challenge [10]. One research problem for multimedia retrieval systems aiming to make large collections accessible is Ad-hoc Video Search (AVS), where a description of a segment is given and users are tasked with finding as many matching segments as possible. Relevance feedback, which allows users to return feedback to the results retrieved for the given query and reflects it to the subsequent retrieval, is particularly well-suited for this kind of task since the

© Springer Nature Switzerland AG 2020
S. Satoh et al. (Eds.): SISAP 2020, LNCS 12440, pp. 379–386, 2020.
https://doi.org/10.1007/978-3-030-60936-8_30

aim is to find more candidates similar to the initially retrieved positive results rather than narrowing down the search to a single correct result.

Interactive Retrieval based on relevance feedback has been an active area of systems research and used by participants of evaluation campaigns such as the Video Browser Showdown (VBS) in the past [7–9]. Two prominent examples include the winner of the 2020 VBS campaign SOMHunter [6] and Exquisitor [5].

vitrivr [3] is an open-source full-stack content-based multimedia retrieval system, which has also participated successfully in VBS in previous years [11]. One limitation of vitrivr has always been that, in contrast to the aforementioned systems, it did not offer support for the exploration of collections and sophisticated relevance feedback.

In order to overcome the limitations of vitrivr, and inspired by these systems, we have extended vitrivr by adding support for guided content-based exploration that uses a combination of deep features and self-organizing maps. The result, vitrivr-explore, allows users to provide relevance feedback to better steer the retrieval process.

In this paper, we present vitrivr-explore, elaborate on the mechanisms used for training set selection, discuss semantic features we employed, and present results as to how effective this approach proved to be in AVS tasks compared to the regular system. In our evaluation, we have mimicked the VBS competition with queries of all types from VBS 2020, and compared the new SOM features of vitrivr-explore to a plain vitrivr instance as well as a complete system encompassing all vitrivr features plus the new SOM-based relevance feedback. We also discuss lessons learned for future distributed evaluations.

The remainder of this paper is structured as follows: Sect. 2 describes the theoretical foundation of our approach. Section 3 discusses the implementation and user interface of vitrivr-explore. In Sect. 4 we present the evaluation results and Sect. 5 concludes.

2   Methods

2.1   Architecture

vitrivr-explore builds upon the vitrivr architecture, which is depicted in Fig. 1. It makes additions to all three components of vitrivr. The database Cottontail DB [1] has newly implemented API calls for random sampling, which are required for SOMs without user initialization, and supports batched k-Nearest Neighbors (kNN) queries, which is a very useful performance optimization.

Cineast has received new feature modules, which we will cover in the next section, and newly implemented relevance feedback and training processes. Additionally, there is a new user interface to facilitate the exploration, which will be discussed in Sect. 3.

2.2   Available Features

All vector-based feature modules of Cineast, such as color- or edge-based ones can be used for both relevance feedback and training of the SOM. Additionally, we

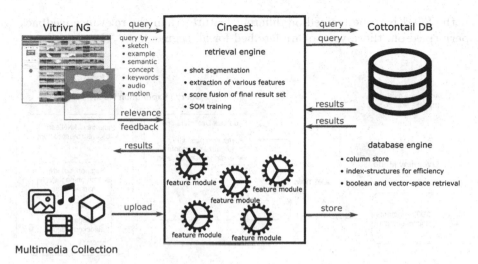

Fig. 1. System overview for vitrivr and its three major components (adapted from [2])

are using deep features which are based on MobileNet V1[1] [4]. Usually, semantic features are used for relevance feedback, while the SOM is then trained with either visual or semantic features.

Feature extraction was performed on the most representative element per shot, which was pre-processed by first cropping the image to square and then resizing it to 192×192 pixels. The features are extracted from the last hidden layer resulting in a feature vector with 512 components.

2.3   Relevance Feedback

Figure 2 provides an overview of the flow of relevance feedback in vitrivr-explore. In vitrivr-explore, users can mark segments either as positive (i.e., "more-like-this"), negative (i.e., "less-like-this"), or choose to not mark them at all. Additionally, users can exclude entire videos, as vitrivr tends to over-segment, which can be a problem for long videos with homogeneous content. Based on the list of positively marked segments P and negatively marked segments N, the training set of vectors for the SOM is generated as follows: Two kNN queries are performed, the first one returning $k = \frac{n}{|P|}$ elements per positively marked segment $p \in P$, and the second one returning $k = \alpha$ elements per negatively marked segment, where n is the upper bound for the number of desired elements α is a user-defined threshold that should be chosen based on collection size and n. This results in two sets of vectors; R and I, consisting of the relevant and irrelevant vectors, respectively. Subsequently, T, the set of vectors for training, is calculated by the formula $T = R \setminus (I \cup B)$, where B is the set of blocked elements.

---

[1] https://tfhub.dev/google/imagenet/mobilenet_v1_050_192/quantops/feature_ve ctor/3.

If the SOM is to be trained on different features than the relevance feedback is performed on, those vectors are fetc.hed for all segments $t \in T$.

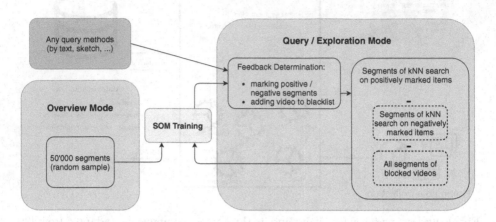

Fig. 2. Interaction of a user with vitrivr-explore

3   Implementation

In the backend, we have made improvements to Kohonen4j[2], which avoid completely imbalanced clusters and overfitting, to train the SOM. Users are able to choose multiple parameters such as the size of the SOM, which vectors should be used for relevance feedback and SOM training, and the degree α that indicates how much exploration the user is willing to do. We set α to a fixed value of 1,000 empirically, while the upper bound n is defined as $n = \min(m \cdot |P|, l)$. m is a factor depending on the selected exploration depth and provides some initial amount of nearest neighbours that will be used per segment. There is an upper limit l of segments used for training. Multiple modes available which differ in how much results are retrieved and how wide exploration should be, ranging from *query close* ($m = 1,000, l = 6,000$) to *explore wide* ($m = 10,000, l = 40,000$). All of these parameters are then sent to the backend, which selects the new segments the SOM should be trained on, requests the feature vectors for those segments, and performs the training. Figure 3 depicts a sample screenshot containing a resulting map with two churches and a church tower that were positively marked, a negatively marked crane as well as a video that should be entirely excluded. Users can inspect temporal context by clicking on the thumbnail, which plays the video from the start of the shot.

---

[2] https://github.com/dashaub/kohonen4j.

Fig. 3. Screenshot of a SOM based on relevance feedback in vitrivr-explore

4   Evaluation

4.1   Setup

For the evaluation, we performed a VBS-like competition with 6 participating teams. Two teams used the current iteration of vitrivr [3]; two teams used vitrivr-explore, meaning no possibility for initialization of positive examples using, e.g., tag-based retrieval; two teams were allowed to combine vitrivr-explore and regular vitrivr. We used the Distributed Retrieval Evaluation Server (DRES[3]), which is designated as the successor to the VBS-Server to perform the competition.

DRES is currently still under development; its motivation is twofold: i) it stems of the desire to improve participant experience at VBS and ii) due to the fact that the current vbs-server[4] does not enable remote participation, which would have made an evaluation impossible due to COVID-19 related restrictions.

Multiple AVS tasks from the 2020 iteration of VBS (VBS20) were used. In order to provide a level playing field, we selected participants that were not part of VBS20. However, all participants received an introduction to the system beforehand and some had prior experience in using vitrivr.

In VBS, there is no groundtruth for AVS tasks, but submissions are judged. In our case, participants and judges of VBS20 served as judges and were given the task description, the submitted shot and additional context video material before and after the submitted frame. Shot segmentation was taken from the dataset [12]. Based on this information, they had to assess a submission as either

---

[3] https://github.com/lucaro/DRES.
[4] https://github.com/klschoef/vbsserver.

correct, if there was *any frame* matching the task description, or *incorrect*, i.e., not matching the task description. Additionally, judges could mark a submission as *unsure*—an enhancement provided by DRES—in case of doubt or inability to reach a verdict. Such submissions neither increase nor penalize the total score and are treated as if they had never been submitted.

There are some limitations to the evaluation results. We show results only for retrieval performance, and not retrieval time, memory usage or other system metrics due to the challenge of collecting and analyzing meaningful data from the heterogeneous hardware used by participants. Additionally, it would have been ideal to benchmark vitrivr-explore also against other systems such as SOMHunter. With the development of DRES, we are working on making it easier and more reproducible to compare different systems with a larger number of participants. Third, it would be interesting to gain qualitative feedback from participants. All of these limitations are being actively tackled in the context of designing future iterations of evaluation campaigns.

4.2 AVS Results

The results from the evaluation have been analyzed using four metrics: precision, videos from which segments were correctly submitted, uniquely submitted correct videos, and score. The reason for reporting on videos instead of segments is that vitrivr tends to over-segment, and therefore users often submit segments very close to one another, which makes numbers of uniquely submitted segments less meaningful. As already discussed in analyses of AVS tasks at VBS [8], measuring the actual recall is not possible due to the lack of a ground truth.

Figure 4 shows all metrics per task and team. Metrics were combined per system category for readability. The maximum score achievable per task is 50. Scoring for AVS Tasks at VBS is relative to the performance of other teams.

We think vitrivr is a reasonable baseline for AVS tasks as it has been very competitive in past evaluation campaigns. The teams which were able to use a combination of vitrivr and vitrivr-explore had the highest score in three out of four tasks, showing that the SOM interface adds to AVS performance in this case. The results also show that without initialization, vitrivr-explore has difficulties in finding positive elements and that the more specific the query was, the harder the challenge. For the first task, the description was *Find shots of a kid smiling, with no adults visible in the shot* and for the last task *Find selfie shots (i.e., showing face or at least head of the person filming) of someone on a paraglider*. These are both tasks with important restrictions, where in addition to the initialization challenge, the lack of tag-based querying was also conceived as a difficulty. This is reflected in the score, with no shots being found for Task 1 and only segments out of two videos for Task 4. Additionally, as the deep feature extraction is trained a general classification dataset, these features can miss out on information present in the video frame that was not in the training set, making relevance feedback harder.

Our evaluation has similar problems as VBS had in the past: Due to different approaches by users with the same level of knowledge, results can vary widely between tasks. Additionally, due to the small number of participants and the vast collection size, it is relatively easy to find unique parts of a video that have not been submitted by other teams and performance can vary between users of the same system.

Task	Metric	Teams		
		vitrivr	combination	vitrivr-explore
AVS 1	precision	0.5	0.85	-
	correct	9	12	0
	unique	7	10	0
	score	35	45	0
AVS 2	precision	0.82	0.73	0.8
	correct	6	6	6
	unique	3	3	4
	score	45	45	42
AVS 3	precision	0.86	0.85	0.9
	correct	5	6	7
	unique	3	4	6
	score	47	48	49
AVS 4	precision	0.71	0.86	0.5
	correct	15	9	2
	unique	13	6	1
	score	41	47	33

Fig. 4. Competition metric for all AVS tasks per team. The best metric is always highlighted in boldface font.

Due to DRES being in early development, the advanced interaction logging of VBS was not applied. As a consequence, we do not have access to more detailed information such as whether correct segments would have been present in the displayed result sets, but overlooked by users.

5 Conclusion

In this paper, we have introduced vitrivr-explore, a first implementation for content-based multimedia collection exploration in vitrivr. We have shown that using relevance feedback in combination with SOMs retrieves different results than regular vitrivr, which increases performance for AVS tasks. While performance varies across tasks due to both task definitions and search strategies by users, vitrivr-explore has shown its effectiveness and competitiveness for certain (types of) tasks. This is in line with lessons learned from past evaluation campaigns which have shown that a diversity in retrieval approaches is key for interactive multimedia retrieval.

For future work, we plan to perform an evaluation with more participants of vitrivr and vitrivr-explore against other systems using DRES. One focus should be to make sure that participants have approximately the same level of experience and finding ways to account for varying performance between team members. Additionally, the training set selection is currently kNN-based and might benefit from interactive learning.

References

1. Gasser, R., Rossetto, L., Heller, S., Schuldt, H.: Cottontail DB: an open source database system for multimedia retrieval and analysis. In: Proceedings of the 28th ACM International Conference on Multimedia (2020)
2. Heller, S., Amiri Parian, M., Gasser, R., Sauter, L., Schuldt, H.: Interactive lifelog retrieval with vitrivr. In: Proceedings of the Third Annual Workshop on Lifelog Search Challenge, LSC 2020, p. 1–6 (2020)
3. Heller, S., Sauter, L., Schuldt, H., Rossetto, L.: Multi-stage queries and temporal scoring in vitrivr. In: 2020 IEEE International Conference on Multimedia Expo Workshops (ICMEW), pp. 1–5. IEEE (2020)
4. Howard, A.G., et al.: MobileNets: efficient convolutional neural networks for mobile vision applications. arXiv preprint arXiv:1704.04861 (2017)
5. Jónsson, B.Þ., et al.: Exquisitor: interactive learning at large. arXiv preprint arXiv:1904.08689 (2019)
6. Kratochvíl, M., Veselý, P., Mejzlík, F., Lokoč, J.: SOM-Hunter: video browsing with relevance-to-SOM feedback loop. In: Ro, Y.M., et al. (eds.) MMM 2020. LNCS, vol. 11962, pp. 790–795. Springer, Cham (2020). https://doi.org/10.1007/978-3-030-37734-2_71
7. Lokoč, J., Bailer, W., Schoeffmann, K., Münzer, B., Awad, G.: On influential trends in interactive video retrieval: video browser showdown 2015–2017. IEEE Trans. Multimedia **20**(12), 3361–3376 (2018)
8. Lokoč, J., et al.: Interactive search or sequential browsing? A detailed analysis of the video browser showdown 2018. ACM Trans. Multimedia Comput. Commun. Appl. (TOMM) **15**(1), 29 (2019)
9. Rossetto, L., et al.: Interactive video retrieval in the age of deep learning-detailed evaluation of VBS 2019. IEEE Trans. Multimedia (2020). https://ieeexplore.ieee.org/abstract/document/9037125/
10. Rossetto, L., Giangreco, I., Heller, S., Tănase, C., Schuldt, H.: Searching in video collections using sketches and sample images – the Cineast system. In: Tian, Q., Sebe, N., Qi, G.-J., Huet, B., Hong, R., Liu, X. (eds.) MMM 2016. LNCS, vol. 9517, pp. 336–341. Springer, Cham (2016). https://doi.org/10.1007/978-3-319-27674-8_30
11. Rossetto, L., Amiri Parian, M., Gasser, R., Giangreco, I., Heller, S., Schuldt, H.: deep learning-based concept detection in vitrivr. In: Kompatsiaris, I., Huet, B., Mezaris, V., Gurrin, C., Cheng, W.-H., Vrochidis, S. (eds.) MMM 2019. LNCS, vol. 11296, pp. 616–621. Springer, Cham (2019). https://doi.org/10.1007/978-3-030-05716-9_55
12. Rossetto, L., Schuldt, H., Awad, G., Butt, A.A.: V3C – a research video collection. In: Kompatsiaris, I., Huet, B., Mezaris, V., Gurrin, C., Cheng, W.-H., Vrochidis, S. (eds.) MMM 2019. LNCS, vol. 11295, pp. 349–360. Springer, Cham (2019). https://doi.org/10.1007/978-3-030-05710-7_29

Running Experiments with Confidence and Sanity

Martin Aumüller[1] and Matteo Ceccarello[2]

[1] IT University of Copenhagen, Copenhagen, Denmark
maau@itu.dk
[2] Free University of Bozen, Bolzano, Italy
mceccarello@unibz.it

Abstract. Analyzing data from large experimental suites is a daily task for anyone doing experimental algorithmics. In this paper we report on several approaches we tried for this seemingly mundane task in a similarity search setting, reflecting on the challenges it poses.

We conclude by proposing a workflow, which can be implemented using several tools, that allows to analyze experimental data with confidence.

The extended version of this paper and the support code are provided at https://github.com/Cecca/running-experiments.

Keywords: Experimental algorithmics · Experimental analysis

1 Introduction

One of the peculiar aspects of *experimental algorithmics* [17] is that the object of the study (an algorithm and its implementation) is often crafted by the same people carrying out the analysis. This has the advantage that the insights obtained from preliminary investigations of early versions of an algorithm can be used to improve the algorithm itself. In fact, the understanding required for an implementation may uncover features of the algorithms that would otherwise go unnoticed [17], giving insights about aspects not easily described by theoretical models of computation [15]. At the same time, this feedback-based process leads to the accumulation of obsolete data, referring to old versions of algorithms and their implementations. Not mixing results from different versions of an algorithm or implementation is an obvious requirement, which however requires some care in practice. In fact, a study often involves different algorithms and datasets, each evolving at a different pace: weeks-old results might be up to date for one algorithm, and obsolete for another.

As we shall see, the literature is mainly concerned with the design and analysis of experiments and with reproducibility. In this position paper, instead, we report on our experience with the day to day tasks that have to be carried out in between those tasks, and the approaches we developed to tackle the perils and frustrations of this often menial work.

© Springer Nature Switzerland AG 2020
S. Satoh et al. (Eds.): SISAP 2020, LNCS 12440, pp. 387–395, 2020.
https://doi.org/10.1007/978-3-030-60936-8_31

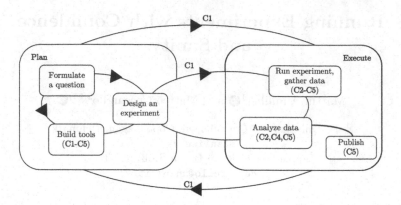

Fig. 1. Overview of the different stages of an experimental study; adapted from [16].

We do not advocate for any specific technology. Rather, we propose a workflow that can be implemented with a variety of tools that can be easily integrated into existing setups. We demonstrate such a setup with a toy project that concerns an efficient implementation of a brute-force nearest neighbor search.

2   Related Work

Moret and Shapiro [17] advocate for the importance of complementing the theoretical analysis of algorithms with their implementation. McGeoch [16] gives several guidelines on how to design and carry out experimental analyses of algorithms. The book [3] collects several contributions on the characterization and analysis of algorithm performance. Earlier, a Dagstuhl seminar was devoted to the discussion of the experimental evaluation of algorithms [10]. More recently, a structured approach to experimental analysis was discussed in [4].

In recent years there has been a discussion about the lack of reproducibility of research findings in several areas, including computer science [8,12]. Much effort has been devoted to finding a solution to this issue. Several contributions have been collected in [19] and [13]. Among the tools to support reproducible research, VisTrails [7] allows to explicitly define reproducible workflows. `knitr` and `Jupyter` take a *literate programming* approach, allowing experiment's code, analysis, and text to be interleaved in a single "executable" document. To solve the issues deriving from software dependencies, some tools aim at capturing the execution environment at runtime [9,11,18], while others such as Docker [5] and Singularity [14] follow a *declarative* approach, where the description of the execution environment is part of the code base.

3   Challenges in Large Scale Experimental Evaluation

We define the following challenges of running large-scale experiments:

(C1) **Feedback Loops-by-Design.** Implementations and tools support the iterative nature of an experimental study.

(C2) **Economic Execution.** Exactly those experiments that change through code changes have to be re-run, but nothing else. Moreover, only the changing parts of the experimental evaluation should be recomputed.

(C3) **Versioning.** The ability to go back in time and compare old results to more recent ones, finding regressions or bugs; the workflow is *append-only*.

(C4) **Machine Independence.** Code and tools are designed in a way that allow them to run in a general setting.

(C5) **Reproducibility-by-Design.** We strive for an automatic workflow that processes an experimental setup into measurements used for evaluating the experiment. Results to be included into a publication should not require manual work to transform these measurements into tables and plots.

Typically, an experimental evaluation spans several weeks, if not months. An overview of a typical experimental evaluation is given in Fig. 1. During this time, the experiments being run have different meanings: early on during initial development, experiments are useful to find out the most appropriate parameter ranges, find bugs, and check assumptions; later on, experiments collect the results of the study. This is not a process that proceeds linearly from start to finish. Rather, the analysis of the results might prompt the modification of an algorithm or dataset, or the introduction of new algorithms and datasets into the study, followed by a new round of experiments. Together with the algorithms and the datasets, also the parameterizations and the quantities being measured are subject to evolution during the lifetime of a project (C1).

For the analysis to be sound, it is of paramount importance not to mix results related to different versions of the algorithms and datasets (C3), in particular when experiments are run on a set of different machines (C4). The simplest solution would be to re-run the entire experimental suite whenever something is modified. This usually takes a very long time, and a change might affect only a small part of the results, making this solution wasteful of time, energy, computational resources and money if computing resources are rented (C2). A potential solution might be to divide the experimental suite in smaller components, each investigating a particular aspect, re-running only those affected by a change. While this works in the short term, as the experimental study progresses the subdivision of the experimental suite will evolve with it, leading to the need of re-arranging the results. On the other hand, manually re-running only parts of an experimental suite, while reusing results from old runs, requires much care in order to exclude obsolete results from the analysis, undermining the confidence in the soundness of the whole analysis. The situation worsens in the rushed final days preceding a submission: some last minute changes are made, there is no time to re-run all the experiments, the possibility of erroneously mixing results is very concrete. Additionally, reviewers will often demand running a new set of experiments, reporting on some other quality measures, or experimentation using different computer architectures (C2, C3, C4). *Reproducibility-By-Design* (C5) requires that such wishes can be accommodated since the whole process from starting at an experimental design to a published table or figure is automated.

As for the analysis itself, it is usually executed on a machine different from the experimental code, using a different programming language (C4). The input of the analysis is the set of results produced by the experimental suite, which is usually quite large due to the fact that many parameter combinations need to be evaluated. While the analysis code may not need the computational resources of the experimental code, it still needs to execute reasonably fast, in order to be able to examine the results interactively. This implies that the results produced by the experiments need to be stored in a convenient format that is at the same time easily manageable, convenient to transfer, and efficient to access (C2, C4).

4   Case Study: Engineering a Linear Scan

As our toy project, we engineer a nearest neighbor search algorithm that just carries out a linear scan over the dataset[1].

Formally, we are given a dataset $S \subset \mathbb{R}^d$ of n points in a d-dimensional space with a distance measure dist: $\mathbb{R}^d \times \mathbb{R}^d \to \mathbb{R}$, such that given a query $q \in \mathbb{R}^d$ we want to return a point $p \in S$ that minimizes $\text{dist}(p', q)$ over all $p' \in S$. Solving this problem via a linear scan is a straight-forward exercise in an introduction to programming class: Compare all points $p' \in S$ one by one to q, and keep track of the point that is closest to q. This results in a running time $O(nd)$ per query.

To make this problem more interesting, we consider engineering choices to speed up a linear scan under inner product similarity $\text{dist}_{\text{IP}}(p, q) = \sum_{1 \le i \le d} x_i y_i$ on unit vectors, similar to Cosine similarity. For the purpose of this project, we consider (i) input representation, (ii) parallelization, and (iii) saving distance computations as factors of the experiment.

Input Representation. A vector in $\mathbb{R}^d$ is traditionally represented as d 64-bit floating point values (`double`) or 32-bit floating point (`float`). Since we guarantee $0 \le x_i \le 1$ for normalized vectors, we also consider a 16-bit representations of the value $\lceil x_i \cdot 2^{16} \rceil / 2^{16}$ (which could of course affect the accuracy of the result).

Parallelization. Naïvely, the CPU has to carry out d multiplications and $d - 1$ additions to compute the distance of two vectors. However, we notice that the structure is inherently parallel because the multiplications are data independent. This is an ideal setup for using so-called SIMD instructions (single instruction multiple data). We split up each vector into blocks of size B, and carry out d/B parallel multiplications, d/B parallel additions to aggregate terms in a register of size B, and one horizontal sum. Depending on the CPU architecture used in the experiment, B is usually 128, 256, or—very recently—512 bits.

---

[1] We would like to thank Michael Vesterli for the many code optimizations that we are using, that he developed for PUFFINN [2].

Saving Distance Computations. Computing the distance between two vectors is certainly the most expensive operation in our linear scan. Hence, if we could decide for a data point p' that it probably is not the nearest neighbor faster than carrying out a distance computation could increase the performance of our linear scan. We include experiments with SimHash sketches with probabilistic quality guarantees in our experiments (see the extended version for details).

We consider this toy project representable for an experimentation task in a similarity search setting. The different choices of input representation, parallelization, and distance filter methods provide an evolutionary setting in which we start with a standard linear scan and add features to the code base one by one. From starting with a measurement of running time, we quickly end up focusing on the quality of the achieved result when using a low-precision input representation, or analyzing the effectivity of the sketch by counting distance computations. The experiment has to be carried out on different machines because of the hardware dependencies, which might mean to rent cloud instances to carry out measurements on recent hardware with $B = 512$ bit AVX512 support.

Fig. 2. Dimensions for running large-scale experimental evaluations.

For the scope of this paper, we consider the *support code* that takes care of handling the setup as the main contribution. For the interested reader, the evaluation of the toy project is given in the extended version of this paper.

5   Approaches to Experimental Evaluation

We now describe a workflow we developed to address the challenges outlined in the previous sections, demonstrating it with our case study. We remark that this workflow does not directly deal with issues like parameter tuning, prediction of runtimes, and trade-off evaluations, which are the topic of *experimental design* [16]. Rather this workflow aims to ease and aid all the aforementioned tasks, along with the analysis. We split up the discussion into different dimensions of running a successful experimental study. These dimensions are summarized in Fig. 2. In the following, each dimension will be introduced with general guidelines and a discussion of our actual solution.

5.1   Manage the Datasets and Workloads Efficiently

- Dataset download and preprocessing should be automated as much as possible, ideally with a single script responsible to manage all the datasets. This makes reproducibility easier, allows to share preprocessing steps between similar datasets, makes it easy to relocate the experiments on a different machine, and makes all the decisions about datasets explicit. Furthermore, it enables the community to change the datasets to observe how these changes are reflected in the experiments.
- It must be possible to create all datasets locally, but the preprocessed datasets should also be shared, for instance using plain `http` or a service such as `S3`. This makes it easier for collaborators, reviewers, and the community to re-run the experiments without incurring the set-up cost of the datasets.
- Datasets should be annotated with meta-data necessary in the evaluation, such as workloads and the related ground truth answers.
- To ease debugging, a small dataset of random data that can be created in a few seconds should also be included. This dataset can be run via Continuous Integration (CI), and results on it can be stored to enable regression testing.

In our code, the main C++ code calls a Python script (`datasets.py`) that takes care of preprocessing datasets in a well-defined manner. It checks for the existence of a shared dataset (of the same version) and computes it locally if such a dataset is not available. It supports the creation of tiny random datasets that allows to run all parts of the workflow on actual data. The query set that is later used for experimentation is created in this process as well, and data is stored as an `HDF5` file for efficient processing in many different programming languages.

5.2   Manage the Experimental Configurations Clearly

- Never run experiments from the command line. Direct command line execution should be limited to testing.
- Experiments should be described in one or more files. This makes it easier to reproduce the entire experimental suite. There are several options, which we both demonstrate in the associated code:
  - Files in a declarative language such as YAML listing all the combinations of parameters to be tested. These files are then interpreted by a script that spawns the appropriately configured experimental code. This approach has the advantage of being declarative, and the disadvantage of requiring some additional software.
  - Shell scripts that directly invoke the experimental code using the appropriate parameters. This is a more procedural approach, which however has the advantage of requiring very little setup.
- All the aforementioned experimental files should be tracked with version control along with the code. Before running the experiments, any pending changes should be committed.

- There should be a mechanism allowing to skip already-run configurations. This allows both to save time (C2) without having to continuously edit the configuration files to remove the configurations that do not need to be run.

Example files can be seen in the code repository. While a direct experimental file written in Bash is straight-forward, the YAML structure gives a much more structured overview. The YAML file is run through an additional Python script that invokes the main implementation with the correct parameters. Using versioning and the result database (Subsect. 5.5) the code can decide whether an algorithm has to be rerun.

5.3   Infrastructure Management

Any implementation will likely depend on many different environmental settings, such as the correct versions of libraries/compiler/OS. To allow a machine independent workflow, we suggest to:

- Provide a containerized development environment[2].
- Consider different container formats for running experiments [1].
- Use continuous integration to test all parts of the workflow.

Our code provides a `Dockerfile` that installs a well-defined Linux environment and sets up the correct compilers and libraries. Each component is run from the local system via a `dockerrun` script that will run the intended process within the container.

Sometimes an implementation relies on system dependent libraries (such as CUDA or MPI): in such cases tracking the version of these libraries helps in handling updates to the underlying system.

5.4   Version Everything

To address challenges (C2) and (C3), version control systems might not be sufficient, since source code revisions lack both a semantic meaning and a total order. Furthermore, different components of a project might evolve independently, thus needing independent versioning to address challenge (C2). Therefore, we suggest to keep track of the versions of individual components of the project, including datasets, algorithms, and database schemas, alongside the versioning provided by the version control system.

In our code, each dataset and database schema provides their own version number. Additionally, components of the implementation such as input representation type or SIMD definitions are versioned. This allows us to map each parameter set for the linear scan to a unique identifier. As an example, during continuous integration we found a bug that only affected the AVX2 inner product computation with floating point numbers. An update of the version number

---

[2] Recently, such environments are included in programming IDE such as https://code.visualstudio.com/docs/remote/containers.

of this part of the code led to a re-run of all parameter configurations that used that particular combination. Each measurement obtained is versioned with its *git identifier*, the algorithm version in question, and the dataset version.

5.5 Manage the Experimental Results Thoughtfully

As for the management of experimental results, structured text file formats like CSV address challenge (C4), but are expensive to parse and require to be fully loaded in main memory prior to the analysis, even when only a subset is needed. Moreover, it is hard to evolve the structure of these files together with the project.

- Use a database to store the results: it presents data conveniently indexed and removes the need for expensive parsing.
- Using schema migrations the database can evolve along with the rest of the project (C3), as demonstrated in our case study's code.
- For simple projects, an embeddable database like SQLite addresses challenge (C4): the results are stored in a single file which can be easily moved between machines, and many languages used for the analysis (like Python and R) provide facilities to access it, as shown in our code. For larger projects, where experimental code is executed on different machines, a database with a client-server architecture (such as PostgreSQL) might be more suitable.
- The experimental code can query the database to detect whether an experiment has already been run with the current version (C2).
- Track the *provenance* [6] of each result, by storing alongside the parameters also the configuration file (and its version) that generated the result (C5).
- By means of database views we can enforce that the analysis code has access only to the most recent results related to each algorithm/dataset (C3): our code demonstrates how to embed a query in the database so to present only the results related to the most recent version of algorithms and datasets.

References

1. Arango, C., Dernat, R., Sanabria, J.: Performance evaluation of container-based virtualization for high performance computing environments. arXiv:1709.10140 (2017)
2. Aumüller, M., Christiani, T., Pagh, R., Vesterli, M.: PUFFINN: parameterless and universally fast finding of nearest neighbors. In: ESA 2019 (2019)
3. Bartz-Beielstein, T., Chiarandini, M., Paquete, L., Preuss, M. (eds.): Experimental Methods for the Analysis of Optimization Algorithms. Springer, Heidelberg (2010). https://doi.org/10.1007/978-3-642-02538-9
4. Bartz-Beielstein, T., Preuss, M.: Experimental analysis of optimization algorithms: tuning and beyond. In: Borenstein, Y., Moraglio, A. (eds.) Theory and Principled Methods for the Design of Metaheuristics. NCS, pp. 205–245. Springer, Heidelberg (2014). https://doi.org/10.1007/978-3-642-33206-7_10
5. Boettiger, C.: An introduction to docker for reproducible research. Oper. Syst. Rev. **49**(1), 71–79 (2015)

6. Buneman, P., Khanna, S., Wang-Chiew, T.: Why and where: a characterization of data provenance. In: Van den Bussche, J., Vianu, V. (eds.) ICDT 2001. LNCS, vol. 1973, pp. 316–330. Springer, Heidelberg (2001). https://doi.org/10.1007/3-540-44503-X_20

7. Callahan, S.P., Freire, J., Santos, E., Scheidegger, C.E., Silva, C.T., Vo, H.T.: VisTrails: visualization meets data management. In: SIGMOD 2006 (2006)

8. Collberg, C.S., Proebsting, T.A.: Repeatability in computer systems research. Commun. ACM **59**, 62–69 (2016)

9. Davison, A.P., Mattioni, M., Samarkanov, D., Telenczuk, B.: Sumatra: a toolkit for reproducible research. In: Implementing Reproducible Research. CRC Press (2014)

10. Fleischer, R., Moret, B., Schmidt, E.M. (eds.): Experimental Algorithmics: From Algorithm Design to Robust and Efficient Software. LNCS, vol. 2547. Springer, Heidelberg (2002). https://doi.org/10.1007/3-540-36383-1

11. Guo, P.J.: CDE: a tool for creating portable experimental software packages. Comput. Sci. Eng. **14**(4), 32–35 (2012)

12. Hutson, M.: Artificial intelligence faces reproducibility crisis **359**(6377) (2018)

13. Kitzes, J., Turek, D., Deniz, F.: The Practice of Reproducible Research: Case Studies and Lessons from the Data-Intensive Sciences. Univ of California Press (2017)

14. Kurtzer, G.M., Sochat, V., Bauer, M.W.: Singularity: scientific containers for mobility of compute. PLoS One **12**(5), e0177459 (2017)

15. McGeoch, C.C.: Experimental algorithmics. Commun. ACM **50**(11), 27–31 (2007)

16. McGeoch, C.: Experimental methods for algorithm analysis. In: Kao, M.Y. (ed.) Encyclopedia of Algorithms. Springer, Boston (2008). https://doi.org/10.1007/978-0-387-30162-4_135

17. Moret, B.M.E., Shapiro, H.D.: Algorithms and experiments: the new (and old) methodology. J. UCS **7**(5), 434–446 (2001)

18. Rampin, R., Chirigati, F., Shasha, D.E., Freire, J., Steeves, V.: ReproZip: the reproducibility packer. J. Open Source Softw. **1**(8), 107 (2016)

19. Stodden, V., Leisch, F., Peng, R.D.: Implementing Reproducible Research. CRC Press (2014)

Doctoral Symposium

Temporal Similarity of Trajectories
in Graphs

Shima Moghtasedi(✉)

Dipartimento di Informatica, Università di Pisa, Pisa, Italy
shima.moghtasedi@di.unipi.it

Abstract. The analysis of similar trajectories in a network provides useful information for different applications. In this study, we are interested
in algorithms to efficiently retrieve similar trajectories. Many studies
have focused on retrieving similar trajectories by extracting the geometrical and geographical information of trajectories. We provide a similarity
function by making use of both the temporal aspect of trajectories and
the structure of the underlying network. We propose an approximation
technique offering the top-k similar trajectories with respect to a query
in a specified time interval in an efficient way. We also investigate how
our idea can be applied to similar behavior of the tourists, so as to offer
a high-quality prediction of their next movements.

Keywords: Temporal-structural similarity · Graph · PoI prediction

1 Introduction

Studying and presenting efficient methodologies for querying similar trajectories
is important as brings the potential applicability to support different applications. Trajectories are complex objects with useful information. Most of the
existing works who study the behavior of network constrained trajectories, they
map trajectories into a spatial network assigned to geographical information,
which can impose high-dimensional indexing challenges and large overall size of
data. In this study, we present each network constrained trajectory as a sequence
of nodes with no spatial information. Determining the similarity between trajectories on networks is a difficult task, and the naive exact solutions do not
scale to large applications. In this study we specify the similarity between two
trajectories with any arbitrary length, taking into account the temporal aspect
of trajectories and the location of trajectories on graphs, in a single linear-time
function [3]. While, the existing functions in the literature (e.g., [10,12,14,15])
are quadratic by computing temporal and spatial distances in a linear combination, which can impose some unnecessary computations.

Goal. In this paper for a given query trajectory as a sequence of nodes on a
graph with the corresponding time intervals, we aim at answering two following

© Springer Nature Switzerland AG 2020
S. Satoh et al. (Eds.): SISAP 2020, LNCS 12440, pp. 399–404, 2020.
https://doi.org/10.1007/978-3-030-60936-8_32

research questions: RQ1: finding k most similar trajectories to query in a specified time interval in an efficient way; RQ2: identifying the node (i.e. Point of Interest (in short, PoI)) that the trajectory (i.e. a tourist) will visit with the highest probability in the future. Usually, this problem is solved by using machine learning-based techniques [1,4,7,8,16]. The overall performance of these techniques depends on the effectiveness of the proposed feature set. In particular, given a trajectory dataset, this model needs to extract a set of useful features for prediction, which is a challenging task. Thus, the objective of the RQ2 task is to reflect the structural-temporal similar behavior of the past tourists to provide an effective prediction.

Contributions. In this study our main contributions can be summarized as follows: (1) we propose a low complexity similarity function considering temporal aspect and the location of trajectories on the graph, directly in a single function; (2) we propose a storage scheme to speed up searching for similar trajectories [6]; (3) we use the Network Voronoi Diagram to accelerate the query processing by computing the minimum number of shortest path distances [3]; (4) we propose a technique to predict the next movement of a tourist, by taking into account the structural-temporal similarities between tourists, outperforming state-of-the-art machine-learning based competitors by providing at least twice more accurate results [13].

2   Preliminaries and Definitions

Definition 1 (Structural-Temporal Trajectory). *Given a graph $G(V, E)$, a Structural-Temporal trajectory T is defined as a sequence of pairs of the form (v_i, t_i), where v_i is a node and t_i a time interval, i.e., $T = \langle (v_1, t_1), (v_2, t_2), \cdots, (v_l, t_l) \rangle$ such that, $1 \leq i \leq l - 1$, we have $(v_i, v_{i+1}) \in E$ and t_i and t_{i+1} are two consecutive time intervals. Letting $t_1 = [s_1, e_1]$ and $t_l = [s_l, e_l]$, we refer to s_1 and e_l as the starting time and ending time of T.*

We restrict the given trajectory T in a specified time interval t as a sequence of nodes are traversed by T in t as $T[t]$. Thus, we can compute the distance between a given node on the graph and a trajectory in a given time interval as $dist(v, T, t) = \dfrac{\min_{(v_i, t_i) \in T[t]} d(v_i, v)}{D_G}$, where D_G is the diameter of the graph. This distance always is a value between 0 and 1. Having this distance we determine our linear similarity function between two query trajectory Q and target trajectory T in a specified time interval t in an exponential way as $Sim(Q, T, t) = \dfrac{\sum_{(v_i, t_i) \in Q[t]} |t_i| \times e^{-dist(v_i, T[t_i])}}{|t|}$.

We define K-MsTraj query, as the K trajectories with maximum similarity score to the query with respect to the proposed function.

Managing the Structural-Temporal Trajectories. Facing with a new type of data as the Structural-Temporal trajectories, how to efficiently organize this data to process queries is a challenge. As the existing indexing methods [2,5,9,17], which typically designed for geo-located (spatial) trajectories, and none of them consider the time intervals spent by objects in the nodes of the network, could be either infeasible or inefficient on structural-temporal trajectories. In this study, we build an indexing structure (i.e. NTRAJI) based on the *interval tree* to store the trajectories on graphs. We model our structure in the main memory to support short response time, although it is possible to extend the methods considering the external memory interval tree. In NTRAJI, we build an Interval Tree IT_u assigned to each node $u \in V$. Each IT_u stores the time intervals of the trajectories in dataset spent in either u or the neighbors of u, and maintains the corresponding trajectory ids.

The Baseline Method (BASE). Due to the lack of competitors, since there is no linear time similarity function considering structural-temporal trajectories, we design a pruning technique as the baseline method to quickly find exact K-MsTraj. First, BASE finds a set of trajectories are close to the query by searching through NTRAJI within the corresponding time interval. Then, by computing the similarity score for each discovered trajectory, reports top-K ones. We use this method as a baseline in the experimental evaluation of our methods.

3 RQ1: Approximate Computation of K-MsTraj

To accelerate the searching process, we propose an approximated method with two main steps.

- First, by making use of Voronoi Diagrams for graphs, we partition G into disjoint groups of nodes, precomputing the distances among the centers of each group in $O(m \cdot n^{1/2})$ time, when n, m are the number of nodes and edges of G, respectively. In comparison to the state-of-the-art [11,12] that precomputes the all-to-all pairwise node shortest path distances.
- Second, we represent each trajectory in the dataset with the Voronoi centers as the SHRUNK trajectory. Therefore, we need to adjust the NTRAJI indexing to support the shrunk trajectories called VoTRAJI. In VoTrajI, for each center node, we build an interval tree maintaining the time intervals are spent by trajectories through the nodes of the corresponding Voronoi group.

In the query processing phase, we search through VoTRAJI and find the trajectories are most likely to be top-K similar trajectories, with respect to the query. We have two variants to proceed with the discovered trajectories: SHQ: estimating similarity scores of the trajectories by considering shrunk query in place of the exact query; SHQT: estimating the scores of the trajectories considering shrunk query and shrunk trajectories. We evaluate the effectiveness of both variants in comparison with the baseline method in terms of the time needed to process K-MsTraj and the precision of the produced results.

Table 1. The average time for answering a query for each method (in sec.)

DATASETS	BASE	SHQ	SHQT
Facebook	1.09	0.74	0.43
Milan	380.03	376.15	85.48
Rome	26.19	19.42	15.83

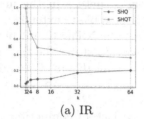 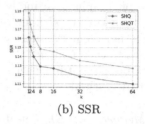

(a) IR   (b) SSR

Fig. 1. The quality of the results in terms of IR and SSR ratio vs κ in Rome.

3.1   Experiments

We conduct the experiments on real-world datasets: a set of synthetic trajectories on Facebook,[1] a real set of trajectories of private cars in Milan[2] and a real set of Flickr geotagged photos provided by [8], containing tourist trajectories in Rome. Regarding the time needed for query processing, Table 1 reports our results, showing the average query time over 100 queries in each dataset. As it can be seen, both SHQ and SHQT outperform BASE, while the running time by SHQT dramatically is less than BASE and SHQ. This is due to that SHQT uses precomputed distances between Voronoi centers. We evaluate the precision of the produced solution by SHQ and SHQT with respect to BASE. The effectiveness of the methods is assessed by means of two metrics. Let γ_1 and γ_2 be two output sets containing top-κ trajectories, e.g., the exact and approximated solutions. We define the *similarity score ratio* as $\mathrm{SSR}(\gamma_1, \gamma_2) = \dfrac{\sum_{T \in \gamma_1} Sim(Q,T,t)}{\sum_{S \in \gamma_2} Sim(Q,S,t)}$, and the *intersection ratio* as $\mathrm{IR}(\gamma_1, \gamma_2) = \dfrac{|\gamma_1 \cap \gamma_2|}{\kappa}$. In both ratios, the values close to 1 are more desirable. Due to the lack of space, we present only the results of the experiments in terms of the SSR and IR ratios on Rome as a function of κ. As shown in Fig. 1(a), SHQT has the best performance for smaller values of κ and outperforms SHQ. However, we observe that the lower values of IR correspond to SSR values that are close to 1. Indeed, Fig. 1(b) shows SSR which is almost always very close to 1 for both variants SHQ and SHQT, which confirms the effectiveness of the proposed model.

---

[1] https://snap.stanford.edu/data/ego-Facebook.html.
[2] https://sobigdata.d4science.org/catalogue-sobigdata?path=/dataset/gps_track_milan_italy.

4 RQ2: Predicting the Next-PoI

Given a trajectory $T = \langle(v_1, t_1), \cdots, (v_l, t_l)\rangle$ on graph, we aim at finding the next node will visit by T in future. To this end, we solve K-MsTRAJ for the query T within the time interval t_l, when K= 1. Then after, we specify the recently visited node by the most similar trajectory to T in t_l (i.e. 1-MsTraj), as the next position of T. To evaluate our method we use three different datasets provided in [8] containing tourist movements in three Italian cities Pisa, Rome, Florence. We compare our method with a probability baseline PROB [8] and LEARNEXT [8]. Although there are two important state-of-the-art techniques WHERENEXT [7] and Random Walk [4], we do not include them in our experimental study, since LEARNEXT outperforms them. We follow the same evaluation strategy adopted in [8] over the three aforementioned datasets, which is a standard training/test evaluation strategy. For each city, we consider 80% of trajectories as a training set and 20% of trajectories as the test set. The effectiveness of the methods is assessed by means of Success@k (i.e., the percentage of times that the correct answer is in the top-k ranked PoIs). Specifically, we will use $k = 1$ in our evaluations, which is the topmost PoI. Results of the experiments are provided for our proposed method MsTRAJ along with two methods (PROB and LEARNEXT). Table 2 shows the results of the experiments, where we show that our method outperforms the competitors. As we can observe, MsTRAJ provides almost twice more accurate results than LEARNEXT in terms of Success@1 in each city. While MsTRAJ provides almost six times more accurate results than PROB for Pisa and Rome, and almost ten times more for Florence, confirming the effectiveness of the proposed method.

Future Direction. Finally, we would like to present an exact solution for the proposed query by taking benefit of the dominant relationship between trajectories taking into account the time duration trajectories spent close to the query and their location on the underlying graph. In this way, we are able to eliminate unnecessary computations and consecutively reduce the total running time. Moreover, we would like to conduct the case study for validation of our model in other applications: real-time detection of trajectories of people, news, events, and virus outbreaks.

Table 2. Effectiveness in terms of Success@1 of the proposed method (MsTRAJ) along with the competitors.

Predictors	Pisa	Rome	Florence
PROB	15.57	12.59	4.96
LEARNEXT	40.70	30.95	37.56
MsTRAJ	**67.33**	**77.96**	**53.57**

References

1. Boukhechba, M., Bouzouane, A., Gaboury, S., Gouin-Vallerand, C., Giroux, S., Bouchard, B.: Prediction of next destinations from irregular patterns. J. Ambient Intell. Hum. Comput. **9**, 1345–1357 (2017)
2. De Almeida, V.T., Güting, R.H.: Indexing the trajectories of moving objects in networks. GeoInformatica **9**(1), 33–60 (2005)
3. Grossi, R., Marino, A., Moghtasedi, S.: Finding structurally and temporally similar trajectories in graphs. In: 18th International Symposium on Experimental Algorithms (SEA 2020). Schloss Dagstuhl-Leibniz-Zentrum für Informatik (2020)
4. Lucchese, C., Perego, R., Silvestri, F., Vahabi, H., Venturini, R.: How random walks can help tourism. In: Baeza-Yates, R., et al. (eds.) ECIR 2012. LNCS, vol. 7224, pp. 195–206. Springer, Heidelberg (2012). https://doi.org/10.1007/978-3-642-28997-2_17
5. Luo, W., Tan, H., Chen, L., Ni, L.M.: Finding time period-based most frequent path in big trajectory data. In: Proceedings of the 2013 ACM SIGMOD International Conference on Management of Data, pp. 713–724. ACM (2013)
6. Moghtasedi, S.: Time-based similar trajectories on graphs. In: ICTCS, pp. 82–86 (2018)
7. Monreale, A., Pinelli, F., Trasarti, R., Giannotti, F.: WhereNext: a location predictor on trajectory pattern mining. In: Proceedings of the 15th ACM SIGKDD International Conference on Knowledge Discovery and Data Mining, pp. 637–646. ACM (2009)
8. Muntean, C.I., Nardini, F.M., Silvestri, F., Baraglia, R.: On learning prediction models for tourists paths. ACM Trans. Intell. Syst. Technol. (TIST) **7**(1), 8 (2015)
9. Popa, I.S., Zeitouni, K., Oria, V., Barth, D., Vial, S.: PariNet: a tunable access method for in-network trajectories. In: 2010 IEEE 26th International Conference on Data Engineering (ICDE 2010), pp. 177–188. IEEE (2010)
10. Shang, S., Chen, L., Wei, Z., Jensen, C.S., Zheng, K., Kalnis, P.: Trajectory similarity join in spatial networks. Proc. VLDB Endow. **10**(11), 1178–1189 (2017)
11. Shang, S., Ding, R., Yuan, B., Xie, K., Zheng, K., Kalnis, P.: User oriented trajectory search for trip recommendation. In: Proceedings of the 15th International Conference on Extending Database Technology, pp. 156–167 (2012)
12. Shang, S., Ding, R., Zheng, K., Jensen, C.S., Kalnis, P., Zhou, X.: Personalized trajectory matching in spatial networks. VLDB J. **23**(3), 449–468 (2013). https://doi.org/10.1007/s00778-013-0331-0
13. Shima, M., Cristina Ioana, M., Franco Maria, N., Roberto, G., Andrea, M.: High-quality prediction of tourist movements using temporal trajectories in graphs (under submission)
14. Tiakas, E., Papadopoulos, A.N., Nanopoulos, A., Manolopoulos, Y., Stojanovic, D., Djordjevic-Kajan, S.: Trajectory similarity search in spatial networks, pp. 185–192. IEEE (2006)
15. Tiakas, E., Rafailidis, D.: Scalable trajectory similarity search based on locations in spatial networks. In: Bellatreche, L., Manolopoulos, Y. (eds.) MEDI 2015. LNCS, vol. 9344, pp. 213–224. Springer, Cham (2015). https://doi.org/10.1007/978-3-319-23781-7_17
16. Ying, H., et al.: Time-aware metric embedding with asymmetric projection for successive poi recommendation. World Wide Web **22**(5), 2209–2224 (2019)
17. Zhu, J., Jiang, W., Liu, A., Liu, G., Zhao, L.: Effective and efficient trajectory outlier detection based on time-dependent popular route. World Wide Web **20**(1), 111–134 (2016). https://doi.org/10.1007/s11280-016-0400-6

Relational Visual-Textual Information Retrieval

Nicola Messina[(✉)][iD]

Institute of Information Science and Technologies,
National Research Council, Pisa, Italy
nicola.messina@isti.cnr.it

Abstract. With the advent of deep learning, multimedia information processing gained a huge boost, and astonishing results have been observed on a multitude of interesting visual-textual tasks. Relation networks paved the way towards an attentive processing methodology that considers images and texts as sets of basic interconnected elements (regions and words). These winning ideas recently helped to reach the state-of-the-art on the image-text matching task. Cross-media information retrieval has been proposed as a benchmark to test the capabilities of the proposed networks to match complex multi-modal concepts in the same common space. Modern deep-learning powered networks are complex and almost all of them cannot provide concise multi-modal descriptions that can be used in fast multi-modal search engines. In fact, the latest image-sentence matching networks use cross-attention and early-fusion approaches, which force all the elements of the database to be considered at query time. In this work, I will try to lay down some ideas to bridge the gap between the effectiveness of modern deep-learning multi-modal matching architectures and their efficiency, as far as fast and scalable visual-textual information retrieval is concerned.

Keywords: Cross-media retrieval · Deep features · Neural networks

1 Introduction

Image-text matching has shown impressive results on many image-sentence retrieval benchmarks, where the objective consists in retrieving images given a sentence as a query, or vice versa. The image-sentence retrieval task has been used to evaluate the network's ability to correctly match together relevant images and sentences. However, the image-sentence retrieval problem is interesting in itself, as it lays the basis for efficient search engines working with multi-modal data.

Search engines must be fast and scalable, as they need to process queries on huge databases in few milliseconds. However, state-of-the-art image-sentence matching approaches usually employ cross-attention mechanisms in the early stages of the data pipeline that makes it impossible to separately forward the

© Springer Nature Switzerland AG 2020
S. Satoh et al. (Eds.): SISAP 2020, LNCS 12440, pp. 405–411, 2020.
https://doi.org/10.1007/978-3-030-60936-8_33

visual and the textual information (Fig. 1). This separation is needed to disentangle the offline indexing phase, usually very expensive, from the online query processing, that instead should be completed in milliseconds.

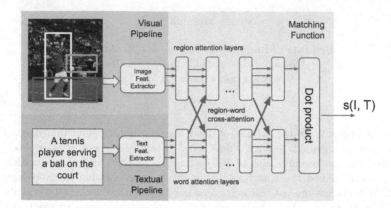

Fig. 1. The overall architecture of state-of-the-art proposals concerning image-text matching. Cross attention in early stages makes the matching function outputting image-text similarity score very complex. It is not possible to extract separately visual and textual features.

Despite the overall loss in efficiency, the use of cross-attention produces an effective multi-step reasoning process that is highly beneficial for producing good matches. As shown in previous works such as Relation Networks [16], trying to infer a relational bias between the basic building blocks of the visual and textual inputs helps in developing abstract links between multi-modal concepts to gather a relational view of the world.

Furthermore, it is possible that the optimal image-sentence representation for good indexing is not a fixed-sized vector, but a variable-length set of vectors describing the images and the texts as sets of concepts. This poses new challenges as far as the indexing structures are concerned.

In this work, I will try to pave the way towards the use of effective relational multi-modal descriptions obtained from state-of-the-art self-attentive architectures in scalable retrieval contexts, where efficiency is a key requirement.

2   Related Work

Many works in computer vision and natural language processing works introduced high-level complex reasoning mechanisms [16,19], mainly addressing Visual Question Answering. More recently, the basic ideas behind these reasoning schemes have been implemented in self- and cross-attentive modules [18], and employed in many language-vision tasks [3,7,15]. These works achieved state-of-the-art results on image-sentence retrieval. However, they do not consider efficiency aspects.

On the other hand, many works tackled the problem of indexing visual features coming from deep architectures. As far as content-based image retrieval (CBIR) is concerned, [1,2] addressed the indexability problem of deep visual features coming from Convolutional Neural Networks (CNNs), like [17]. In particular, [1] showed the performances achieved with the quantization of RMAC features, and they compared this methodology to the deep permutation approach [2], using an inverted file as an indexing structure. Standard CNN features do not embed complex relational biases. Furthermore, multi-modality is not addressed in these works.

3   Explored Approaches

In previous publications [10–12] I explored the effectiveness of a relational visual descriptor extracted from a relation-aware architecture reasoning on a scene with multiple objects. This relational descriptor obtained best results in the introduced Relational-CBIR task, which consists in finding all the images having objects in similar spatial relationships, given an image as query. The proposed relational feature defeated common CBIR deep features such as RMAC [17] on this task. This work tackled the problem of producing a compact and effective visual descriptor that could carry very complex scene information, including inter-object relationships, and that could be indexed using already-existing CBIR frameworks.

Given the increasing interest in multi-modal relational information processing, my research is now focused on complex cross-modal retrieval scenarios. The excellent results obtained by recently introduced self- and cross-attentive models made me concentrate on the transformer architecture [18] for processing visual-textual data using multi-step reasoning pipelines.

Although current efficient retrieval methods assume fixed-sized descriptors (e.g., RMAC [17]), the latest works in cross-modal analysis treat images and texts as sets of basic interconnected elements (image regions and words) processed using attentive mechanisms. The native representation available becomes therefore a variable set of features, called *concepts*, for every image or sentence.

One of my recent key contributions in this direction is the introduction of the Transformer Encoder Reasoning Network (TERN) [14]. TERN employs self-attentive mechanisms to produce both a global fixed-size deep feature and sets of fine-grained concepts that are independent of their source modality. Unlike most works in this field, TERN lacks cross-attentive links. Doing so, two well-distinguished pipelines, a visual and a textual one, are created and can be used separately in the online search and in the offline indexing phases. The produced cross-modal representations are compared with simple dot-products so that the similarity search can be very efficient by employing already existing indexing schemes working on standard metric spaces (Fig. 2).

Concerning the global fixed-size description of images and sentences, in [13] we applied the scalar quantization or deep permutation approaches to multi-modal global features as explained in [1]. These representations can be then easily used in inverted lists without further modifications to the indexing structure.

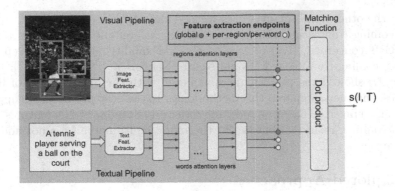

Fig. 2. An high-level overview of the proposed TERN architecture.

On the other hand, TERN can also treat images and sentences as sets and sequences of basic interconnected concepts, coming from regions and words respectively. In this case, TERN does not output a compact global description of images and sentences, but a variable-sized set of features, one for every concept, in the same abstract common space.

The concepts can be clustered to create a dictionary. Following this direction, in [13] we also introduced a model similar to the Bag of Words, that we called Bag of Concepts, for producing image and sentence representations for efficient indexing using inverted lists.

3.1 Early Results

In [11] we were able to obtain a good relation aware descriptor, that reached a Spearman-Rho correlation value of 0.28 against −0.15 of the RMAC features on the Relational-CBIR benchmark built on the CLEVR dataset [5]. We thus showed the efficacy of a relational architecture in producing a fixed-size relation aware image descriptor.

In the recent work on visual-textual information retrieval [13] we used the proposed TERN architecture [14] as a multi-modal feature extractor, both for global fixed-sized descriptors and for the variable-sized set of concepts. In these first experiments, we tested the stability of the extracted features by simulating strong sparsification, as this is the key element for the production of efficient inverted indexes.

On the visual-textual MS-COCO dataset [6], the scalar quantization and the deep permutation approaches on the fixed-sized global feature behaved very similarly (72.7 Recall@10 in the sentence-to-image retrieval and 81.3 in image-to-sentence). When the sparsity rate achieves 99% (only 20 dimensions out of 2048 are not zero), the deep permutation approach loses around 27% on the Recall@10 metric, while scalar quantization loses 23%. At the same very high sparsification rate of 99%, the Bag of Concepts model shows better results than the scalar quantization approach during the re-ranking phase of the sentence-to-image retrieval scenario. The reranking using the non sparsified vectors can be performed efficiently using GPUs on the subset of results selected by the initial approximate search, therefore it defines an overall good compromise between efficiency and effectiveness.

4 Conclusions

In this work, I tried to pave the way towards efficient and effective multi-modal retrieval using state-of-the-art technologies from computer vision and natural language processing worlds. The emphasis is placed on attentive architectures. They are able to implement a multi-step high-level relational reasoning procedure, gaining a lot in effectiveness but creating efficiency problems when scalable information retrieval is addressed.

In my research, I first tried to produce a fixed-sized relational visual descriptor that defeated RMAC on the Relational-CBIR task. Then, considering the interesting cross-modal retrieval problem, I tried to extract powerful fixed- and variable-sized features using the proposed TERN architecture, using existing methods (scalar quantization, deep permutations) when addressing global fixed-sized features and proposing the Bag of Concepts model for producing indexable representations out of the variable-sized sets of multi-modal concepts.

4.1 Next Steps

In the near future, I manage to extensively evaluate the efficiency of the proposed approaches when implemented in inverted indexes structures. To do so, a large multi-modal dataset containing matching images and sentences is needed. MS-COCO can be augmented with Flickr30k [20], obtaining a total of around 36k images annotated with 180k sentences. For further validating these approaches, a whole set of distracting images can be added from available huge image datasets such as MIRFlickr1M[1].

Further experimentation is needed as far as the Bag of Concepts is concerned. An extension of the TERN architecture that I am implementing involves the fine-grained alignment of regions and words at training time. In this case, a precise similarity matrix between every image region and every word is available. It is therefore possible that a custom indexing structure can be built using the region-word alignment matrix. It is furthermore possible to learn this indexing structure imposing some sparsification constraints directly at training time.

Another line of research can be derived by the approaches by [8,9]. In these works, the transformer encoder in the BERT architecture [4] is split in an offline and an online processing stages by partitioning the attention links so that they do not create cross-connections between the two pipelines. In this way, the complex activations produced from the offline pipeline can be stored during the indexing

---

[1] http://press.liacs.nl/mirflickr/.

phase and efficiently retrieved at query time. They applied this methodology for textual document retrieval. However, this approach can be directly applied to state-of-the-art visual-textual processing architectures based on the BERT model, such as [3,7,15].

References

1. Amato, G., Carrara, F., Falchi, F., Gennaro, C., Vadicamo, L.: Large-scale instance-level image retrieval. Inf. Process. Manag. 102100 (2019)
2. Amato, G., Falchi, F., Gennaro, C., Vadicamo, L.: Deep permutations: deep convolutional neural networks and permutation-based indexing. In: Amsaleg, L., Houle, M.E., Schubert, E. (eds.) SISAP 2016. LNCS, vol. 9939, pp. 93–106. Springer, Cham (2016). https://doi.org/10.1007/978-3-319-46759-7_7
3. Chen, Y.C., et al.: UNITER: learning universal image-text representations. arXiv preprint arXiv:1909.11740 (2019)
4. Devlin, J., Chang, M., Lee, K., Toutanova, K.: BERT: pre-training of deep bidirectional transformers for language understanding. In: NAACL-HLT 2019, pp. 4171–4186. Association for Computational Linguistics (2019)
5. Johnson, J., Hariharan, B., van der Maaten, L., Fei-Fei, L., Lawrence Zitnick, C., Girshick, R.: CLEVR: a diagnostic dataset for compositional language and elementary visual reasoning. In: Proceedings of the IEEE Conference on Computer Vision and Pattern Recognition, pp. 2901–2910 (2017)
6. Lin, T.-Y., et al.: Microsoft COCO: common objects in context. In: Fleet, D., Pajdla, T., Schiele, B., Tuytelaars, T. (eds.) ECCV 2014. LNCS, vol. 8693, pp. 740–755. Springer, Cham (2014). https://doi.org/10.1007/978-3-319-10602-1_48
7. Lu, J., Batra, D., Parikh, D., Lee, S.: ViLBERT: pretraining task-agnostic visiolinguistic representations for vision-and-language tasks. In: NeurIPS 2019, pp. 13–23 (2019)
8. MacAvaney, S., Nardini, F.M., Perego, R., Tonellotto, N., Goharian, N., Frieder, O.: Efficient document re-ranking for transformers by precomputing term representations. arXiv preprint arXiv:2004.14255 (2020)
9. MacAvaney, S., Nardini, F.M., Perego, R., Tonellotto, N., Goharian, N., Frieder, O.: Expansion via prediction of importance with contextualization. arXiv preprint arXiv:2004.14245 (2020)
10. Messina, N., Amato, G., Carrara, F., Falchi, F., Gennaro, C.: Learning relationship-aware visual features. In: Leal-Taixé, L., Roth, S. (eds.) ECCV 2018. LNCS, vol. 11132, pp. 486–501. Springer, Cham (2019). https://doi.org/10.1007/978-3-030-11018-5_40
11. Messina, N., Amato, G., Carrara, F., Falchi, F., Gennaro, C.: Learning visual features for relational CBIR. Int. J. Multimed. Inf. Retrieval (2019). https://doi.org/10.1007/s13735-019-00178-7
12. Messina, N., Amato, G., Falchi, F.: Re implementing and extending relation network for R-CBIR. In: Ceci, M., Ferilli, S., Poggi, A. (eds.) IRCDL 2020. CCIS, vol. 1177, pp. 82–92. Springer, Cham (2020). https://doi.org/10.1007/978-3-030-39905-4_9
13. Messina, N., Amato, G., Falchi, F., Gennaro, C., Marchand-Maillet, S.: Cross-media visual and textual retrieval using transformer-encoder deep features. In: SISAP 2020 (2020, submitted)

14. Messina, N., Falchi, F., Esuli, A., Amato, G.: Transformer reasoning network for image-text matching and retrieval. arXiv preprint arXiv:2004.09144 (2020)
15. Qi, D., Su, L., Song, J., Cui, E., Bharti, T., Sacheti, A.: ImageBERT: cross-modal pre-training with large-scale weak-supervised image-text data. CoRR abs/2001.07966 (2020)
16. Santoro, A., et al.: A simple neural network module for relational reasoning. In: Advances in Neural Information Processing Systems, pp. 4967–4976 (2017)
17. Tolias, G., Sicre, R., Jégou, H.: Particular object retrieval with integral max-pooling of CNN activations. In: Bengio, Y., LeCun, Y. (eds.) 4th International Conference on Learning Representations, ICLR 2016, San Juan, Puerto Rico, 2–4 May 2016, Conference Track Proceedings (2016)
18. Vaswani, A., et al.: Attention is all you need. In: NeurIPS 2017, pp. 5998–6008 (2017)
19. Yang, Z., He, X., Gao, J., Deng, L., Smola, A.: Stacked attention networks for image question answering. In: Proceedings of the IEEE Conference on Computer Vision and Pattern Recognition, pp. 21–29 (2016)
20. Young, P., Lai, A., Hodosh, M., Hockenmaier, J.: From image descriptions to visual denotations: new similarity metrics for semantic inference over event descriptions. Trans. Assoc. Comput. Linguist. 2, 67–78 (2014)

Author Index